ADVANCES IN HIGH TEMPERATURE SUPERCONDUCTIVITY

ADVANCES IN HIGH TEMPERATURE SUPERCONDUCTIVITY

Torino, Italy 18 – 20 November 1991

Edited by: I.S.I.

D. Andreone, R. S. Gonnelli & E. Mezzetti

INSTITUTE FOR
SCIENTIFIC INTERCHANGE

World Scientific
Singapore • New Jersey • London • Hong Kong

Published by

World Scientific Publishing Co Pte Ltd
5 Toh Tuck Link, Singapore 596224
USA office: 27 Warren Street, Suite 401-402, Hackensack, NJ 07601
UK office: 57 Shelton Street, Covent Garden, London WC2H 9HE

British Library Cataloguing-in-Publication Data
A catalogue record for this book is available from the British Library.

ADVANCES IN HIGH TEMPERATURE SUPERCONDUCTIVITY

ISBN-13 978-981-02-1297-1
ISBN-10 981-02-1297-6

FOREWORD

This volume contains a number of reviews and original contributions on outstanding selected topics in the field of high-temperature superconductivity collected at the first Workshop on "Advances in High-T_c Superconductivity". The Workshop took place in Torino, Villa Gualino, in November 17-20, 1991 and was organized by the Institute for Scientific Interchange Foundation (ISI).

The about hundred participants coming from 11 different countries enjoied the particular stimulating atmosphere of scientific "full immersion" provided by the Villa Gualino facility. The style was set up in such a way that an overview of important recent developments in the selected topics was given, through a limited number of invited oral presentations of outstanding scholars from eight different countries.

In order to stimulate the exchange of ideas on particular "hot" questions, a guided discussion was introduced at the end of each session, in which an expert chairperson summarized the talks and triggered the informal contributions of all the audience. The Organizing Committee considered these extended discussions as one of the focal points of the Workshop.

I believe that the success of the meeting, demonstrated by the presence of outstanding scientists and by the positive comments of all the participants, will induce the Organizing Committee to consider the possibility to repeat this proficuous initiative in the near future.

The workshop is highly indebted to all my colleagues of the Organizing Committee and, in particular, to Dr. Domenico Andreone, Dr. Renato Gonnelli, Prof. Enrica Mezzetti and Ing. Franco Pavese for planning and organizing the meeting and to Dr. Giorgio Castagno and Ing. Carlo Ferrero for arranging the social program. Special thanks are due to Prof. Fedele Abbattista, Prof. Emilio Acerbi, Dr. Roberta Cimberle and Prof. Mario Vallino for the continuous help during the organization. We are very grateful to Prof. Carlo Rizzuto for taking the time to give us the brilliant opening lecture on the state of art of high-T_c superconducting materials.

We acknowledge the ISI Scientific Coordinator Dr. Tiziana Bertoletti and her assistant Mrs. Enza Palazzo, as well as Dr. Maria Gabella of the Associazione per lo Sviluppo Scientifico e Tecnologico del Piemonte (ASP) and Mrs. Rita Aimone of the Istituto Nazionale di Fisica Nucleare (INFN) for the precious help in the organization and management of the Workshop.

Finally the Organizing Committee warmly acknowledge the financial support of: ASP, INFN, Politecnico di Torino, Istituto Elettrotecnico Nazionale Galileo Ferraris, Europa Metalli LMI S.p.A., Zanon S.p.A., Cassa di Risparmio di Torino.

Emilio Olzi

PREFACE

The structure of this volume essentially follows the organization of the Workshop which was divided in three sessions devoted to:

1 *Preparation and characterization of consolidated and new high-T_c materials*
2 *Critical currents* (flux motion and flux pinning enhancement). *Magnetic shielding*
3 *Weak superconductivity: junctions and devices* (fabrication and tunneling spectroscopy).

The first paper by A. Barone is an introductory review on the particular features of high-T_c superconductors which make these materials unique. The book is then divided in three parts which follow the above mentioned main topics.

The first part, concerning the synthesis and the characterization of high-T_c superconductors, is opened by the comprehensive contribution of C.W. Chu which deals with the present status and challenges of high temperature superconductivity research and development from the side both of basic science and of materials for small and large current applications. Three papers follow (F. Licci et al., R. Masini et al. and V.P. Seminozhenko et al.) which are devoted to the preparation of pure and doped superconducting materials in the form of single crystals, Ag-sheathed tapes and textured ceramics and coatings on metallic wires. The remaining papers (B. Raveau et al., F. Abbattista et al., G. Calestani et al. and F.C. Matacotta et al.) approach the problem of the characterization of HTS mainly from the chemical point of view including the optimization of lead and bismuth cuprates, the study of phase diagrams and structural modulation and the analysis of the effect of substitutions in bismuth compounds.

In the second part, the crucial questions of how to optimize the current density in a controlled manner is revised in view of the two existing method to enhance the properties: chemical precipitation and controlled radiation damage. Certainly, the optimization depends on synthesis, shaping and processing, so that the investigation of the microstructure of high-temperature superconductors, as well as, of the pinning-centers structure is the basis for understanding the critical currents. In this respect also the application of high-resolution methods, in particular high-resolution transmission electron-microscopy has brought much insight in to the field. As a consequence some contributions are devoted to the study of flux lattice relaxation, dissipative phenomena, intergranular fluxons and anisotropy and dimensionality in HTS (V. Calzona et al., D. Fiorani et al., R. Vaccarone et al. and M. Giura et al.) while the others (H.W. Weber, R.B. van Dover et al. and F. Abbattista et al.) concern the effects of proton and fast neutron irradiation on critical currents and flux pinning in 123 compounds.

For what concerns magnetic shielding, it is well known how attractive this topic is from the point of view of the applications of HTS. The two papers appearing in the volume demonstrate that small scale applications are already a reality while the possibility of shielding large volumes at room temperature is under investigation (Mineo Itoh and F. Pavese et al.).

The third part of the volume is devoted to weak superconductivity from the side both of the preparation and characterization of the junctions and of tunneling spectroscopy of high-T_c materials. The first three contributions (R. Gross et al., G.J. Gerritsma and C. Camerlingo et al.) indicate that the technology of grain-boundary and multilayer-type Josephson junctions is mature and high-T_c devices as SQUIDs and magnetometers have been already produced, but a great amount of work has yet to be done to understand the basic physics of junctions in HTS. The study of microwave properties of superconducting films has been an important tool for the characterization of the materials and for the study of the fundamental physics of superconductivity. The application to high-T_c superconductors is described in the paper of C. Attanasio et al.. Finally there are four papers (J.F. Zasadzinski et al., A.M. Cucolo et al., A. Di Chiara et al. and R.S. Gonnelli et al.) concerning tunneling spectroscopy studies in high temperature superconductors. Several topics are discussed, among them: Point-contact tunneling spectroscopy of SIN and SIS junctions, planar junctions on chemically etched YBCO surface, the application of the proximity effect model for the interpretation of the tunneling data and of the temperature dependence of the gap and the determination of the Eliashberg function in high-T_c superconductors.

Domenico Andreone
Renato Gonnelli
Enrica Mezzetti

CONTENTS

Part 2
Critical Currents, Flux Motion and Flux Pinning Enhancement. Magnetic Shielding

ADVANCES IN HIGH TEMPERATURE SUPERCONDUCTIVITY

HIGH-T_c SUPERCONDUCTORS: MUCH MORE THAN SUPERCONDUCTORS OF HIGHER T_c

ANTONIO BARONE
Dipartimento di Scienze Fisiche,
Università di Napoli "Federico II", Napoli, Italy

ABSTRACT

Five years after the discovery of the new class of high-Tc superconductor oxides by Bednorz and Müller, achievements and open questions legitimate the belief that this topic represents one of the most important and challenging chapters of condensed matter physics. The search for new explanations, the difficulty of subtle implications, the perspective of interesting applications, are all witnessing in these years a dramatic development of superconductivity.
This stimulating and timely Workshop will cover some of the important aspects of this evolution. Leaving more specific and complete accounts to the following lectures, I shall give here a brief outline of just a few of the recent advances whose choice will obviously reflect, besides a lack of space, both my personal views and limits.

1. - Introduction

The message underlying the title of this introductory lecture although quite obvious has been, in my opinion, not always fully considered.

The avenue opened by Bednorz and Müller[1] five years ago is indeed fascinating to the extent that it is complex and difficult. It started from a great discovery, it crosses areas of deep interest, it reaches points of unexpected features, it certainly goes toward new surprises.

The lack of theory or, if you prefer, the existence of several theories represents, by itself, one of the greatest challenges of condensed matter physics of our days. This is a physiological consequence of a real breakthrough. Indeed, superconductors with high critical temperature (say T_c>30K) would be almost unconceivable according to previous beliefs.

Together with underlying theory, new aspects of the intriguing phenomenology and the perspective of both near term and potential applications keep this subject in the limelight of the worldwide scientific community.

In the next Section is given a brief sketch summarizing some of the properties of superconductor cuprate oxides and a bird's eye view of theoretical proposals.

Section 3 will discuss some aspects of weak superconductivity in various high-T_c structure configurations.

2. Superconducting Properties

The five years history of high-Tc superconductivity has produced a variety of interesting results and posed new questions.

As far as the superconducting nature of high-T_c oxides is concerned, the situation is still quite puzzling and perphaps far from being completely understood. Even more elusive appear to be the properties of the normal phase whose understanding would very likely represent the key to explain also the superconducting state of cuprate oxides. In these materials, basic issues such as the shape of the Fermi surface are themselves subject of debate. The electronic band structure has been intensively investigated by different experimental tools such as PES (Photoemission Electron Spectroscopy), specific heat and spin susceptibility measurements. Magnetic aspects have also strongly attracted the attention of both theoretical and experimental investigators on properties such as those concerning spin dynamics, (see general accounts given in ref. [2] and references reported therein).

2.1 - Layered structure.

One of the most peculiar features of cuprate superconductors is the marked anysotropy which stems from the layered structure of these materials. As a consequence it has been observed a resistivity ratio $\rho_c/\rho_{a,b}$ (ρ_c and $\rho_{a,b}$ being the resistivity perpendicular to the CuO planes and along such planes respectively) which ranges between 10 (for YBCO) and 10^5 (for BSCCO). Moreover other parameters, such as the characteristic superconducting lengths are characterized by a marked anisotropy. Indeed using the same subindex notations a, b and c to label quantities referred to the directions parallel and perpendicular to the planes, it is found for the London and coherence lengths the relation

$$\lambda_c/\lambda_{a,b} = \xi_{a,b}/\xi_c = (m^*/m)^{1/2}$$

where m^*/m is the large effective mass ratio.

As a starting point for a phenomenological theory of such layered superconductor structures it can be adopted the suitable modification of the Ginzburg-Landau theory given by Lawrence and Doniach[3]. The consequences of the layering is of relevant interest in different ways. In fact the existence and the nature of vortices in these superconductors appear to be strongly related to their layered structure and has been subject of great attention during these years.

Depending on whether $\sqrt{2}\xi/d >> 1$ or $\sqrt{2}\xi/d < 1$ (d being the interlayer distance) the character of a 3D or a 2D system will dominate. In the former case the superconductor can be considered as a homogeneous anysotropic system and vortices of Abrikosov type are expected. In the latter case it is commonly accepted a picture of such a layered superconductor as consisting of planes which are linked together by Josephson coupling and, accordingly, Josephson coreless vortices can also occur. YBCO can be regarded as 3D, while bismuth and thallium compounds should belong to the 2D case.

A great attention has been paid in these years to the pinning properties which ultimately determine the achievable value of the maximum current density[4] As we know in microscopically homogeneous samples, regardless of whether isotropic or anysotropic, flux pinning is due to extrinsic defects present in the structure. In high-T_c superconductors for an applied magnetic field H_o parallel to the layers an additional pinning mechanism has been proposed which is due to the layering itself[5-7]. Such an intrinsic pinning stems from the order parameter modulation, namely the assumption of an order parameter Δ_k for each kth layer[7].

The corresponding expression of the free energy for $T \sim T_c$ is given by:

$$F = vd \int dx^2 \sum_k \left[\frac{a}{d^2} \left| \Delta_k - \Delta_{k+1} \, exp\left(-2ie \int_{kd}^{(k+d)d} A_z dz \right) \right|^2 + b \left| \left(\frac{\partial}{\partial x} - 2ieA_x \right) \Delta_k \right|^2 - \tau \left| \Delta_k \right|^2 + \frac{C}{2} \left| \Delta_k \right|^4 \right] + \frac{1}{8\pi} \int \left[(rot\mathbf{A})^2 - 2 H_o \, rot\mathbf{A} \right] dr^3$$

with

$$\xi_{ab} = \left(\frac{b}{\tau}\right)^{1/2}, \qquad \xi_c = \left(\frac{a}{\tau}\right)^{1/2}, \qquad \lambda_{ab}^{-2} = 32\,\pi v e^2 |\Delta|^2 b,$$
$$\lambda_c^{-2} = 32\,\pi v e^2 |\Delta|^2 a, \qquad |\Delta|^2 = \frac{\tau}{C}$$

where **A** is a vector potential, a is a measure of the interlayer coupling, $\tau = 1-T/T_C$ and ν is the density of states at E_F. The behavior of the system strongly depends on the adimensional parameter $\gamma = a/\tau d^2$. For $\gamma >> 1$, in the zero order approximation, the result is closely the same as that of anisotropic superconductors whereas in the next order approximation we have:

$$F = \int_{-\infty}^{+\infty} dy \left[\frac{J_c \Phi_o d}{2\pi} \cos\left(\frac{2\pi U}{d}\right) - J_c \Phi_o U + \frac{A}{2}\left(\frac{\partial U}{\partial y}\right)^2 \right]$$

with $\qquad A = (\Phi_o{}^2/8\pi^2)\,(\lambda^2{}_c/\lambda^2{}_{ab})\ \ln(\lambda_c/\xi_{ab})$

J_c results to be exponentially small (see ref.7). In this frame (order parameters defined only on the layers and H_o parallel to the layers) the vortex core lies between the superconducting layers. It has been found[8] also the critical current value for a magnetic field **B** not alligned with the a,b planes. Such a result has been obtained by assuming strong pinning centers and using the London expression of the free energy for an anisotropic superconductor:

$$F = \frac{1}{8\pi} \int dr^3 \left[B^2 + rot(\mathbf{B} \wedge rot\,\mathbf{B}) \right]$$

where Λ is the angle dependent penetration depth matrix.

Of great importance is also the role of the precise match of the vortex lattice (commensurability) with the periodicity of the interlayer structure[9].

The pinning mechanisms in connection with the observed magnetic flux creep has been subject of intensive investigations[10,11]. Indeed the resulting current-voltage characteristics drastically depend upon the properties of the vortex system, in particular, whether it is liquid (V~I) or not[12]. Following Anderson's theory of flux creep[13] it has been widely recognized the occurrence of "flux bundles" jumps over the pinning sites at a rate $v=v_o\exp(-U/kT)$. The consequences of such thermal activation are several (decreasing of J_c, broadening of the resistivity vs. temperature curve etc.). In high-T_c superconductors this effect is more relevant due to the combination of low pinning energy and higher operation temperature. A large number of papers has been devoted to this specific topic both for the understanding of underlaying physics and the fundamental importance for applications[14].

Thermally activated flux creep at low temperature can turn to the quantum one[15]. A vortex, in this case, can tunnel through a pinning barrier. This is a further manifestation of the Macroscopic Quantum Tunneling in addition to that investigated in small Josephson junctions.

2.2 - A comment on the Pairing Mechanism

As for the charge carriers, experimental evidence has shown unambiguously that, due to some pairing process, an effective charge $e^*=2e$ is present in the superconducting state. The question of which pairing mechanism should be invoked has stimulated various answers which are based on exotic or more conventional models. Theoretical approaches include say phonons, magnons, anyons etc. Although rather arbitrarily, I shall assume as a main distinction, on one side, the Anderson theory[16] as representative of new avenues which, besides providing a consistent description of the normal state, gives a whole "new" explanation for this class of superconductors which appear quite reluctant to be classified as conventional. In the normal state description given by Anderson it is adopted the concept of "Luttinger liquid". A Fermi surface is identified which refers now to "spinons" which are chargeless 1/2 spin excitations. The charge is carried by the "holons", the counterpart excitations without spin. The scheme resulting from such a separation of charge and spin is certainly stimulating and some experimental evidence is claimed to support this theory.

On the other side, a variety of approaches can be collected which are variants of conventional theories rather than intrinsically different in nature. Indeed, several of the proposed schemes[17] are different from the usual BCS governing weak coupling low-Tc superconductors. However these models fall, in any case, within the BCS theory in its more general significance of a phase coherent condensate state, where pairing can be mediated by various types of excitations and the superconductor can be for instance in a non-zero angular momentum state, such as expected for heavy Fermions, rather than being traditional s-wave.

Recently[18], in connection with the thermodynamic stability of the Kondo-lattice and to the extended three band Hubbard model, it has been discussed the role that phase separation can play in the occurrence of superconductivity. Namely it follows from attractive interactions which, in some regions of the phase diagram can also give rise to pair formation. Such a correspondence between phase separation and superconductivity reflects the circumstance that both phenomena stem somehow from a common origin. Indeed near the phase separation region is, for a Fermi liquid, $dn/d\mu \propto m^*/m(1-A^s{}_o)$, $dn/d\mu$ being the compressibility and $A_o{}^s$ the Landau scattering amplitude. The large value of the compressibility and the relatively small mass enhancement in high-T_c upper oxides ($m^*/m \sim 2$-5 say) implies a negative and large value of $A_o{}^s$. This indicates the presence of relevant attractive forces in the particle hole channel which can in turn lead to phase separation but can also give a contribution to the particle-particle channel.

As far as the high T_c values of cuprate oxides and even more of the recent fullerene (C_{60}) compounds the attention has been addressed to the circumstance that for such materials the condition of a Fermi velocity much larger than the sound velocity is no longer satisfied. This implies the violation of the Migdal theorem which assumes the necessary condition $\omega_D/E_F \rightarrow 0$ in order to neglect vertex renormalization and higher order diagrams in the self consistent gap equation.

In Ref.19 it is shown how the T_c enhancement in the new classes of superconductors can follow from a generalization of the many body of superconductivity in the nonadiabatic limit (ω_D/E_F 0).

Obviously a parameter of great interest in the superconducting state is represented by the energy gap, Δ, since the parameter $\gamma=2\Delta/kT_c$ can give a great insight into the nature of the superconductors. Since the beginning of the history of high-T_c superconductors, a large variety of experiments has been concentrated on the measurements of Δ. These include photoemission, optical and Raman scattering spectroscopies as well as tunneling spectroscopy, using a variety of junction structures. As far as concerns the determination of the energy gap through superconductive tunneling spectroscopy, the experiments have shown, since the beginning, a number of ambiguities in the interpretation of the data[20]. In the case of bulk pellets this was partly connected to the "morphology" rather than to the actual gap-like structure, as appeared since the early point contact measurements. There has been a variety of experiments in the direction of realizing a clear tunnel structure and a higher degree of reproducibility[21]. The interpretation is however still open to a debate (see also lectures by A. M. Cucolo and by R. Gonnelli at this workshop and quoted references). In Fig. 1 a large variety of results concerning the determination of the parameter γ is reported (the data are taken from reference[22]).

3. High T_c Josephson Junctions

The development of reliable procedures for film fabrication such as the laser ablation, the increased capabilities in film growth and desired orientation, the realization of crystal lattice match with excellent epitaxial multilayer configurations, the increased control of grain boundaries and improved substrates quality and dimensions, are all factors which allowed, in the last years, the realization of an impressive and encouraging variety of weakly coupled superconductor structures. As stated in the previous section, however, the realization of Josephson tunnel junction of the conventional S-I-S type employing high-T_c superconductors remains a puzzling topic. As we know the unsuccess is mainly related to the problem of the small coherence length (Fig. 2).

Quite explicitly, Rowell[23], among a selection of "significant advances" of these years includes "the decision to stop trying to make Josephson tunnel junctions". I'm inclined to share this view, though some possibilities may deserve further investigations. The c-axis on plane junctions such as that discussed by Camerlingo et al. in this volume, (and quoted references) allow to deal with the more favorable (larger) coherence length at the junction interface. This configuration combined with a proper barrier growth may offer further perspectives. As far as non-cuprate high-T_c superconductors, symmetric Josephson tunnel junctions of the type $Ba_{1-x}K_xBiO_3/Ba_{1-x}K_xBiO_3$[24] have shown good I-V characteristics with reasonable values of the I_cR_n product (critical current times normal tunneling resistance).

As pointed out at the beginning of this Section high-Tc junctions of good characteristics have been obtained. These are confined to structures exhibiting an S-N-S behaviour. Indeed, different ways to circumvent the difficulty of making "archetype" sandwich tunnel junctions have been explored. A variety of configurations is summarized in Fig. 3 where also some sort of figure of merit is given. It is worthwile recalling that some of these junctions can be employed in laboratory applications and that, for instance, bicrystal type structures have

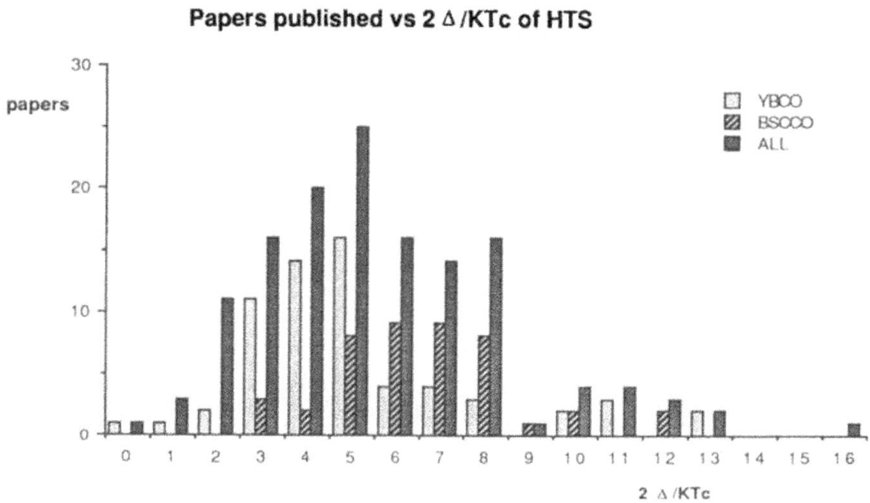

Fig. 1
Data distribution for the determination of the parameter $\gamma = 2\Delta/k\ T_C$ (the data are taken from ref. 22)

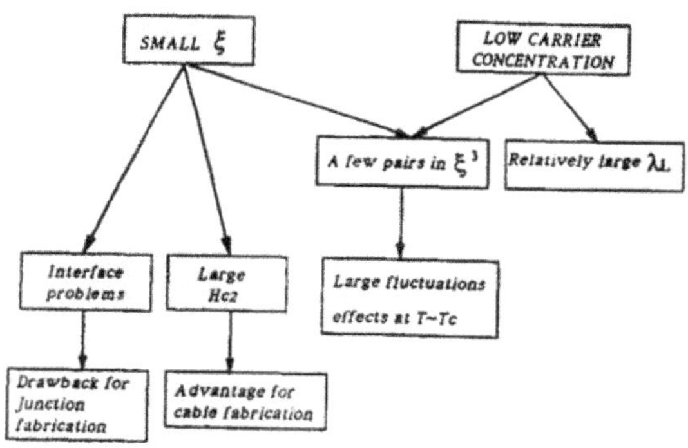

Fig. 2
Qualitative sketch of some of the physical properties encountered in high T_C superconductors and related implications

already left the laboratory being commercialized by CONDUCTUS for didactical demonstration in undergraduate laboratories.

To guarantee good performances in a variety of applications, the junction parameters have to satisfy some essential prescriptions. A sufficiently large critical current density J_c is one of the desirable properties. The value of the product $I_c R_n$ is quite an important parameter. It reaches values as large as tens of milivolts for low-T_c junctions, whereas for high-T_c samples rarely exceeds a few millivolts.

The values of R_n have also to be large enough to allow best matching with other circuit components and possible ibrydization with conventional electronics. The lack of reliable Josephson high-T_c tunnel junctions rules out the possibility of realizing those logic circuits which require the hysteretic character (double valued) of the current-voltage characteristics. On the other hand, should such a device exist, the relative large values of the energy gap Δ would allow a faster switching time ($\Delta t \Delta E \sim h$) at the expenses however of the dramatic drawback of a higher power consumption.

Looking at the scheme of Fig. 3 let us try to make some of the most essential comments on this wide zoology of junction configurations. Point contacts, in spite of their rather naive structure, can supply some information as tunneling spectroscopy, providing a cautious and careful "reading" of the results (mainly dI/dV vs. V measurements). Indeed, though far from an actual device, point contacts, by a proper contact pressure, could explore a more "representative" region of a bulk sample whereas sandwich junction configurations which reproduce the archetype Josephson junction, can, on the other hand, suffer of interface effects due to degradation (oxygen deficiency or other defects) of the surface layer whose "uniformity" has to compete with the prohibitive short coherence length of high-T_c superconductors. Step-edge junctions somehow replicate the low-T_c counterparts giving also interesting results. Moreover Josephson effect has been observed in coplanar devices employing a metal (e.g. Au) bridging a narrow gap between "banks". Also strips, properly damaged in a small region, realized a superconducting weak-link. Other very interesting structures with edge and/or coplanar coupling have recently been proposed[25].

As far as sandwich junctions, great attention deserve full epitaxial SNS junctions. In these structures two epitaxial films electrodes are devided by a barrier layer which can correctly match the lattice. In this context YBCO/YBCO with PrBCO barrier has been successfully realized[22]. Similarly, Bismuth junctions consisting of 2122 Bi phase,

			Junction type	Reproducibility	Difficulty of fabrication	Fully high T_c	Physical properties investigation	Device applications
BULK			Point Contact	POOR	LOW	YES	YES (a)	NO
BULK			Break	POOR	LOW	YES	YES	NO
BULK	PLANAR		Bulk	MEDIUM	MEDIUM	NO	NO (b)	MAYBE
FILM	PLANAR		Sandwitch	GOOD	HIGH	YES	NO (f)	YES (g)
FILM	PLANAR	INTERGRAIN	Natural	POOR	MEDIUM (c)	YES	NO (d)	MAYBE (e)
FILM	PLANAR	INTERGRAIN	Bi-epitaxial	GOOD	MEDIUM	YES	YES	YES
FILM	PLANAR	INTERGRAIN	Bi-crystal	GOOD	MEDIUM	YES	YES	MAYBE
FILM	PLANAR		Step Edge	GOOD	MEDIUM	YES	YES	MAYBE

Remarks

(a) In paricular STM studies on single crystals
(b) Surface degradation
(c) Worse with good filma
(d) No control on grain interface
(e) Only SQUIDs
(f) Surface degradation. Possible study of proximity effect and anisotropy
(g) Require further improvement

Fig 3
High-T_C junction zoology

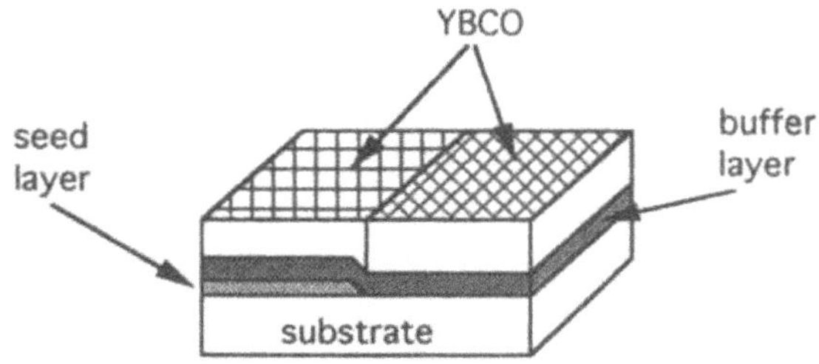

Fig. 4
By-epitaxial junction structure.

superconductive electrodes, separated by 2201 Bi phase barrier layer gave also excellent results[26]. As stated above great interest deserves the investigation of junctions with the c-axis on the plane which allows a larger coherence length normal to the barrier[27].

Furthermore, a stimulating variety of junctions which are somehow more tightly connected to the intrinsic nature of high-T_c superconductors is that based on intergrain coupling. Natural occurring grains are by definition out of control in their realization, however interesting results and projections of SQUID configurations have been proposed.

Intergrain deliberately realized have been accomplished with success by growing films on substrates having tilted crystal boundaries[28]. Very recently Josephson junctions realized by YBCO films following a by-epitaxial process have proved to be of great interest as a single junction and in a SQUID configuration system which is leaving the laboratory toward market applications[29]. In this case, as shown schematically in Fig. 4, a "seed layer" leads to two different epitaxial relations between the "base layer" and the "buffer layer'" leading thereby to the by-epitaxial growth. For advanced technology of grain boundary junctions and multilayer type configurations extensive accounts are given at this Workshop by R. Gross and G. L. Garritsmaa respectively.

4. Conclusions

Five years after the discovery of high-T_c superconductivity there has been a great advance both in the understanding of many aspects of the basic phenomenoogy as well as in the development of technology which legitimate more optimism for the applications. However some of the important problems concerning a complete description of the superconducting mechanism are yet to be solved and further investigation is also required to circumvent some of the peculiar material constraints shown by this class of superconductors.

Short coherence length, marked anysotropy, possible large energy gap etc., all contribute to the puzzling character of this research. This certainly affects also the possibility of a reliable tunneling spectroscopy in these superconductors. As far as it concerns the objective of realizing good junctions, the actual possibilities are fixed by the quality of the interfaces which, in turn, is set by the severely short coherence length.

Several alternative solutions have been proposed which, far from being conclusive, allow the possibility of realizing reliable devices in spite of many intriguing material issues.

Partially supported by the Italian National Project *"Superconductive and Cryogenic Technologies"* of the CNR.

5. References

1. J.G. Bednorz and K. A. Müller, Z. Phys. B 64, 189 (1986); K. A. Müller, M2S-HTSC III, Kanazawa Japan, July 22-26 (1991) J.M. Rowell, Supercond. Sci. Techn. 4, 692 (1991)
2. B. Battlogg, Physics Today 44 no.6 (1991); R. Simon ibid pag. 64
3. W. E. Lawrence and S. Doniach - in Proc. 12th Intern. Conf. on Low Temperature Physics, E. Kanda Ed. (Academic Press of Japan, Kyoto (1971)
4. e. g. see review paper by A. I. Larkin and Yu. N. Ovchinnikov, in Nonequilibrium Superconductivity (A.I. Larkin and D.N. Langerberg Eds.) Elsevier Science Publ. B.V. 1986
5. M. Tachiki and S. Takahashi, Solid State Comm. 70, 291 (1989)
6. B. I. Ivlev and N.B. Kopnin, J. Low Temp. Phys. 77, 419 (1989)
7. A. Barone, A.I. Larkin and Yu. N. Ovchinnikov, J. of Superconductivity 3, 155 (1990)
8. Yu. N. Ovchinnikov and B. I. Ivlev, Phys. Rev. B 43, 8024 (1991)
9. See also B;I. Ivlev and N.B. Kopnin, Phys. Rev. Lett. 64, 1828 (1990)
10. K. A. Müller, M. Takashige and J. G. Bednorz, Phys. Rev. Lett. 58 1143 (1987)
11. Y. Yeshrun and A. P. Malozemoff, Phys. Rev.Lett. 60, 2202 (1988)
12. P. H. Kes, J. Aarts, J. van den Berg, C. J. van der Beek and J. A. Mydosh, Supercond. Sci. Technol. 1, 242 (1989); and R. H. Koch, V. Foglietti, W. J. Gallegher, G. Koren, A. Gupta and M.P.A. Fisher, Phys. Rev. Lett. 63, 1511 (1989)
13. P. W. Anderson, Phys. Rev. Lett. 9, 309 (1962)
14. See the reviews M. Tinkham in Proc. 19th Inter. Conf. on Low Temp. Phys., Brighton Aug. 1990; E. H. Brandt, Int. J. of Modern Phys. B5, 751 (1991); B. I. Ivlev - Addison Wesley Publ. Co. 1991 (in press) and references reported therein.
15. A. I. Larkin, Yu. N. Ovchinnikov and A. Schmid, Physica B152, (1988); A. V. Mitin, Sov. Phys. JETP 66, 335 (1987); A. Mota, A. Pollini, G. Juri, P. Visani and B. Hilti, Physica A 168 298 (1990); G. Blatter, V. B. Geshkenbein and V. M. Vinokur, Phys. Rev. Lett. 66, 3297 (1991)
16. P. W. Anderson, Science 235, 1196 (1987)
17. See for instance, M. L. Cohen, Mat. Res. Soc. 169, 13 (1990)
18. C. Di Castro and M. Grilli to be presented at the EPS Meeting 1992 (and referenes cited therein)
19. L. Pietronero and S. Strässler, Europhys. Lett. (to be published).
20. e.g. A. Barone Physica C 153-155, 1712 (1988)
21. For an extensive account of the tunneling spectroscopy on high-Tc superconductors, see J. R. Kirtley, Int. J. of Modern Physics B4, 201 (1990)

22. T. Hasegawa, H.Ikuta and k.Kitazawa, "Physical Properties of High Temperature Superconductors III" (D. M. Ginsberg Editor) World Scientific Publishing (to be published)
23. J. M. Rowell, Supercond. Sci.Techn. 4, 692 (1991)
24. A. N. Pargellis et al. , Appl. Phys. Lett. 58, 95 (1991)
25. For a review see for instance, T. Van Duzer in Proc. 3rd FED Work shop on High-Temperature Superconducting Electron Devices, Kumamoto, Kyushu (1991); and M. A.M. Gijs in High Tc Superconductors Materials Science Forum, (1992 (to be published)
26. K. Mizuno, H.Higashino, K. Setsune and K. Wasa - Proc. 2nd ISS'89 Tsukuba 1989
27. H. Akoh, C. Camerlingo and S. Takada, Appl. Phys. Lett. 56, 14 (1990)
J. B. Barner, C. T. Rogers, A. Inam, R. Ramesh and S. Bersey, Appl. Phys. Lett. 59, 742 (1991)
28. For an extensive account see R. Gross and P. Chaudari in "Principles and Applications of Superconducting Quantum Interference Devices (A. Barone Editor) World Scientific (1992)
29. K. Char, M. Colclough, L. Lee and G. Zaharchuk CONDUCTUS, Sunnyvale.

Part 1

Preparation and characterization of high-T_c superconductors

THE PRESENT STATUS AND CHALLENGES OF HIGH TEMPERATURE SUPERCONDUCTIVITY

C. W. Chu
Texas Center for Superconductivity and the
Department of Physics at the
University of Houston
Houston, Texas 77204-5932
U.S.A.

1. ABSTRACT

The present status (prior to March 1991) of high temperature superconductivity research and development is reviewed. The challenges for scientists and engineers in the field to face are briefly summarized and discussed. This talk represents the view of only one of the scientists in the trenches and I have chosen to examine the problem more from a macroscopic than microscopic point of view. Inevitably, therefore, much worthy work in the field cannot be included in the discussion. To minimize the impact of such incompleteness, I have chosen to mention only a few references in the text.

2. INTRODUCTION

Superconductors have interested us for a long time because of their scientific challenges and technological promises. A superconductor displays three main features: (1) the zero resistivity that enables us to make more efficient electrical and electronic devices; (2) the magnetic flux expulsion, the so-called Meissner effect, which allows us to build more efficient and powerful magnetic bearings and more efficient magnetic shields; and (3) the so-called macroscopic quantum phenomenon, as a result of which we can use this material to make ultra-sensitive sensors and ultrafast switches. To achieve the superconducting state, we have to overcome three barriers: the critical temperature (T_c), the critical magnetic field (H_{c2}), and the critical current density (J_c).

What is so special for high temperature superconductors (HTS's) is that, in addition to all of the benefits just mentioned, we can achieve the superconducting state with a simpler coolant and can generate a stronger magnetic field using HTS materials, not to mention the scientific challenges they pose to physicists. Until the discovery of HTS's, prevailing theories predicted that superconductivity was not possible above the 30's K. Furthermore, the larger energy gap associated with HTS materials makes it possible to fabricate ultra-sensitive sensors with a wider electromagnetic wave window. This is why scientists, engineers, and the general public were so excited about high temperature superconductivity (HTS) when the liquid nitrogen barrier of 77 K was broken in early 1987. This is evident by, for example, the *Wall Street Journal* article (March 13, 1987), "Gains in Superconductor Signal New Era of Electrical Devices," followed by the cover story in *USA Today* (April 22, 1987), "Research Electrifies Industry: Breakthrough Could Create $36 Billion Market for Materials." The Japanese even envisioned an age of superconductivity by the year 2030, using superconductors for electric cars, magnetic energy storage, power transmission, ship propulsion, microelectronics, jet catapults, *etc.* As more and more studies were made, people started to realize that there existed a wide range of material hurdles to cross to make the application of HTS a reality. The *Wall Street Journal* (May 15, 1987) article, "Enthusiasm Over Superconductor is Tempered by Daunting Problems," appeared because, at that time, HTS materials could not carry large J_c. The most damning message was printed in *Science* (June 1989), "Superconductivity: Is the Party Over?", based on the false alarm that HTS's might be inherently unable to carry high J_c, making applications of HTS's impractical. However, the optimism and perseverance that characterized people in the field finally paid off. The *New York Times* (September 16, 1990) carried the headline, "Superconductors Enter the Marketplace Modestly." The whole situation may be illustrated as a HTS roller coaster ride (Fig. 1). The first peak occurred in the spring of 1987 when liquid nitrogen superconductivity was discovered and the first deep trough in the summer of 1989 when the *Science* article appeared. Now we are out of the valley and on our way up. In the future we certainly expect more ups and downs in the field of HTS. However, one can be sure that the slope of the baseline is positive, representing steady progress. For the last four years, the progress has been impressive and at an unprecedented rate. This is equally true in the scientific front. Experimental data start to converge between experiments and between laboratories

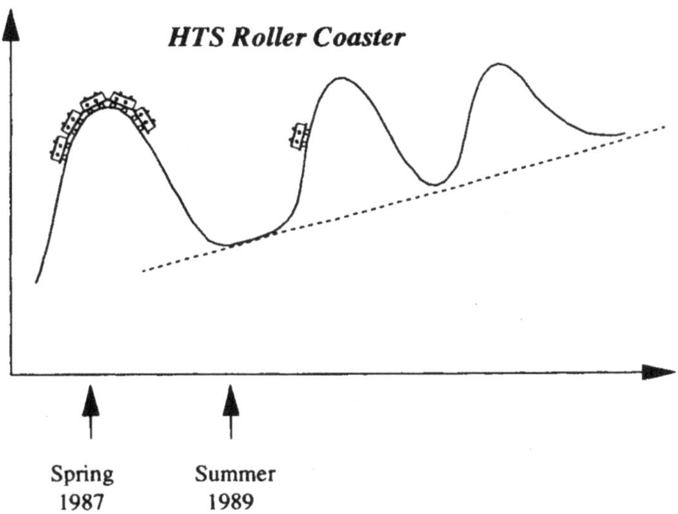

Figure 1. HTS Roller Coaster

following the successful synthesis of quality samples and the successful differentiation of intrinsic properties from parasitic characteristics of the HTS compounds. Now I am going to describe briefly the present status and future challenges in the study of HTS's.

3. CURRENT STATUS

3.1. Materials

- More than 75 HTS's (cuprates, bismuthates, and fullerites) with $T_c > 21$ K
- Quality single crystals of some cuprate and bismuthate HTS's
- Maximum $T_c \sim 125$ K
- Maximum $H_{c2} > 200$ T
- J_c at 77 K already in the application range and better than Nb_3Sn (at 4 K) at high field
- R_s at 77 K reaching that of superconducting Nb and orders of magnitude better than Cu

---

We have about 75 non-intermetallic compounds with a T_c higher than 21 K, the liquid hydrogen boiling point. They belong to the three general groups of cuprates, bismuthates, and fullerites. The confirmed record high T_c is 125 K (high above the liquid nitrogen boiling point at 77 K) in the cuprate $Tl_2Ba_2Ca_2Cu_3O_{10}$; although 127 K has been reported with proper heat-treatment. The highest T_c for bismuthates is ~ 30 K and for the fullerite $(Rb,Tl)_3C_{60}$, ~ 42 K. The recently announced $T_c \sim 57$ K in the iodine-doped C_{60} has been withdrawn due to experimental error.

A record high $H_{c2} > 200$ T is set by HTS's. In other words, in principle a magnet can be made of these materials when they are perfected to generate a field above 200 T near 0°K. A critical factor for most applications and a source of many of the doomsday predictions for HTS's, J_c at 77 K has been moving up impressively from 10 A/cm^2 in 1987 when the compound was discovered to ~ 10^5 A/cm^2 in bulk $YBa_2Cu_3O_7$, today (Fig. 2). The requirements of J_c at different fields (H) for various

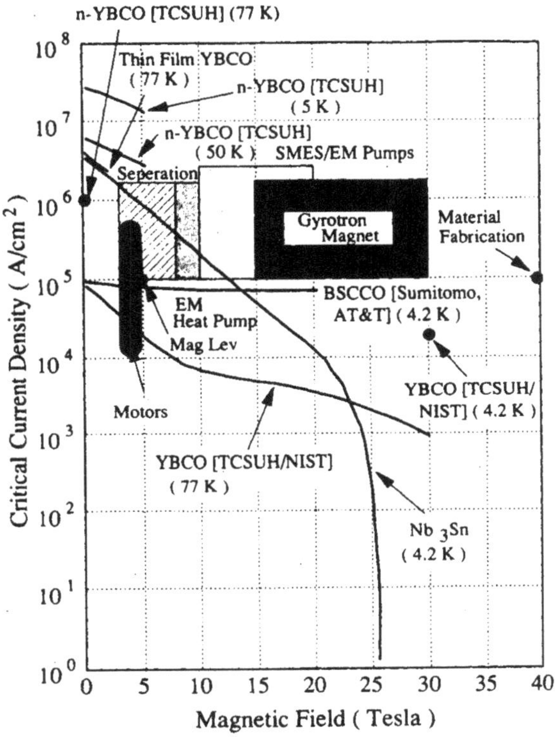

Figure 2. Critical current density (A/cm^2) *vs* magnetic field (Tesla).

applications are also given in the same figure. We have shown that the J_c of $YBa_2Cu_3O_7$ at 77 K under the influence of its own field has reached 6×10^6 A/cm^2 for thin films and ~ 8×10^4 A/cm^2 and ~ 10^6 A/cm^2 for bulk materials before and after fast-neutron irradiation, respectively. As expected, J_c decreases with H. However, it grows and becomes less H-dependent at lower temperatures. At very high H, HTS's at 77 K clearly out-perform Nb_3Sn at 4.2 K which has the highest J_c among the low temperature superconductors (LTS's), as demonstrated by Salama *et al.* Recently, $Bi_2Sr_2CaCu_2O_8$-ribbons have been shown by Sumitomo and AT&T Bell to exhibit a J_c at 4.2 K exceeding that of Nb_3Sn at the same temperature at fields above 13 T. The surface impedance (R_S) in the microwave region of HTS's in both thin film and bulk form at 77 K has come close to that of superconducting Nb, orders of magnitude better than that of Cu (Fig. 3). This clearly demonstrates the important role HTS's will play in future microwave devices and systems.

SURFACE RESISTANCE AS A FUNCTION OF FREQUENCY FOR YBCO THIN FILMS AT 77 K

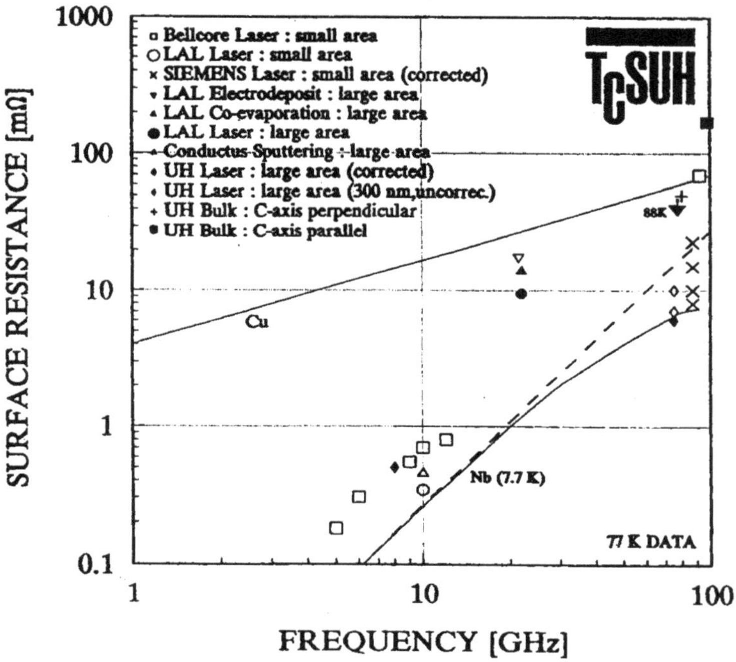

Figure 3. Surface resistance (mΩ) *vs* frequency (GHz).

3.2. Science

- Insulator → Metal (HTS) → Normal Metal transition sequence
- Unusual normal state properties -- possibly not a simple Fermi liquid
- Anomalies at or slightly above T_C
- Complicated magnetic phase diagram
- Complex flux pinning
- Pairing
- Energy gap
- Short coherence-length
- Low carrier concentration (n); large penetration depth
- T_C-n correlation
- BCS (bismuthates and fullerites) or non-BCS (cuprates)
- Fermi surface
- Charge transfer
- Complex spin dynamics and excitations

Because of the great improvement made in the quality of HTS samples, experimental results from different laboratories finally start to converge. While the global picture of HTS science are yet to become transparent, many of the observations crucial to our understanding of HTS have been confirmed. For instance, it is known that each HTS has its respective parent insulator. Through hole- or electron-doping,

HTS's become metals which become superconducting on cooling. Further doping destroys superconductivity but leaves the metallic character of the material intact. Although, at times, structural anisotropy was considered to be essential for the occurrence of HTS based on the highly anisotropic, layered-structure of the cuprates, superconductivity up to 30 K and in the 40's K has been observed in the isotropic cubic-structured bismuthates and fullerites, respectively.

Unusual normal-state properties (such as the linear-temperature-dependent resistivity, the temperature-dependent Hall constant, the drastic variation of carrier concentration with doping, *etc.*) have been detected, strongly suggesting that HTS's cannot be easily described in terms of the simple Fermi liquid picture applicable to ordinary metals. On the other hand, T^2-dependent resistivity, characteristic of electron-phonon scattering in a metal, has also been detected in the electron-doped and some hole-doped HTS's with a $T_c \leq 30$ K. Anomalies in the acoustic, structural, channeling, Raman, thermal expansion, positron-lifetime, and Doppler-parameter-broadening measurements have been observed at or slightly above T_c. Some of these anomalies indicate a stiffening of the lattice and some suggest a change in the electron-energy spectrum. Yet others, such as the thermal expansion anomaly can occur at the same temperature even when the sample is made non-superconducting by hole-doping.

LTS's are grouped as either Type I or Type II superconductors. There exists only one phase-boundary $H_c(T)$ in the H-T diagram separating the superconducting Meissner state from the normal state for Type I superconductors. However, for Type II superconductors, there are two phase boundaries, *i.e.* $H_{c1}(T)$ separating the Meissner state at low T from the vortex state at higher T's, and $H_{c2}(T)$ partitioning the vortex state from the normal state. A superconductor is expected to carry an electric current without loss below $H_{c2}(T)$. However, an irreversibility line $H_i(T)$ was found in the HTS's by the IBM group based on magnetic relaxation, ac magnetic susceptibility, and resistivity measurements (Fig. 4). The region between $H_i(T)$ and $H_{c2}(T)$ is dissipative although superconducting. Recently, yet another boundary, shown as $H_n(T)$ below $H_i(T)$ (Fig. 4), was found by us in detailed studies of the magnetic relaxation in various fields and temperatures. In contrast to the original belief that such a complex magnetic phase diagram was uniquely associated with the

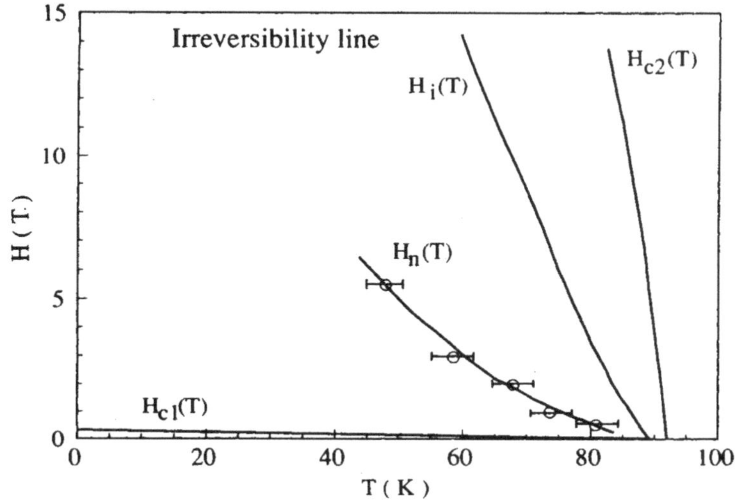

Figure 4. The magnetic phase diagram: $H_{c1}(T)$, $H_{c2}(T)$, $H_i(T)$ and $H_n(T)$ (schematic).

weak pinning of HTS's, both $H_n(T)$ and $H_i(T)$ have also been observed recently in LTS's by us and others.

It is known that motion of fluxoids in a HTS in the presence of an electrical current results in a loss of energy. Defects of a size comparable to the coherence length are needed to pin the fluxoids to avoid such energy loss. The conventional metallurgical technique to introduce fine dispersion of a non-superconducting second phase to the LTS's to act as pinning centers is no longer applicable to HTS's, because the dispersion so introduced is usually too large to be effective. Effective pinning centers of 20 to 60 Å in HTS's have been successfully introduced by fast-particle or heavy-ion irradiation, leading to a large increase in J_c. Recently, such a J_c-enhancement was also achieved by W. K. Chu *et al.* by irradiating a Li- or B-doped HTS with thermal neutrons as shown in Fig. 5. The thermal neutron is captured by the Li- or B-nucleus which then undergoes a nuclear reaction, generating fast particles to induce defects in a HTS. Thermal-neutron irradiation is much better than fast-

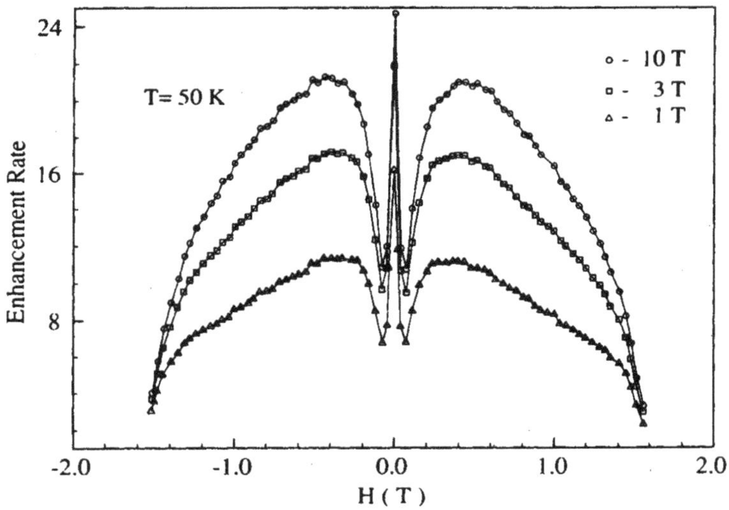

Figure 5. J_C-enhancement induced by thermal neutron irradiation of an HTS slightly doped with Li or B.

neutron irradiation in terms of the availability of neutron sources and thus in economic terms. *in situ* chemical decomposition of $YBa_2Cu_4O_8$ to $YBa_2Cu_3O_7$ and fine non-superconducting dispersions of atomic scale to enhance J_C were also demonstrated by Jin *et al.*

Charge carriers were shown to form pairs below T_C by microwave and tunneling experiments shortly after the discovery of HTS's. An energy gap has been detected, but it is not clear if it follows the universal gap equation predicted by the BCS theory (except the bismuthates). HTS's are found to possess a short coherence length (ξ), low carrier concentration (n), and a large penetration depth. The results of optical and muon-relaxation experiments by Tanaka *et al.*, and Uemura *et al.* seem to suggest a universal linear T_C-(n/m*) relationship with m* being the effective mass, if one neglects the deviation from such linear behavior by the different HTS families as shown in Fig. 6. Preliminary data from tunneling and other experiments suggests that bismuthates are BCS-superconductors but the situation of cuprates is not yet clear.

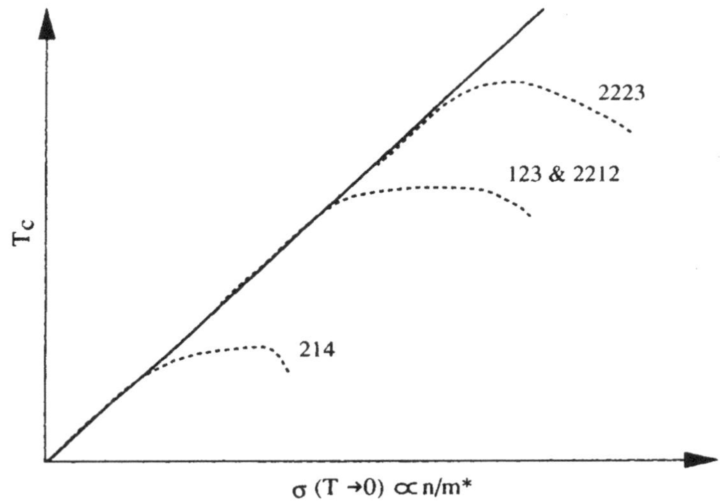

Figure 6. T_c-n (schematic) [Unemura's data]

The once-heated debate on the presence or absence of a Fermi surface was put to rest recently by careful angular correlative positron and de Haas-van Alphen experiments, by West *et al.* and Müller *et al.*, respectively.

Earlier, we mentioned that HTS's can be induced from their respective parent insulators by chemical doping. An accompanying change in the chemical bond length is therefore expected through doping. Indeed, this was first observed by Magliori *et al.* in $YBa_2Cu_3O_{7-\delta}$ near the metal-insulator transition. We later found that such a sudden change is universal for all HTS's and occurs when the boundary is crossed by chemical doping (Fig. 7). The positron lifetime (τ) undergoes a drastic change when the properly-doped HTS is cooled to below its T_c as shown by us in Fig. 8. Equally true is the fact that t varies with pressure. All these suggest that a charge transfer in HTS's can be induced by chemical doping, cooling, or pressure. The results are consistent with the picture that electrons are transferred from the CuO_2-layer to their surrounding for the hole-doped HTS's and to the CuO_2-layers for the electron-doped HTS's.

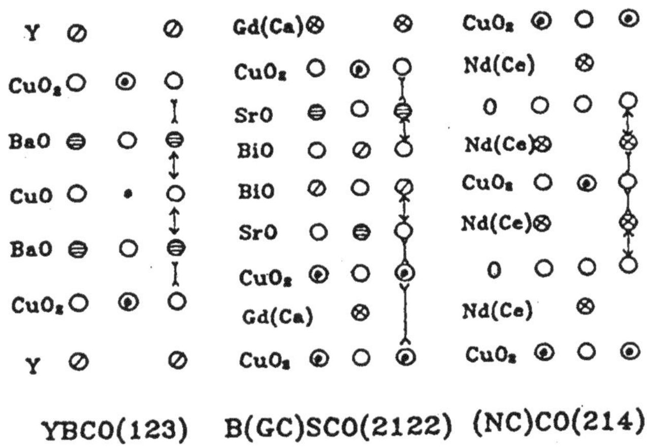

Figure 7. Drastic variations of inter-layer distances as the compounds are made metallic and thus superconducting by chemical doping: →← stands for contraction and ←→ stands for expansion.

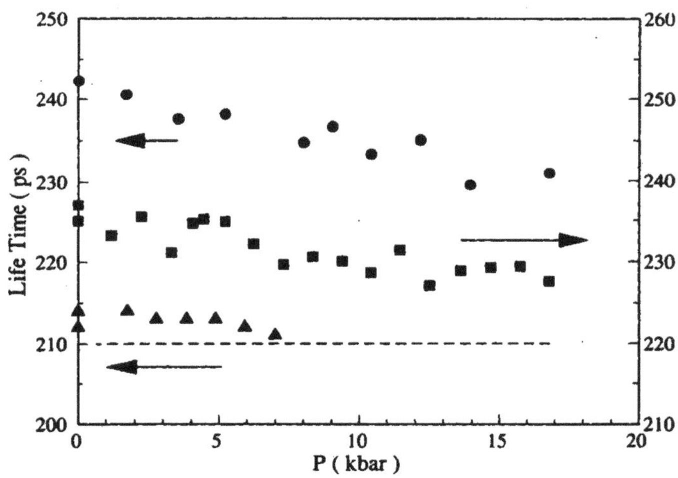

Figure 8. t *vs* T for 214, 123 and Tl2223

3.3. Applications

3.3.1. Small Current Applications

- Thin film processing: reproducibility, growth rate, orientation, grain growth, film size, superlattice, overlayer
- Patterning and lithography
- SQUID sensors, IR bolometers, rf resonators, rf filters, rf delay lines, *etc.*
- Monolithic integrated SQUID magnetometers
- Flux flow transistors
- Electric field effect transistors

Immediately after the discovery of HTS above 77 K in 1987, the IBM group succeeded in sputtering HTS thin films with a $J_c \sim 10^5$ A/cm^2 at 77 K, demonstrating the technological potential of HTS's in microelectronics and sensor applications. Subsequent rapid progress in thin film processing by sputtering as well as laser ablation techniques led to the attainment of high quality films with a J_c reaching the theoretical limit of $\sim 10^7$ A/cm^2 at 77 K, and with atomic surface smoothness enabling the synthesis of superlattices of HTS's by the Geneva group, for instance. Grain growth was controlled to such an extent that superconducting bridge-junctions have been fabricated by various techniques. Uniform high quality thin films of large area ~ 1.5"x1/5" have been prepared by Wolfe *et al.* Both wet and dry lithography and patterning techniques were developed for HTS thin films. Microstrips of a width ~ 0.25 μm with a $J_c \sim 10^6$ A/cm^2 were obtained by Wolfe *et al.* in $YBa_2Cu_3O_7$. Many nonhysteretic Josephson junctions, IR bolometers, and rf resonators, filters, and delay lines, have been fabricated and their functions have been demonstrated. The recent success in depositing over-layers with different electrical properties and the precise control of the orientation of the thin films enable Conductus to fabricated the monolithic integrated SQUID magnetometer with a structure consisting of up to 15 layers. This opens up many possibilities for the preparation of many other sophisticated HTS microelectronic devices.

3.3.2. Large Current Applications

- Bulk processing: grain alignment, size, shape, mechanical strength
- Passive magnetic bearings (~ 15 psi)
- Hybrid magnetic bearings (~ 60 psi)
- Magnetic replicas
- Demonstration motors
- rf resonators, antennas, *etc.*
- Disks: EM projectiles, dent pullers, rivet guns
- Rods: large current leads
- BiSrCaCuO ribbons
- HTS wire magnet at 4.2 K

Since the cuprate HTS's are highly anisotropic, large electrical current flows without loss only along the a-b plane or in the CuO_2-layers. The direction of the grain growth is therefore of utmost importance for achieving large J_c. We found that it can be controlled by adjusting the temperature gradients along different directions parallel or perpendicular to the a-b plane during the melt-texturing process. Many of the superior superconducting properties of HTS's mentioned earlier were originally achieved only in selected small pieces of large samples, which are basically of no significant consequence to applications. We have therefore developed a continuous processing technique for the preparation of monolithic bulk HTS materials with large J_C. The technique involves the transferring of a bulk HTS through a tubular furnace with a prescribed special temperature profile at a predetermined rate. Quality samples of various shapes and dimensions so obtained are shown in Figs. 9a and 9b. One of

Figure 9a. Quality bulk HTS's of various shapes.

Figure 9b. The SEM data for a long sample. The dark end was caused by the heat generated at the contact after the lead was burnt.

these rod ~ 4 cm long showed no loss when our maximum current of 600 A was passed through it at 77 K. If one wants to take advantage of the record H_{c2} of HTS's for ultrahigh field applications, the mechanical strength of the material must also be improved. The relative fracture toughness of sintered $YBa_2Cu_3O_7$ has been obtained by White *et al.* to approach that of alumina after doping with Ag, but the J_c remains low.

Magnetic bearings have been considered to be the simplest application of bulk HTS's in a non-wire form. This was demonstrated immediately after the discovery of HTS by floating a magnet over a HTS chilled to 77 K. The maximum magnetic pressure so obtained with the best HTS cannot exceed 15 psi. For some applications, such as pump bearings for pumping liquid hydrogen for space rockets, a much greater levitation pressure is required. In principle, two opposing magnets can generate a very large repulsive force when the two magnets are brought close to each other. However, instability, making the two opposing poles fly apart sideways, sets in as the repulsion grows. By taking advantage of the flux-trapping characteristics of HTS's, the instability has been removed by inserting a slab of HTS material between the poles as shown in Fig. 10. As a result, a magnetic pressure greater than 60 psi has been obtained by W. K. Chu *et al.*

Recognizing the difficulties inherent in making HTS wires, a parallel approach has been adopted to make use of HTS's in their present synthesizable forms. Magnetic replicas in plate-form have been made by Weinstein *et al.* to trap a field up to 1.5 T at 77 K, which is greater than that generated by the best permanent magnet presently available. It is believed that magnetic replicas of simple geometry can be made to trap a stronger magnetic field with a complex field-profile with high fidelity. A demonstration motor using a HTS magnetic replica has been built as shown in Fig. 11 and operated at 77 K, although at a low power level. With the availability of high-quality bulk HTS materials, concepts of HTS disks (for electromagnetic projectiles, dent pullers, and rivet guns) and rods (for low-loss large-current leads between 4.2 and 77 K) are being actively pursued by us. Very low surface impedance at 77 K in the microwave region has been achieved in melt-textured HTS materials by Salama *et al.* HTS rf resonators and antennas have thus been demonstrated.

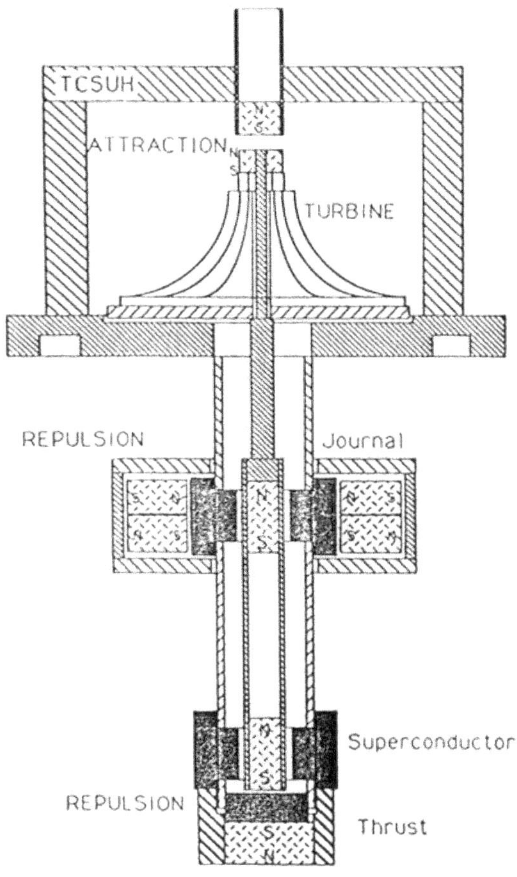

Figure 10. The basic principle for a hybrid magnetic bearing with a high stability and large levitation pressure.

Because of the easy texture-growth of the $Bi_2Sr_2CaCu_2O_8$ near the Ag-surface, Ag-sheathed ribbons of this compound has been successfully fabricated with a large J_c by Sumitomo, AT&T Bell, and others, including us. Their J_c at 4.2 K has reached ~ 10^5 A/cm^2 and is rather magnetic-field insensitive. In other words, above 14 T, Bi-based HTS ribbons out-perform Nb_3Sn, the best LTS we have at this time.

Hybrid-Electric Motor using High Tc SC Magnetic Replica
IBPD/TCSUH, 9/90

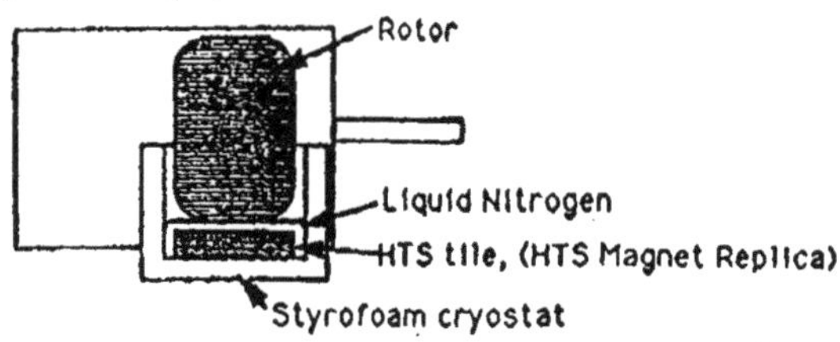

Emerson Electric Motors with HTS Magnet Replica
IBPD/TCSUH & Emerson Electric, 3/91

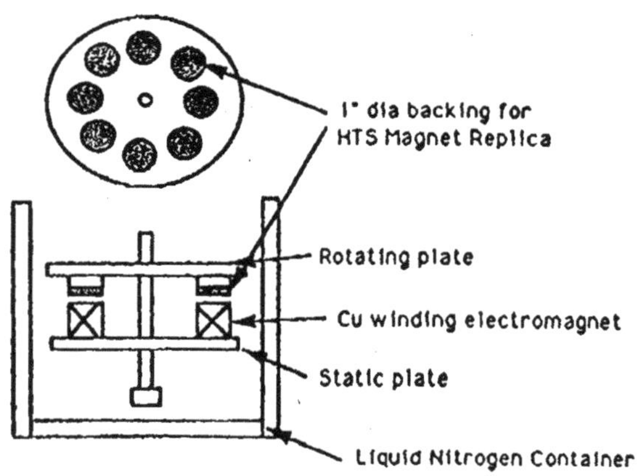

Figure 11. Demonstration motors using HTS magnet replicas.

Using this kind of ribbon, a HTS magnet was constructed to generate a field of ~ 1 T at 4.2 K.

In short, advances in the processing of quality HTS materials and in the demonstration of basic concepts for HTS applications have reached the level that design engineers can now start to consider HTS materials for functional small and large current devices.

4. CHALLENGES

In spite of the impressive progress made in the last four years in all areas of HTS research, great challenges still lie ahead of us to unravel the mystery of HTS and to make the dream of HTS applications come true.

4.1. Science

- HTS mechanism: phononic, electronic, and magnetic
- BCS or non-BCS
- Pairing state
- HTS *vs* anomalies
- HTS *vs* charge transfer
- Cuprates *vs* bismuthates *vs* fullerites
- Antiferromagnet → HTS → normal metal
- Dimensionality
- Flux dynamics
- Universal rule
- $T_C > 125$ K

One of the most challenging HTS-science problems concerns the mechanism responsible for superconductivity in such an unusual class of materials and at such unusually high temperatures. Many mechanisms have been proposed. They include excitations, such as phonons, excitons, polarons, bipolarons, demons, spinons, holons, anyons, even frustrati-ons. It appears at this moment that anything with an "-on" at the end of it is operative. More experiments covering a wide range of HTS's are needed to determine which of the above is the dominant factor. Until now, most of the reliable data was obtained only in the Y123 compound system because high quality samples can be made from it. Therefore, one should try to improve the sample quality of other existing HTS systems from which intrinsic properties can be determined to broaden the data base.

Assuming that reliable data are finally obtained on all existing HTS's, the question remains whether the existing HTS's are diverse enough to form a material base sufficiently broad to pose enough constraints on the comprehensive theory for HTS. To shed some light on this question, we decided to focus on the types of Cu-O layers in the cuprate HTS's. Clearly, there are only three types: CuO_6 octahedrons, CuO_5 pyramids, and CuO_4 squares. The first type is the basic building block of the 214-T superconductor; the second type is the basic building block of the 123 superconductor; and the third type is the basic building block of 214-T' superconductor. By examining the structures of all existing HTS's, we have suggested earlier that the known cuprate HTS's may be no more than mere modifications of 214-T, 214-T', 123 or a combination of the last two. To look for further evidence, we have examined Y123, Y247, and Y124, which all possess two CuO_5-pyramid layers per unit formula but with different T_c's at ambient pressure due to the different charge transfers between the CuO_5-layers and the other intervening layers. If the suggestion above is correct, all the different T_c's should merge into one when the charge transfer can be properly tuned, *e.g.* by the application of pressure. Indeed, this was the observation, shown in Fig. 12. Similar T_c-behavior under pressure was also detected in $YBa_2Cu_3O_7$ and $YBaCu_2O_{6.7}$ which had ambient T_c's at 90 and 55 K, respectively. Therefore, I feel that the HTS material base is far narrower than the large number of HTS's suggests and that more, truly different types of HTS's are needed in order to understand better the mechanism of superconductivity.

The extremely short coherence length of HTS's, particularly that perpendicular to the CuO_2-planes which is shorter than the atomic cell dimension, puts these materials in a different perspective, as far as superconductivity is concerned. The possibility arises of the existence of atomic-scale chemical and structural orders which affect the HTS properties. The determination of such orders or the atomic structure/properties relationship will be very interesting. We always find that the

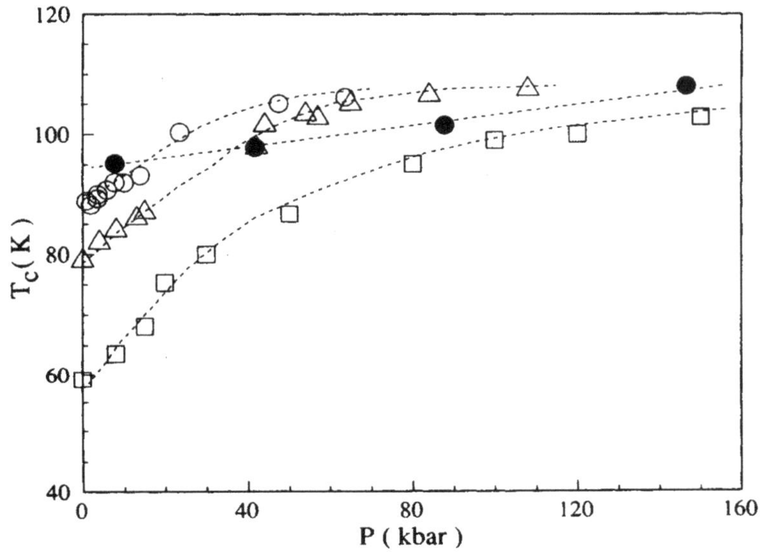

Figure 12. T_c vs P for
● $YBa_2Cu_3O_7$, □ $YBa_2Cu_3O_{6.7}$, ○ $YBa_2Cu_4O_8$, △ $Y_2Ba_4Cu_7O_{15}$.

volume fraction of an HTS decreases when one moves always from the optimal doping condition. I happen not to believe claims of 100% superconducting volume fraction in their HTS's with different dopings provided they were "properly synthesized." I wonder if HTS materials can ever be made microscopically homogeneous and how this microscopically heterogeneous nature will affect the T_c. A positive affect on T_c due to this microscopic heterogeneity seems to be consistent with the higher on-set T_c one usually finds in materials with a broad transition.

One major question is whether HTS's can be described by the conventional BCS Theory, a modified version of the BCS Theory, or a non-BCS theory. In some of the non-BCS models, bosons already form above T_c, undergo a Bose-Einstein condensation at T_c and enter into a superconducting state on cooling. Presently, it is believed that bismuthates and (perhaps) fullerites fit the BCS-framework, but the jury is still out on the cuprates. For the cuprates, there exists almost as much experimental evidence for BCS as for non-BCS. High resolution spectroscopy experiments as well as high resolution neutron-scattering experiments on large, quality HTS crystals will no doubt help resolve the problem. To design an experiment to determine unambiguously the pairing states of the electrons is very important, and several preliminary experiments have already been done in this respect. Unfortunately, no definitive conclusion has been reached yet.

As mentioned earlier, many anomalies occur in HTS's at or slightly above T_c, in strong contrast to the LTS's. More studies about their relation to HTS are clearly needed. These anomalies cannot be entirely accidental. I strongly believe that these anomalies carry with them a great deal of insight to the occurrence of HTS. No HTS theory can be complete without explaining these anomalies.

The antiferromagnetic insulator → HTS metal → normal metal transitions through doping a cuprate is very intriguing. Various experimental evidence, *e.g.* in inelastic neutron scattering, has been obtained to suggest that the HTS mechanism may be magnetic in nature. In view of the complex chemical nature and thus the possible accompanying microscopically heterogeneous characteristics of the compounds, the experimental situation appears to be more cloudy than some would like to believe. It should be noted that the general correlation between T_c and the long-range antiferromagnetic order in cuprate HTS's far from the insulator → HTS metal phase-boundary is very similar to that observed in LTS's. The interplay between superconductivity and magnetism in LTS's has equally been masked by the complication in materials near the magnet → superconductor phase-boundary.

The cuprate HTS's have a layered structure and as a result, many HTS theoretical models have been developed based on a two-dimensional charge system. However, the exact dimensionality of cuprate HTS's remains unsettled. This is particularly true in view of the high T_c's (although lower than 42 K) in the three-dimensional bismuthates and fullerites, and the strong fluctuations associated with low dimensional systems that tend to suppress T_c. It will be extremely important, in this respect, to determine if a single layer sample of the 214-T compound or $(Sr,Ca)CuO_2$ is superconducting with a high T_c. Uniaxial high-pressure experiments should shed significant light on this problem.

The complicated phase diagram of HTS's (and of LTS's) provides a fertile ground for the study of flux dynamics and phase transition. For example, the nature of $H_i(T)$ is still not clear. It has been suggested that it is associated with a depinning of the fluxoids and thus is a viscosity change and not even a phase transition; it might be a melting of the fluxoids and hence a first-order phase-transition; or it might be a vortex-glass transition. Experimental evidence for each of the above has been reported. Based on our own data, it appears that the character of $H_i(T)$ depends on the type of pinning centers introduced. The nature of $H_n(T)$ shown in Fig. 4 detected by us is yet to be determined. All we know at this moment is that the magnetic relaxation behavior is rather different above the boundary than it is below the boundary. This is also reflected in the different current-voltage characteristics below and above. Although magnetization-relaxation measurements have been used to test the various theoretical models on the flux dynamics below the $H_i(T)$-line, the narrow time-window, for the experiments makes it extremely difficult to differentiate one model from the others. However, recently we have developed a pulse magnetometer which enables us to extend the time-window for the experiment by at least four orders of magnitude from 10^{-4} to 10^{5} sec. The results seem to support the vortex-glass model although the details of the excitations of fluxoids are more complicated than the theory predicts. A systematic study of the magnetic relaxation using the pulse magnetometer on samples with different types of defects, *e.g.* induced by fast-neutron or heavy-particle irradiation will be very fruitful.

Muon relaxation and optical experiments did show a possible universal linear T_c *vs* n/m^* behavior for cuprate HTS's as schematically shown in Fig. 6, if one neglects the deviation near the T_c-peak of each HTS family. Another universal relation between T_c and the Madelung energy difference ΔV_{AC} has also been proposed by Maekawa *et al.* as shown in Fig. 13. According to a simple rigid band model, both predict a positive pressure effect on the T_c of HTS's. However, a negative pressure effect on T_c has also been observed by us as well as by Mori *et al.* both for the electron- and hole-HTS's. We found that the electron- and hole-HTS's fall into two distinct regions in the $\partial \ln T_c/\partial P$ *vs* T_c plot, shown in Fig. 14, suggesting that the two may belong to two different groups. The fact that $\partial \ln T_c/\partial P$ for cuprate hole-HTS's can be either positive or negative and that all cuprate hole-HTS's fall into one general band in $\partial \ln T_c/\partial P$ *vs* T_c leads us to the proposition that there may exist a

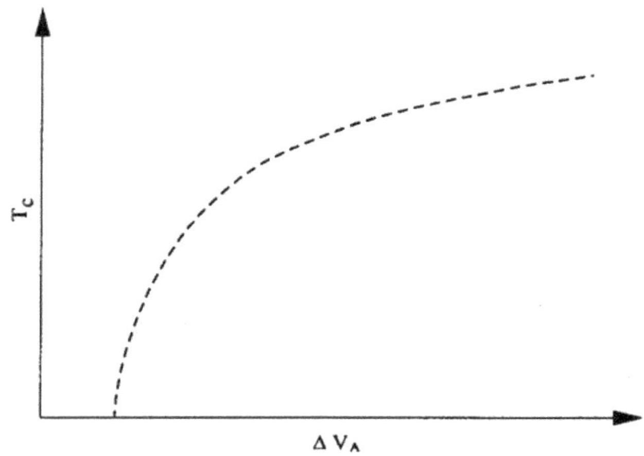

Figure 13. T_c *vs* ΔV_{AC}.

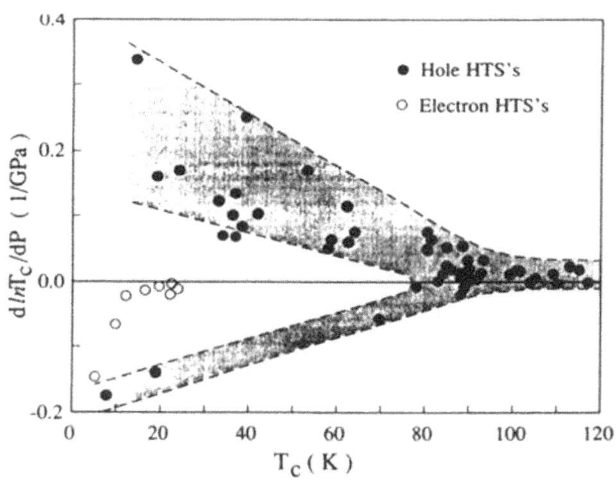

Figure 14. $\partial \ln T_c/\partial P$ *vs* T_c.

universal T_c variation with a single parameter X as shown in Fig. 15. X is closely but not completely related to the carrier concentration. If this is proven, one must look for a compound system other than the layered-cuprate system in order to achieve a T_c greater than ~ 150 K.

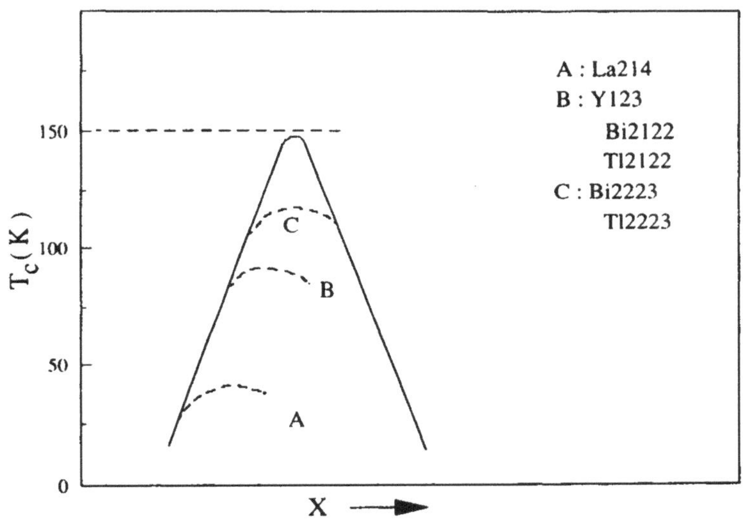

Figure 15. T_c *vs* X (our general T_c-behavior).

4.1. Materials and Applications

4.2.1. Bulk

- Quality large, single crystals
- Chemical phase equilibrium diagrams
- Chemical stability
- More truly new HTS's and more improved HTS's
- Improved bulk and wire processing
- Improved mechanical strength without sacrificing superconducting properties
- Material compatibility
- Fullerites

Let us now first consider the area of bulk HTS. We need large, quality single crystals of HTS's for the definitive determination of much of the spectroscopic data crucial to unraveling the HTS science. Until now quality single crystals were available only for the Y123 and 214-T compounds in small sizes. One of the important steps in overcoming this dilemma is to carry out a systematic investigation on the chemical-phase equilibrium diagrams and to try to grown large single crystal growth from seeds. Apparently, the critical parameters dictating the formation and decomposition of the various HTS's phases are still not known. Not only will this information help control the compound growth to achieve quality, large crystals, it will also enable us to determine the chemical stability of HTS's. The life-span of HTS devices depends on the chemical stability of the HTS materials under different operating conditions. It becomes all the more important for large-current applications for the utility industry. As already pointed out earlier, the available HTS material-base is much narrower than the number of existing HTS's suggests. More, truly different HTS's are needed to test the generality of the theoretical models. While searching for truly new HTS's, one should continue to look for HTS's with improved properties for scientific study and for practical applications. We all know that many of the HTS material problems stem from the highly anisotropic character of HTS's. One of the immediate challenges is to find an isotropic HTS with a $T_c > 90$ K and with a large J_c, through the formation of a new phase, appropriate doping, or better processing.

I have mentioned the many advances in the preparation of bulk HTS materials. However, I have not yet told you that the yield and production rate of those impressive HTS items are still very low. The main reason lies in the difficult control of grain growth, the occurrence of microcracks, the appearance of chemical inhomogeneity, and the slow grain-growth rate near the peritectic recombination temperature. Our preliminary results show that part of these difficulties can be overcome. For instance, by varying the ratio of the longitudinal to perpendicular temperature gradients with respect to the sample traveling direction in the continual processing of HTS, one can control the direction of grain-growth; through doping of Y_2O_3 and Ag, one can drastically reduce the microcracks and enhance the rate of grain-growth. Y_2O_3-doping improves the dispersion of the 211 insulating phase, which is believed to generate stacking faults near the 123/211 interfaces and thus to increase J_c. To prepare large-sized HTS-materials, the continuous processing technique was developed and used. The inherent zone-refining nature of the technique (by transporting a bulk HTS of specific shape through a predetermined temperature profile which includes an narrow hot-zone near the peritectic temperature of HTS) poses a dilemma in the chemical homogeneity of the large, bulk sample prepared. Again, we found that proper doping of Y_2O_3 can significantly improve the uniform chemical distribution. By properly applying all of these findings to HTS bulk processing, I believe the yield and rate will be greatly improved.

The most desirable form for large-current applications is wire or ribbon. Great progress has been made in Japan in making Ag-sheathed Bi2212 ribbon. At 4.2 K, its J_c remains high even above ~ 13 T, in contrast to Nb_3Sn. Again, the yield is rather poor. Presently, there are very few labs where BSCCO ribbons of this quality can be produced. The greater challenge now is to fabricate quality ribbon of long length. By examining carefully the texturing and the stability of the HTS in a Ag-HTS interface under different processing conditions, one should be able to improve the reproducibility. Our recent data shows that texturing along the ribbon direction appears only near the interface. This may set a limit of ~ 60 μm on the maximum thickness of well-textured BSCCO ribbon one can make. For Y123, texturing occurs only at a high temperatures, near 1050° C, the peritectic decomposition temperature. Until very recently, no material could retain its own integrity without interacting with Y123, is ductile, and at the same time is permeable to oxygen at this high temperature. As a result, no wire or ribbon with well-textured Y123 has been made, even though Y123 has a superior J_C to that of Bi2212 at 77 K. However, we just found that Pd-doped Ag, with a melting point above 1050° C, depending on the doping level, does not react with remains permeable to oxygen Y123 and enables texturing of Y123 at the interface. High J_C is yet to be achieved in this Pd/Ag-sheath Y123-wire due to the formation of voids in the wire. Also, rolling, which is used for BSCCO ribbon forming, does not appear to be suitable for Y123.

Material compatibility needs to be thoroughly examined, since it is critical to the performance and lifetime of HTS devices. The study can also provide interesting challenges to surface scientists.

The newly-discovered HTS in fullerites A_3C_{60} with A being the Alkaline elements represents a very interesting new opportunity for people in the field. Although the extreme chemical activity of these HTS fullerites may pose serious problems for applications, their isotropic nature remains attractive if their T_C can be raised to above 77 K. While people are actively studying C_{60} by doping it with various elements, it will be important to see if C_{60} can be used to modify the properties of existing materials or to form new inorganic compounds with interesting properties.

Along this direction, preliminary evidence of partial melting has been detected by us in C_{60} under pressure, which might be a possible channel toward the formation of new inorganic compounds.

4.2.2. Thin Films

- thin film processing
- SIS junction
- Grain boundary effect
- Noise
- Substrates
- Ultra-small dimension effect
- Novel superconducting devices
- Novel HTS/semiconductor hybrid devices

While truly impressive progress has been made in HTS thin-films, HTS commercialization still requires (1) better understanding and hence better controlling the thin film growth process on various substrates under different conditions, (2) lowering the deposition temperature without sacrificing J_C, (3) enhancing the deposition rate, (4) increasing the size of the thin film deposited, and (5) searching for low-cost substrates with better lattice matching, no twins, low dielectric constant, and low loss tangent.

It is not yet clear if the absence of Josephson tunneling in cuprate HTS's is a genuine characteristic of the material or merely an imperfection in junction preparation. More understanding of the grain-boundary effect on the properties of HTS thin films is in order, since they are directly affecting the performance of the devices. The real source of noise in HTS films remain a mystery. As the HTS devices become smaller in the future, we have to know the effect of small dimensions on the properties of HTS's. Most of the efforts in HTS thin film applications appear to be based on concepts developed over the last 75 years. It may be fruitful to spend some time on the development of novel device-concepts and to build devices unique to HTS materials.

5 . CONCLUSION

As we can see from the above, great progress has indeed been made at an unprecedented rate in HTS materials. The advances have sufficiently demonstrated that HTS's hold great technological promise. However, they are not yet commercially viable materials. The material advances made so far have reached the level needed to attract the attention of design engineers to consider the material appropriate for functional devices and systems. HTS's materials are ripe for device construction and demonstration but not yet ready for immediate system fabrication. The experimental data obtained are reliable enough for us to develop and test theoretical models, although a broader material base would be much more helpful. Larger, quality HTS-crystals are indispensible for spectroscopy studies, crucial for the determination of the microscopic mechanism of HTS.

One of the most asked questions is regarding the relative importance of the various HTS material families that we have. I believe the answer depends on the specific applications. For instance, BSCCO has an advantage over 123 HTS's for wire-applications at 4 K and high field. However, Y123 is irreplaceable at this moment for other bulk applications, particularly above 77 K. For monolayer thin film devices, Tl-based HTS's appear to be better than Y123 in terms of its higher T_c and ease in preparation. However, Y123 is indispensable for multilayer devices.

Over the last four years, HTS research has not only rejuvenated the field of superconductivity, it has also had a profound impact on the way we do material science. It shows how important a multidisciplinary approach is. I believe that the recent discovery of HTS in fullerites would have been delayed without the expertise and experience accumulated over the last four years for HTS research. The effects of HTS will be long lasting. I strongly believe that, with determination and perseverance guided by vision, one day in the not too distant future, we will be able to reach the promised land of HTS.

6. ACKNOWLEDGMENT

I would like to thank my colleagues P. H. Hor, Y. Y. Xue, R. L. Meng, K. Matsuishi and J. Kulick, and students Q. Xiong, G. Lin, Z. J. Huang, L. Gao and C. Diaz for useful discussions, comments, and technical help. Research at TCSUH is supported in part by the NSF Low Temperature Physics Program Grant No. DMR 86-126539, DARPA Grant No. MDA 972-88-G-002, NASA Grant No. NAGW-977, Texas Center for Superconductivity at the University of Houston, and the T. L. L. Temple Foundation.

CRYSTAL GROWTH AND COMPOSITION CONTROL OF PURE AND DOPED 123 MATERIALS

F.LICCI, L.RAFFO, C.FRIGERI, T.BESAGNI, P.FERRO
MASPEC Institute of C.N.R., Via Chiavari, 18/A
43100 Parma, Italy

ABSTRACT

Physical properties of high T_c superconductors are strongly depending on the materials quality, including stoichiometry and impurity presence. The development of the crystallization procedures and the availability of single crystals improved the reproducibility and reliability of experiments, but some problems still remain because of the the low thermodynamic stability and the high chemical reactivity of such materials. A few examples concerning crystal growth and composition control of 123-type compounds are reported: they include contamination of YBCO crystals from the crucible, crystal growth and characterization of Fe doped YBCO, and crystallization of specific compounds in the $Nd_{1+x}Ba_{2-x}Cu_3O_{7-\partial}$ system.

1.Introduction

Extensive research on high temperature superconductivity and related materials has emphasised the fundamental role of the materials quality on the physical properties, together with the need to control the sample composition and microstructure accurately, on an atomic scale. The very short coherence length of high T_c superconductors imposes strict requirements of chemical homogeneity and/or defect distribution which, in combination with the high complexity and the low chemico-physical stability of the materials, are difficult to achieve. Specific problems related to the stoichiometric control are determined by some peculiar characteristics: high T_c compounds are non-stoichiometric and thermodynamically metastable[1]; form solid solutions with variable cation ratios; have very elastic structures, able to accomodate a large variety of extraneous ions. The improvement of crystallization techniques and the growth of monocrystalline materials has provided more perfect and reliable specimens for investigation, but even in single crystals, the composition control can be problematic, due to thermodynamic factors or uncontrolled environment contamination.

In the following we review our experience in crystal growth of pure and doped $YBa_2Cu_3O_7$ (123)-type compounds, by emphasising typical effects of stoichiometric factors and impurity contamination on the materials properties. The results we discuss specifically concern:

-crystal contamination from the crucible (Al_2O_3) material;

-intentional doping of YBCO with various amounts of iron;

-preparation of $Nd_{1+x}Ba_{2-x}Cu_3O_7$ crystals, with controlled composition.

2. Crystal Growth and Composition Control

Crystals of $ReBa_2Cu_3O_{7-\delta}$ (123) materials are currently grown by slowly cooling off-stoichiometric melts (Flux method) containing excess of BaO and CuO, in alumina crucibles. Details of the crystal growth procedure, crystal separation from the flux and post-growth oxygen annealing were reported elsewhere[2,3]. Data on phase diagrams, which allowed optimization of the procedure and identification of the primary crystallization field of 123 YBCO phase were reviewed in[4].

The actual cationic composition of the crystals was determined by Energy-Dispersive X-Ray microanalysis (EDX), attached to a Cambridge Instruments scanning electron microscope and equipped with a standard beryllium window detector. The standards used for microanalysis were sintered pellets with chemical composition similar to that expected for the crystals (as far as the main components and the doping impurities were concerned), which were obtained by solid state reaction of oxides and $BaCO_3$. For each sample, at least three measurements were carried out in different areas. Generally the peak intensities were reproducible within 5%, and in a few samples within 10-15%. The errors of single measurements, corrected by the MAGIC IV program, often are smaller. The analyzed volumes were about $3x10^3$ μm^3 as the measured area was $2x10^3$ μm^2, and the approximate penetration depth of the electron beam was ≈ 1.5 μm, as calculated by Monte-Carlo simulations[5].

The corrosion of the crucible and the evaluation of different element concentration in the melts were determined by analysing the off stoichiometric melts for a temperature program similar to that used in crystal growth procedure: after soaking at the maximum temperature, the liquid solutions were quenched to room temperature and the solidified liquid boules were directly used for composition determination by X-ray microanalysis. The distribution coefficient, $K_{eff}(Me)$, of minority components between the liquid and the crystals was calculated according to equation (1):

$$K_{eff}(Me) = \frac{[Me]_{cryst}}{[Me]_{sol}} \times \frac{[\text{Total cations}]_{sol}}{[\text{Total cations}]_{cryst}} \qquad (1)$$

Oxygen content was determined by iodometric titration[6], with a "dead-stop end point" amperometric detection[7]; a modification of the standard procedure was developed in order to determine the Cu valence in presence of interfering Fe[8].

Complementary characterizations consisted with:

-differential calorimetric analysis (DSC), performed in a Netzsch Simultaneous Thermal Analyzer, STA 409, operating under controlled atmosphere between room temperature and 1500°C;

-X-ray diffractograms of powdered specimens and crushed crystals performed on a Siemens D500 Diffractometer using CuKα radiation. Measurements were performed in steps of 0.05° between 20 and 80°, with a counting time of 60 s;

-resistivity as a function of temperature measured between 300 and 12 K. The measurements were performed in a closed-cycle helium cryostat, by applying a pulsed 0.1-1 mA AC current. Electrical contacts of 0.03 mm^2 average area were made with silver paint and 40 μm diameter gold wire. A thermal annealing in air at 300°C for 20 min. improved contact stability and decreased their resistivity to typically 10^{-4} Ωcm at room temperature. The temperature of the specimens was measured by a silicon diode in contact with the crystal. T_c and ΔT values were evaluated from resistivity versus temperature plots; they were defined as the maximum in the first derivative (dR/dT) and the difference between temperatures corresponding to 90 and 10% of the resistivity at 100K, respectively.

3. Aluminum Contamination from the Crucible

At present one of the serious problems which still limit the development of larger and more perfect crystals of 123 compounds is represented by the crucible material and the crucible corrosion. In Table I are summarized a few significant characteristics of different materials which have been proposed as crucible for YBCO crystal growth. Among these alumina remains widely used, mostly for growing mm-size crystals, separated from the flux. Its main features are relatively high thermal shock resistance, availability and price, medium corrosion resistance and creeping behavior of the cuprate melts.

Table I: Proposed crucible for crystal growth of YBCO

	Cor[a]	Sol[b]	Cre[c]	TM[d]	Cost	Yield[e]	Ref.
Pt	h	h	h	h	h	µm-size,in solid	9
Au	m	l	vh	h	h	µm-size,free	10
ZrO_2	m	m	h	l	m	mm size,in cavity	11
Al_2O_3	m	m	l	h	l	mm-size,free	12
SnO_2	na	l	na	na	na	in cavity	13
ThO_2	na	na	na	na	na	na	11
Y_2O_3	na	-	na	na	h	na	14

a=corrosion from the flux; b=solubility in the crystals; c=creeping of the liquid; d=thermomechanical resistance; e=average crystal size and localization.
vh=very high; h=high; m=moderate; l=low; na=not available data

As main drawback, alumina is attacked from the corrosive Ba-Cu-O flux at high temperature, with formation of $BaAl_2O_4$ layers and subsequent dissolution in the flux[15]. This leads to the incorporation of Al in the YBCO crystals, which causes a significant degradation of the superconducting properties[16]. At higher Al concentration secondary phases are also formed[17], which hinder development of larger YBCO crystals[3], or even inhibit the crystallization of YBCO[2].

Our systematic investigations for reducing crucible corrosion and minimizing the aluminum contamination effects in long duration crystal growth experiments showed that the BaO/CuO ratio in the flux influences the corrosion[2]. In Fig.1 are reported the

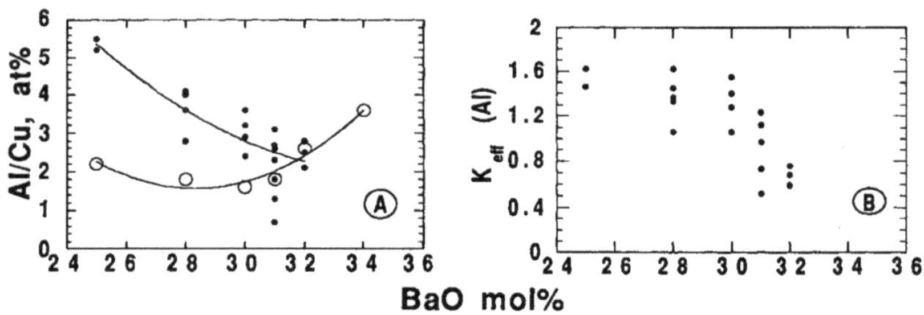

Fig.1 Aluminum content in the crystals (full circles) and in the solution (open circles) (A), and Al distribution coefficient (B) as a function of the flux composition

aluminum content in the crystals and in the solutions and the aluminum distribution coefficient, K_{eff}(Al), between crystals and solution, as a function of the flux composition. The concentration of aluminum in the solution is minimum at a flux composition of about 29 mol% BaO, which is coincident with the BaO/CuO binary eutectic composition[18]. It increases at increasing BaO or CuO concentration with respect to the eutectic value. Incorporation of aluminum in the crystals is maximum at low BaO concentration (CuO-rich side of the eutectic) and decreases with increasing BaO. Also the aluminum distribution coefficient decreases at BaO concentration in the flux higher than the eutectic one. We ascribed the reduced activity of aluminum in the BaO-rich solutions to the formation of stable Ba-Al-O or Ba-Al-Y-O phases[17]: this accounts for both, the reduced segregation of Al in the crystals and the lack of YBCO crystallization from the flux containing 34 mol% BaO[2].

As a consequence of different aluminum content the crystal properties were found to drastically change. In Fig.2 are reported typical resistive characteristics of $YBa_2(Cu_{1-x}Al_x)_3O_7$ crystals with $x \approx 0.01$ and 0.05, respectively. A metallic behaviour with $T_c \geq 90$ K and $\Delta T \leq 1$K was found at low x, while a multistep transition with low T_c occurred at increasing x values.

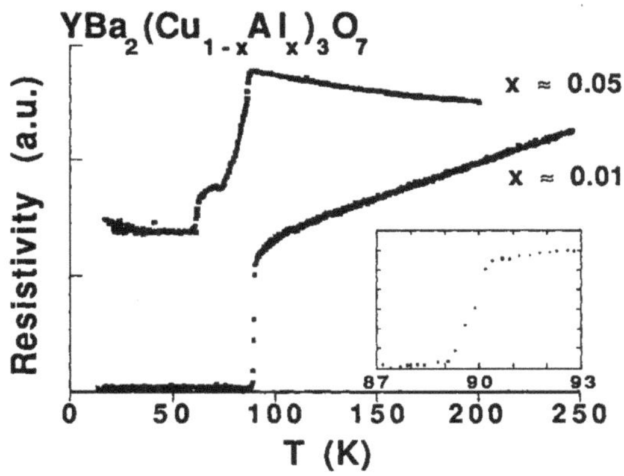

Fig.2:
Resistivity versus temperature of two $YBa_2(Cu_{1-x}Al_x)_3O_{7-\partial}$ crystals with different x
Insert is an enlargement of the lower plot (x=0.01)

4 Iron Doped YBCO Crystals

The effects of iron substitution for Cu in YBCO have been extensively investigated (see for instance[19-23]) under many respects, i.e. microstructure, cation site occupancy, oxygen content and distribution, superconductivity, magnetic and transport properties. However, except for the basic observation that T_c is reduced with increasing Fe content, there is little agreement on both the observed properties and the physical understanding[24,25]. What clearly appears is that preparation methods are strongly affecting the structural[26] and superconducting properties[27] of the samples. For this reason and in order to provide more reliable and better characterized materials for analysis, some attention was devoted to the preparation of single crystal specimens.

Fe-doped YBCO crystals ($YBa_2(Cu_{1-x}Fe_x)_3O_7$) were grown by slowly cooling non stoichiometric melts containing Fe_2O_3 as an additional component, in alumina crucibles. Details on melt composition and Fe_2O_3 concentration are reported in Table II. Experiments with low $FeO_{1.5}$ concentration (≤ $YO_{1.5}$ molar content in the melt) yielded well developed, platelet-like crystals, with dimensions up to 2-4 mm and tickness up to 0.5 mm (Fig.3). When $FeO_{1.5}$ concentration was exceeding twice the $YO_{1.5}$ content no perovskite crystals segregated, and, at intermediate concentration, crystallization of 123 phase was only occasionally obtained. Variation of the relative concentration of BaO/CuO in the melts did not significantly affect the crystallization results. When compared

Table II: Off-stoichiometric melt composition and crystal stoichiometry of Fe doped YBCO

Melt composition (molar ratio[a])				Crystal stoichiometric indexes[a]				
Cu	Ba	Y	Fe	Cu	Ba	Y	Fe	Al[b]
45	18.5	1	0.6	2.73	1.9	1	0.06	0.09
53.4	25	1	0.74	3.18	2.1	1	0.09	0.03
54	25	1	0.9	2.97	2	1	0.18	0.09
52.6	25	1	1.5			*		
54	26	1	1.5			*		
52.6	25	1	1.5	2.79	1.9	1	0.12	0.06
52.4	25	1	1.7	2.85	1.9	1	0.12	0.06
54	26	1	1.7			*		
51.5	25	1	2.5			*		
51	25	1	3.3			*		
54	25	1	3.45			*		

a= normalized to the Y content
b= from the crucible corrosion
*= no crystallization

with previous results[2,28], such data indicate that Fe_2O_3 increases the dissolution rate of alumina crucible and reduce the crystallization field of 123 phase in the BaO-CuO-$YO_{1.5}$ ternary phase diagram, thus limiting the maximum Fe doping obtainable in the crystals.

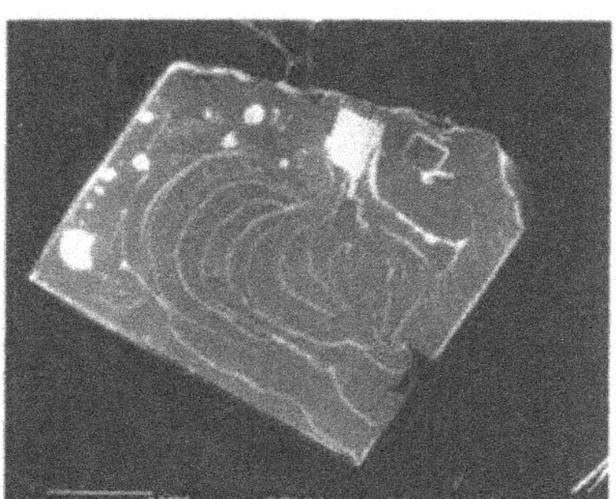

Fig.3: $YBa_{1.95}Cu_{2.85}Fe_{0.09}Al_{0.09}O_7$ crystal

Crystal composition was determined by EDX analysis, and results are summarized in Table II. Crystals from the same batch exhibited almost identical composition and impurity concentration. The distribution coefficient of Fe between liquid and crystals, $K_{eff}(Fe)$, calculated from eq. (1), was found to be ≈ 1.2, thus indicating a preferential segregation of Fe in the crystals. Lattice parameters and crystallographic structures of crystals were determined from X-ray powder diffractograms, obtained on crushed crystals, since the enhanced twinning of the examined specimens made quite problematic the use of single crystal X-ray technique[29]. A few data are reported in Table III. T_c was determined by resistivity measurements as a function of temperature; results are summarized in Fig.4. Data relative to

Table III. Structural data and actual sample stoichiometry (including oxygen content, when measured) of (Fe+Al)-doped compounds

Sample stoichiometry	Lattice parameters a	b	c
$YBa_{1.9}Cu_{2.85}Fe_{0.09}Al_{0.09}$*	3.867(1)	-	11.722(3)
$YBa_{1.9}Cu_{2.79}Fe_{0.12}Al_{0.04}$*	3.866(3)	-	11.729(9)
$YBa_2Cu_{2.97}Fe_{0.18}Al_{0.09}$*	3.871(1)	-	11.722(0)
$YBa_2Cu_{2.91}Fe_{0.06}Al_{0.03}O_{6.94}$	3.864(6)	3.890(0)	11.679(2)
$YBa_2Cu_{2.82}Fe_{0.15}Al_{0.03}O_{6.89}$	3.871(1)	-	11.691(4)
$YBa_2Cu_{2.94}Fe_{0.06}O_{6.89}$	3.848(6)	3.887(6)	11.677(9)
$YBa_2Cu_{2.85}Fe_{0.15}O_{6.92}$	3.870(2)	-	11.696(8)
$YBa_2Cu_{2.925}Fe_{0.075}O_{6.97}$	3.867(4)	3.876(4)	11.668(9)
$YBa_2Cu_{2.85}Fe_{0.15}O_{6.98}$	3.868(2)	-	11.672(7)
$YBa_2Cu_{2.775}Fe_{0.225}O_{7.03}$	3.869(1)	-	11.666(6)
$YBa_2Cu_{2.7}Fe_{0.3}O_{7.02}$	3.873(2)	-	11.660(7)

*Single crystal specimens.

ceramic samples having similar compositions are also shown for comparison in Table III and Fig.4: the oxygen concentration we report was experimentally determined. Our data confirm previous reports: i)the Fe-induced orthorhombic to tetragonal phase transition occurs in fully oxidized compounds at doping levels of about 2.5-3 at% Fe with respect to Cu; ii)T_c decreases with increasing Fe content; iii)the oxygen content per unit formula is higher in the Fe-doped materials than in the undoped 123 compounds. The reasonable agreement obtained among the data relative to the ceramic and monocrystalline specimens with the same Fe content prove the good homogeneity and complete oxydation of crystals. Aluminum contamination at low levels (≤3 at% with respect to Cu) does not affect significantly the resistive and structural characteristics of the Fe-doped crystals, as already proved for the undoped YBCO crystals[2].

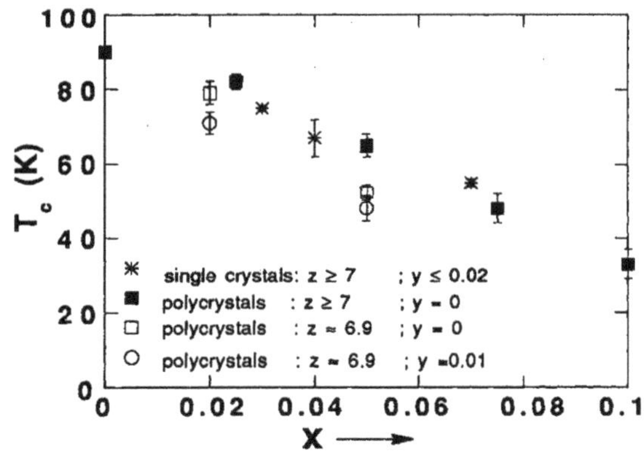

Fig.4:
T_c versus Fe content (x) in $YBa_2(Cu_{1-x-y}Fe_xAl_y)_3O_z$ materials (bars indicate ΔT, as defined in the text).

5. Growth of 123 Nd-Ba-Cu-O crystals.

When substituting for Y in the 123 structure, large rare earth ions exhibit an enhanced tendency to occupy also the Ba sites in the structure[30], thus changing the optimum hole concentration for the superconductivity, and decreasing T_c[31]. In order to clarify such mechanism $Nd_{1+x}Ba_{2-x}Cu_3O_7$ system was widely investigated[31-34], and a few attempts have been also done for obtaining single crystals of the superconducting phase[35-37]. Conventional slow cooling methods of non-stoichiometric melts, containing excess of CuO were used. It was found that, under non-stoichiometric conditions, the thermodynamically stable, first segregating phase is the Nd-rich, non superconducting one, while the 123 superconducting phase is only occasionally obtained as a minority phase[36]. Experiments carried out in our laboratory under condition similar to that used for YBCO compounds[2] confirmed literature results and yielded well-developed, platelet-like crystals (Fig.5A), which were not superconducting down to 10 K and exhibited a tetragonal structure, even after annealing in oxygen, thus suggesting an extra Nd content with respect to the stoichiometric 123 composition[38].

In order to obtain stoichiometric 123 crystals a different approach was then applied. A grain-growth, solid state method was

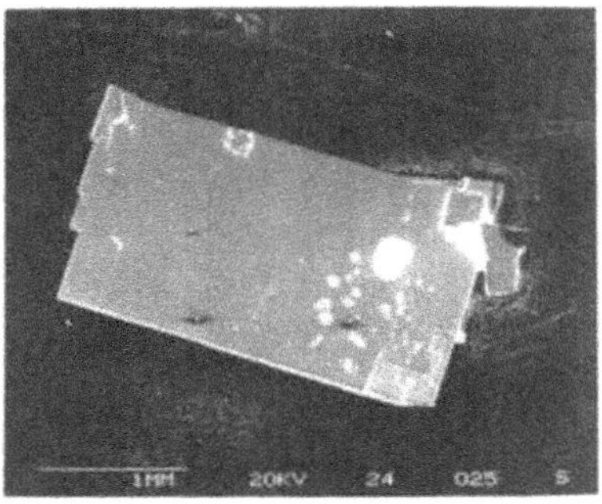

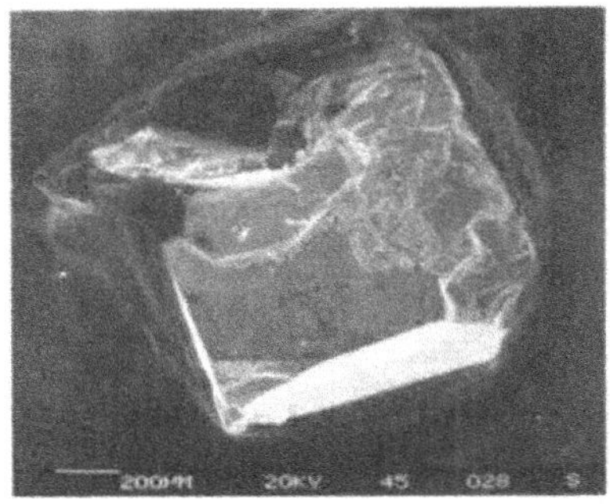

A B

Fig.5: Flux-(A) and solid-state-(B) grown $Nd_{1+x}Ba_{2-x}Cu_3O_{7-\partial}$ crystals.

developed, which is based on the use of stoichiometric mixtures of components. The peritectic decomposition temperature (i.e.1080°C) was determined by DSC analysis of stoichiometric, well-reacted 123 powders and crystal growth was carried out by applying a careful controlled cooling in a narrow temperature range, close to the melting point. Stoichiometric amounts of $BaCO_3$, CuO and Nd_2O_3 materials were mixed and repeatedly reacted at 950°C and ground for producing single phase, polycrystalline 123 perovskite. The powders were compacted as cilinders of about 2 cm in diameter and 3 cm in height, and sintered at usual 950°C. Temperature was increased near the peritectic decomposition reaction and a controlled cooling was applied at 0.5-1°C/hour from about 1078 to 1060°C. Intermediate isothermal steps were allowed, for a total duration of about 48 hours; the furnace was then cooled to room temperature at 50°C/hour. The process was carried out in air and the post-growth annealing in oxygen was applied. Massive crystals of 1x1x1 mm average dimensions were obtained (Fig.5B), sticking in the solid rod, which were mechanically extracted. Resistive measurements indicate superconductive transition at a temperature which depends on the oxygen concentration. DSC analysis evidenced the different composition of the two types of crystals: in Fig.6 are shown the melting characteristics of the solid-state- and the flux-grown crystals[36].

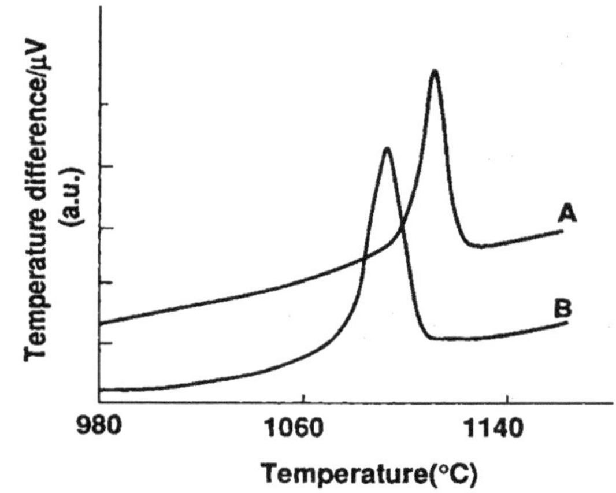

Fig.6:
Melting curves of flux-(A) and solid-state-(B) grown $Nd_{1+x}Ba_{2-x}Cu_3O_7$ crystals.

6. Conclusions

The thermodynamic and chemical characteristics of high Tc superconductors make quite problematic the synthesis of pure compounds with well defined stoichiometry and microstructure. On the other hand, the presence of inadvertently introduced impurities or intentional dopants and the occurrence of small deviations from stoichiometry dramatically affect the physical and superconducting properties of the materials. Such a poor reproducibility of the sample characteristics is probably one of the main causes of the considerable scattering of the experimental results and the slow progresses in the fundamental understanding which still characterize the high T_c superconductivity in spite of the many efforts devoted to the field. In such a situation, accurate controls of the specimen quality and composition are fundamental for a correct evaluation of the results.

The activity was developed with the financial support of the Progetto Finalizzato del C.N.R. "Materiali Speciali per Tecnologie Avanzate"

References

1. A.W.Sleight, *Physica C* **162-164** (1989) 3.
2. F.Licci, C.Frigeri and H.J.Scheel, *J.Crystal Growth* **112** (1991) 606.
3. F.Licci and T.Besagni, in *High Temperature Superconductivity*, eds C.Ferdeghini and A.S.Siri (World Scientific, Singapore, 1990) 26.
4. H.J.Scheel and F.Licci, *Thermochimica Acta* **174** (1991) 115.
5. D.C.Joy, *Inst.Phys.Conf.Ser.* **93**, (Vol.1) (1988) 23.
6. P.Lanza and G.Rossi, *Analyt.Chim.Acta* **244** (1991) 253.
7. A.I.Vogel, *A Text-Book for Quantitative Inorganic Analysis* (Longmans, London, 1964) p.1051.
8. F.Licci, L.Raffo and P.Ferro, in preparation.
9. Z.Z.Wang and N.P.Ong, private communication (December, 1988).
10. D.L.Kaiser. F.Holtzberg, B.A.Scott and T.R.McGuire, *Appl. Phys. Lett.* **51** (1987) 1040,
11. R.A.Laudise, L.F.Scheemeyer and R.L.Barns, *J.Crystal Growth* **85** (1987) 569.
12. See, for instance, *J.Crystal Growth* **85** (1987), and **91** (1988); H.J.Scheel and F.Licci, *Mater. Res. Bull.* **13 (10)** (1988) 56.
13. C.Thomsen, M.Cardona, B.Gegenheimer, R.Liu and A.Simon, *Phys.Rev.B* **37** (1988) 9860.
14. H.J.Scheel, private communication, December 1991.
15. H.J.Scheel, W.Sadowski and L.Schellenberg, *Supercond. Sci. Technol.* **2** (1989) 17.
16. T.Siegrist, L.F.Schneemeyer, J.V.Waszczak, N.P.Singh, R.L.Opila, B.Batlogg, L.W.Rupp and D.W.Murphy, *Phys.Rev.B* **36** (1987) 8365.
17. Th.Wolf, W.Goldacker, B.Obst, G.Roth and R.Flukiger, *J.Crystal Growth* **96** (1989) 1010.
18. F.Licci, H.J.Scheel and P.Tissot, *J.Crystal Growth* **112** (1991) 600.
19. Y.Oda, H.Fujita, H.Toyoda, T.Kaneko, T.Kohara, I.Nakada and K.Asayama, *Jap.J. Appl. Phys.* **26** (1987) L1660.
20. Y.Xu, M.Suenaga, J.Tafto, R.L.Sabatini, A.R.Moodenbaugh and P.Zolliker, *Phys.Rev.B* **39** (1989) 6667.
21. C.Y.Yang, S.M.Heald, J.M.Tranquada, Y.Xu, Y.L.Wang, A.R. Moodenbaugh, D.O.Welch and M.Suenaga, *Phys.Rev. B* **39** (1989) 6683.
22. T.Ohtani, *Mat.Res.Bull.* **24** (1989) 343.

23. B.D.Dunlap, J.D.Jorgensen, C.Segre, A.E.Dwight, J.L.Matykiewicz, H.Lee, W.Peng and C.W.Kimball, *Physica C* **158** (1989) 397.
24. K.Westerholt, H.J.Wuller, H.Bach and P.Stauche, *Phys.Rev. B* **39** (1989) 11680.
25. Y.Xu, R.L.Sabatini, A.R.Moodenbaugh, Y.Zhu, S.-G.Shyu, M. Suenaga, K.W.Dennis and R.W.McCallum, *Physica C* **169** (1990) 205.
26. S.Katsuyama, Y.Ueda and K.Kosuge, *Mat.Res.Bull.* **24** (1989) 603.
27. T.J.Kistenmacher, *Phys.Rev. B* **38** (1988) 8862.
28. F.Licci, T.Besagni, C.Frigeri, C.Paris, F.Aquilano and H.J.Scheel, in *Proceed. Europ.Conf.on Crystal Growth,* (Trans.Tech. Publications, Switzerland, 1992), in press.
29. G.Calestani, unpublished.
30. C.U.Segre, B.Dabrowski, D.G.Hinks, K.Zhang, J.D.Jorgensen, M.A. Beno, and I.K.Schuller, *Nature* **329** (1987) 227.
31. M.W.Shafer, T.Penney, B.L.Olson, R.L.Greene and R.H.Koch, *Phys.Rev. B* **39** (1989) 2914.
32. J.M.Tarascon, W.R.McKinnon, L.H.Greene, G.W.Hull and E.M.Vogel, *Phys.Rev. B* **36** (1987) 226.
33. K.Takita, H.Akinaga, H.Katoh, H.Asano and K.Masuda, *Jap.J. Appl.Phys.* **27** (1988) L67.
34. R.A.Steeman, E.Frikkee, R.B.Helmholdt, D.M.DeLeeuw and C.A.H.A. Mutsaers, *Physica C* **153-155** (1988) 950.
35. A.Katsui, Y.Hidaka and H.Ohtsuka, *Jap.J.Appl.Phys* **26** (1987) L1521.
36. T.Iwata, Y.Tajima and M.Hikita, *J.Crystal Growth* **91** (1988) 274.
37. J.A.Campa, J.M.Gomez de Salazar, E.Gitierrez Puebla, M.A.Monge, I.Rasines and C.Ruiz Valero, *J.Crystal Growth* **91** (1988) 334.
38. F.Licci, T.Besagni and L.Raffo, unpublished.

PREPARATION AND CHARACTERIZATION OF Ag-SHEATHED TAPES IN THE Bi-Ca-Sr-Cu-O SYSTEM

R. Masini, C. A. D'Ovidio and A. Tuissi
Istituto per la Tecnologia dei Materiali Metallici non Tradizionali del Consiglio Nazionale delle Ricerche.
Via Bassini 15, 20133 Milano, Italy

L. Gherardi and P. Metra
Pirelli Cavi Spa Divisione Italia.
V.le Sarca 202, 20126 Milano, Italy

G. Zanella, L. Martini and V. Ottoboni
CISE Spa.
P.O. Box: 12081, 20134 Milano, Italy

ABSTRACT

An extensive investigation on typical 2212 and 2223 BSCCO silver sheathed tapes produced by powder-in-tube method were carried out.

The characterization was focused on critical current measurements, which were carried out at 77 K and 4.2 K in zero and applied magnetic field up to 10 T.

INTRODUCTION

The powder-in-tube method using an Ag sheath is one of the promising techniques to fabricate the composite-type conductors in various forms such as tapes, wires and multifilaments.

Tapes fabricated by this technique appear to be the most likely candidates for power transmission/storage applications, provided that good current carrying properties, comparable to or better than it was recently demostrated on short "championship" samples, are repeatably achieved in long lengths.

Bi-based oxide superconductors show, moreover, more advantages than Y-based ones, since c-axis oriented microstructures can be obtained by mechanical deformation and coupling between grains is much better leading, hence, to high current densities J_c values[1].

EXPERIMENTALS DETAILS

Samples Preparation

Commercially available powders of compositions $Bi_2Sr_2CaCu_2O_x$ (BK924 sample) and $Bi_{1.6}Pb_{0.4}Sr_2Ca_2Cu_3O_x$ (M33L-6 sample) were employed.

The powder-in-tube technique was applied to both types of materials with the same procedure, before a suitable thermal treatment (or cycles of thermal/mechanical treatments, in the case of 2223) is given to the samples[2]. The "final" samples were in both cases tapes with a high aspect ratio, to take advantage of the better orientation and packing of crystallites induced by cold working and by the favourable interaction between superconductor and silver surface.

The phase purity of the starting powders was very different for the two systems: according to the XRD analysis, Bi-2212 was almost single phase, whereas Bi-2223 showed only about 30% of the three-layers phase, with a very high initial fraction of two-layers 2212 phase. Nonetheless, also the latter material reached a phase fraction of more than 90% after full heat treatment (see, for example, in figures 1a and 1b, the X-ray diffraction patterns of Bi-2223 powder before treatment and that of a small slab of the same material extracted from a tape after full treatment. Note also, in the same figure 1b, the high degree of orientation achieved on this sample).

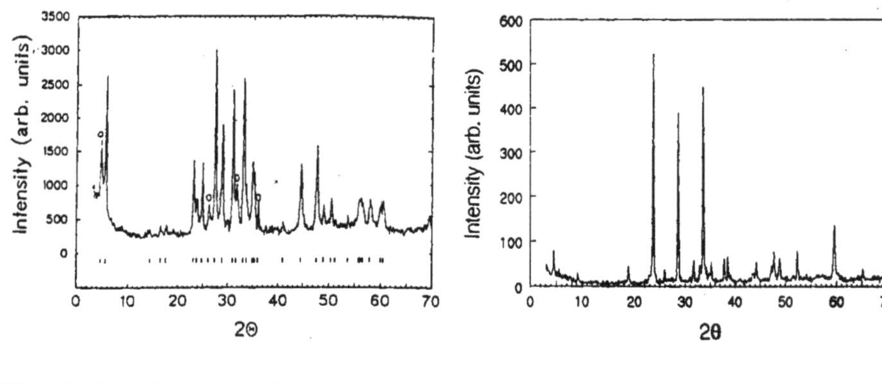

Figure 1: X-ray difractogram for both a) "as supplied" powder and b) after final heat treatment.

Details of treatment conditions, typical dimensions and characteristics of the samples employed are given in Table 1.

Table 1: Samples preparation

	Bi-2212/Ag TAPES	Bi-2223/Ag TAPES
Powder composition	$Bi_2Sr_2Ca_1Cu_2O_x$	$Bi_{1.6}Pb_{.4}Sr_2Ca_2Cu_3O_x$
Phase purity of starting powders	> 90 %	30 %
Detail of "powder-in-tube" method	Filling powders into pure silver pipes, then swageing or drawing to a diameter of about 1.5 mm, then rolling to form tapes.	
Heat treatment	"Melt" processing for a short time at 915°C, annealing at 840°C in air, followed by a short annealing at 500°C in nitrogen.	Repeated (up to 3) treatment cycles including sintering at 830°C in Ar/O_2 mixture (8 % of O_2) and thickness reduction by rolling or pressing.
Typical overall dimensions	2.5 mm * 0.2 mm	5 mm * 0.15 mm
Resistive T_c (zero R)	89 K	106 K
Typical J_c at 77 K	3000 A/cm^2	7000 A/cm^2

Measuring Techniques

Four contact resistive transition of the samples was measured in a liquid nitrogen cryostat. Due to the presence of the silver sheath, the resistive transitions obtained in this way are incomplete: that is, no true onset, but only a "cut" at the level where the resistance of the sample becomes lower than that of the silver sheath can be observed.

Our current procedure during Tc measurements was to ramp the temperature up to beyond transition with a very small current circulating in the specimen, and then to decrease it while adjusting the current value in steps so as to obtain different transition "tails" for the different current densities. This allowed to determine a set of critical current values for different temperatures (see figures 2 and 3).

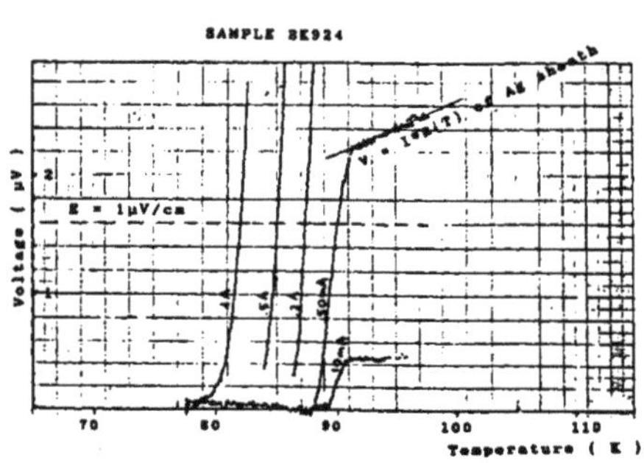

Figure 2: Critical current measurements above 77 K for sample BK924.

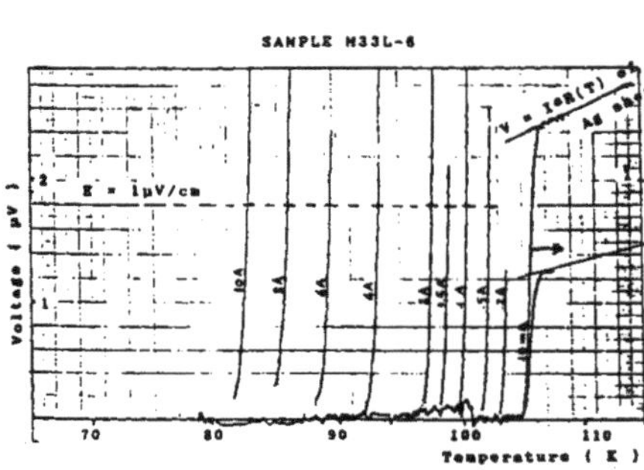

Figure 3: Critical current measurements above 77 K for sample M33L-6.

The determination of the V-I characteristics as a function of applied magnetic field was performed by fitting the sample into a copper winding with an inside field up to 0.2 T (at 77 K) or in the bore of a superconducting magnet (at 4.2 K) with fields up to 10 T. Critical current I_c values for the tapes were determined from standard I-V characteristics at 77 K and 4.2 K as a function of magnetic field H_a under both directions of H_a parallel and perpendicular to the tape surface using the criterion of 1 μV/cm. By means of samples SEM microphotographs the value of their cross sectional area (used to compute our Jc values) was determined.

Magnetic susceptibility versus temperature of tapes was measured either on cooling down the samples to 4.2 K and then warming up them in a magnetic field of 1 mT to above the transition temperature (zero field cooling, ZFC) or cooling the samples from room temperature down to 4.2 K in a 1 mT magnetic field (field cooling, FC).

RESULTS AND DISCUSSION

In figures 4 and 5 are reported the typical ZFC and FC susceptibility curves obtained for the Bi-2212/Ag and Bi-2223/Ag tapes respectively. A sharp diamagnetic transition can be observed starting at 90 K and 110 K for the Bi-2212 and Bi-2223 samples respectively. The susceptibility curves reported in figures 4a and 4b show a high shielding signal for both configurations of magnetic field. A preferential trapping of the magnetic flux when the eccitative magnetic field is oriented parallel to the ab planes (H⊥c axis) can be observed as expected.

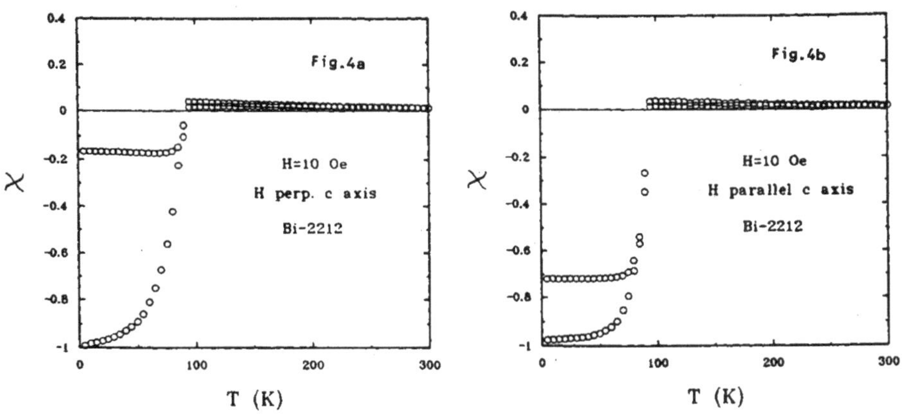

Figure 4: Magnetic susceptibility measurement for different sample orientation for sample BK924.

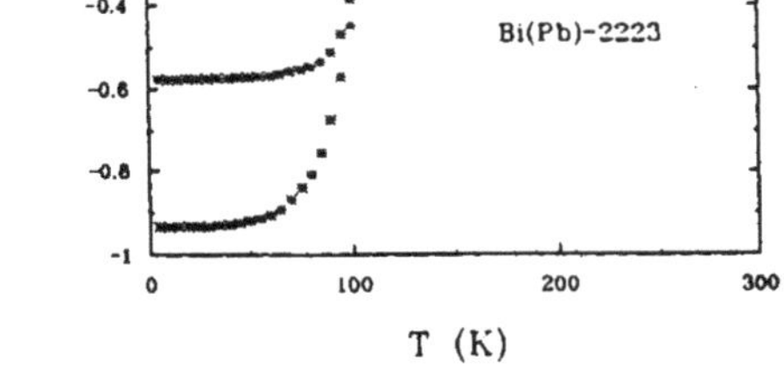

Figure 5: Magnetic susceptibility measurements for sample M33L-6.

The temperature dependence of J_c for conventional metallic superconductors is known to obey an empirical law [3] as

$$J_c(T) \propto (1 - (T/Tc)^2)^m$$

that follows from the temperature dependence of upper critical field $H_{c2}(T)$. The value of the parameter m is believed to reflect the nature of the pinning centers: values of m aroung 1.5 are found in those materials where pinning occurs at planar defects or boundary surfaces, whereas larger values of m (around 4, according to ref. [4]) are expected for point-like pinning centers. Different m values in different temperature ranges are also reported for high Tc superconductors [1], which seems to suggest the possibility of distinct pinning regimes occurring in the same material. This was not the case with our data, that as can observed in figures 6 and 7, shows for both materials, a single slope with m between 1.7 and 2.0 in the whole temperature range.

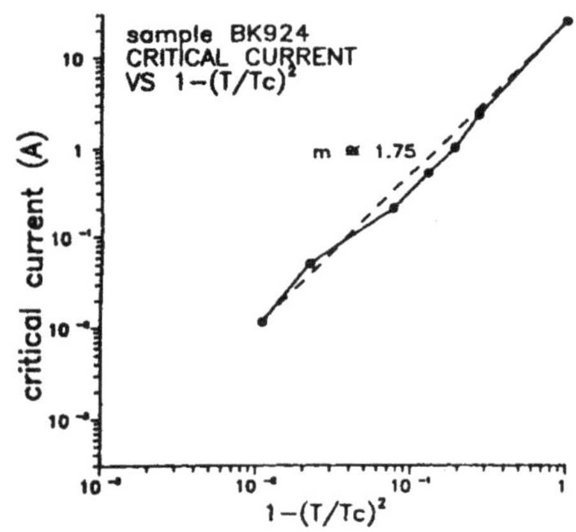

Figure 6: Temperature dependence of J_c for samples MK924.

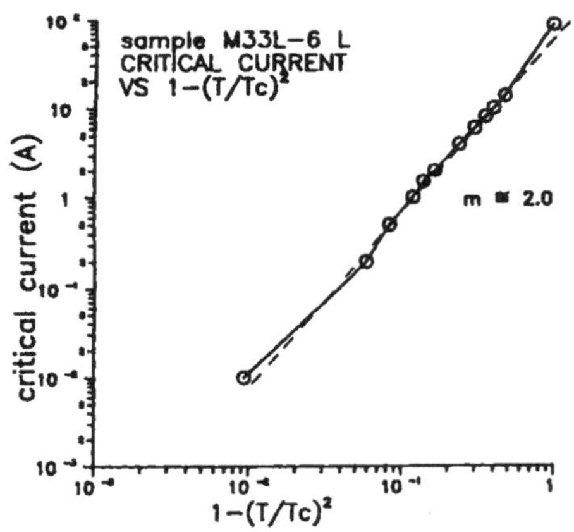

Figure 7: Temperature dependence of J_c for sample M33L-6.

The electrical characterization at 77 K in applied magnetic field which was cycled from zero to 0.2 T and then decreased to zero field, shows a slightly different behaviour of the two samples considered.

In figure 8 is reported a typical J_c/J_{c0} versus B curve for Bi-2212/Ag tape, where J_{c0} is the critical current density at the initial zero field. The experimental data measured increasing and then decreasing the applied magnetic field show a perfectly <u>reversible</u> curve, resulting, hence, in complete absence of flux trapping in this sample at this temperature. This implies the disappearance of the weak links which lead to history effects, as shown in ref 5 and 6, by improvement of the contact at grain boundaries as should be the case for a melted sample. For what concerns the Bi-2223/Ag tape (figure 9), on the other hand, the $J_c(B)$ loop seems, within the experimental errors, to show a hysteretic behaviour. By the way, applied magnetic fields higher than 2 - 3 mT are sufficient to degrade the electrical transport properties of both tapes at 77 K.

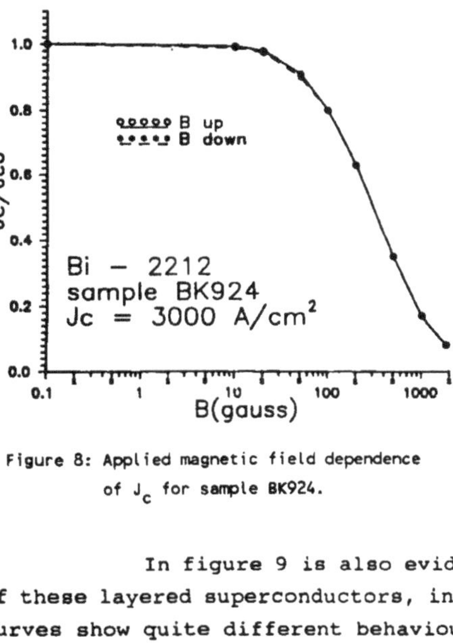

Figure 8: Applied magnetic field dependence of J_c for sample BK924.

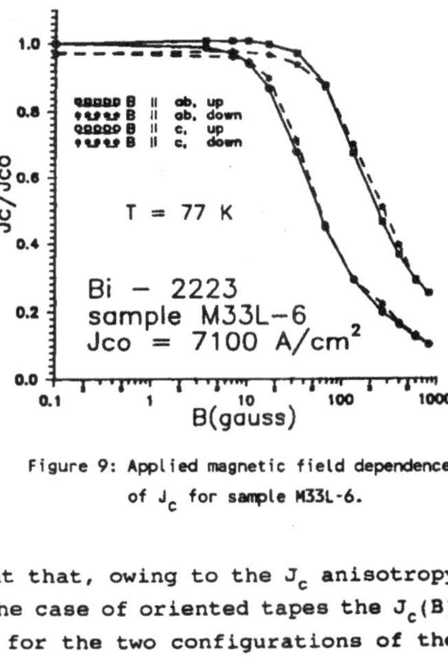

Figure 9: Applied magnetic field dependence of J_c for sample M33L-6.

In figure 9 is also evident that, owing to the J_c anisotropy of these layered superconductors, in the case of oriented tapes the $J_c(B)$ curves show quite different behaviour for the two configurations of the applied magnetic field (H⊥c or H∥c).

In figures 10 and 11 the I-V characteristics measured at 4.2 K for the samples BK924 and M33L-6 respectively are shown. From these data figure 12 is reported for the magnetic field dependence of the obtained values of the transport J_c, normalized by J_{co} in zero magnetic field at 4.2 K. The magnetic field was applied on increasing values perpendicular to the transport current and parallel to the wide plane of the tape. For both samples transport J_c values were also measured on decreasing the applied magnetic field at this temperature. Some hysteretic behaviour has been observed in both cases. Measured J_c values at zero field after the cycle were, for example, of the order of about 50% and 30% of initial J_{c0} for M33L-6 and BK924 sample respectively. Tape

BK924 did not exhibit a history effect at 77 K as shown in figure 8, but exhibits a history effect clearly at 4.2 K. It is considered that such a difference is caused by weak links at grain boundaries whish have their origin by the shortening of the coherence and penetration lenghts compared with those at 77 K, as well as stronger pinning force at 4.2 K.

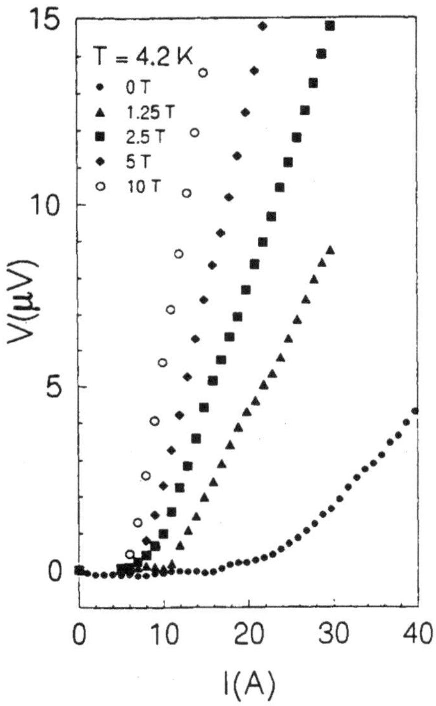

Figure 10: I-V characteristics at 4.2 K for sample BK924.

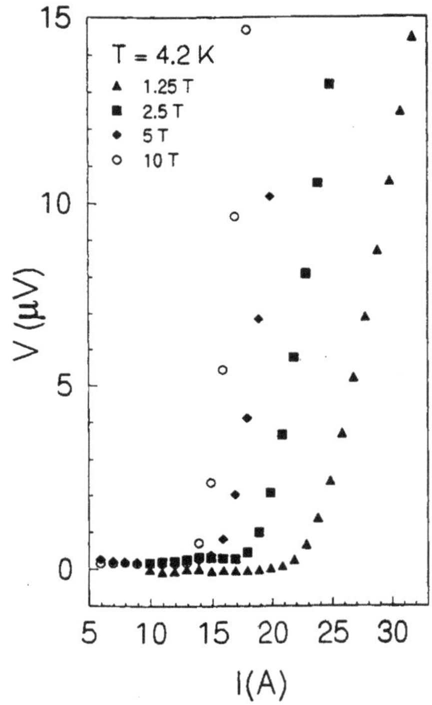

Figure 11: I-V characteristics at 4.2 K for sample M33L-6.

It is also possible to observe (figure 12) that the field dependence for sample BK924, although it shows a strong decrease (~60%) for field values less than 1 T, is then pratically constant in the whole range considered.

For what concerns M33L-6 tape, on the other hand, its field dependence is less pronounced, showing a gradual and monotonic decrease of J_c values up to 10 T.

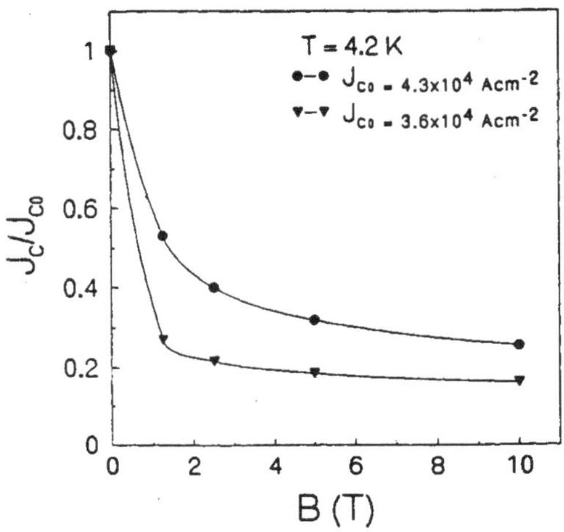

Figure 12: Transport critical current density J_c/J_{c0} at 4.2K: BK924 (▼); M33L-6 (•).

CONCLUSIONS

- An extensive investigation was carried out on typical silver sheathed BSCCO tape samples, both two- and three-layers, produced by powder-in-tube method.

- Melt processing provide Bi-2212 with excellent connectivity: no history effect and granularity, as witnessed by the reversible J_c vs B characteristics, was detected at 77 K.

- Repeated deformation-heat treatment cycles allow to produce Bi-2223 samples with improved performances. Granularity is not overcome, however, and although good characteristics of J_c vs applied field are exhibited at low temperature, substantial improvement is still required for liquid nitrogen operation.

- Better and better wires and tapes are expected to be developed in the near future by continuing the optimization of both materials and manifacturing processes. Based on the present experiencei it is trusted that this will be possible by means of technologies suitable for continuous production.

This work is supported by the National Research Council of Italy (C.N.R) under the "Progetto Finalizzato: Superconductive and Cryogenic Technologies".

REFERENCES

1) Matsushita, T., Otabe, E.S., Ni, B., Hikata, T.H. and Sato, K. *"History effect of critical current density and weak links in superconducting BiPbSrCaCuO tape wires"* Cryogenics 31, 949 (1991).
2) Metra, P., Gherardi, L., Vellego, G., Masini, R., Szabo, G. and Zannella, S. *"Transport properties of different BSCCO wires"* Proc. of ICTPS '90, Rio de Janeiro 29 April - 4 May 1990, R. Nicolsky Ed. , p.747, World Scientific (1990)
3) Campbell, A.M. and Evetts, J.E. *"Flux vortices and transport currents in type-II superconductors"* Adv. Phys 21, 358 (1972)
4) Matsushita, T., Otabe, E.S., Ni, B., Kimura, K.,Morita, M., Tanaka, M.,Kimura, M., Miyamoto, K. and Sawano, K. *"Critical current characteristics in superconducting Y-Ba-Cu-O prepared by the melt process"* Jpn. J. Appl. Phys. 30, L342 (1991)
5) Aomine, T. and Yonekura, A. Phys. Lett 114A, 16 (1986)
6) Watanabe, K., Noto, K., Morita, H., Fujimori, H., Mizuno, K., Aomine, T., Ni, B., Matsushita, T., Yamafuji, K. and Muto, Y. Cryogenics 29, 263 (1989)

PROBLEMS OF OBTAINING TEXTURED HIGH-Tc SUPERCONDUCTING MATERIALS FOR HIGH CURRENT APPLICATIONS

V.P. SEMINOZHENKO, V.M. ISHCHUK, V.A. KHLUS

Institute for Single Crystals of the Ukrainian Academy of Science, Kharkov

1. Introduction

The problem of obtaining high-Tc oxide materials with high critical current densities at liquid nitrogen temperatures has attracted great attention in recent years. The main obstacle on the way to wide-scale applications of HTSC materials is their relatively low critical currents. HTSC materials prepared by various methods usually have polycrystalline structure with different sorts of macroscopic defects such as cracks, voids, etc. Even in the case when we succeed in preparing a good sample with minimal amount of inhomogeneities and close connection between superconducting grains, critical current values still may be low. The reason is that the grains and crystallite boundaries in HTSC materials are weak links with critical current densities much lower than in the bulk material. Another fundamental reason for decreasing critical currents is the high anisotropy of crystal lattice structure of high-Tc oxides. The crystal lattice misorientation in neighbouring grains reduces the critical currents of the inter-grain weak links. Therefore, in addition to close connection between grains, the alignment of crystal lattice axes in neighbouring grains is needed.

These goals may be achieved by preparing HTSC materials with textured structures. Texturing process should form superconducting grains or crystallites with optimum shape and orientation and provide as close connection as possible between them. Furthermore, the maximum of alignment of crystal lattice axes in the grains should be obtained.

The purpose of this report is to review different methods of preparing textured HTSC materials in close connection with the kinetics of phase transformations occurring in the process of growth of HTSC phases.

2. Preparing textured HTSC samples from Bi-Sr-Ca-Cu-O glass

The method, which can be practically important for obtaining applicable HTSC current-carrying structures, is preparing textured coatings on different substrates. This process is usually divided in two stages; at the first stage an amorphous layer with good adhesion to the substrate is formed, and at the second stage the superconducting phases arise in the process of thermal treatment. We have studied the re-crystallization kinetics for the superconducting phase $Bi_2Sr_2CaCu_2O_x$ (2212 phase) starting from the glass precursor system obtained by fast cooling of the melt. The 2212 phase is formed in the temperature region 840-860 °C after 80-90 h of thermal treatment. At the initial stage of the growth the crystal phase $CaCuO_2$ is formed which is later transformed into 2212 phase as a result of diffusion of Bi and Sr from amorphous regions. Textured samples of the size of 5-6 mm have been prepared by moving them with low speed (3-4 mm/h) in a temperature field gradient of about 35 °C/cm. Different plate-like crystallites of 40-50 μm in size were oriented along the temperature gradient. The critical density in the materials obtained by this method is about $(5\text{-}6)\times10^4$ A/cm^2 at 4.2 K and $(2\text{-}3)\times10^3$ A/cm^2 at 77 K.

3. Thermal method of texturing 123 ceramics

For the HTSC compound $YBa_2Cu_3O_{7-x}$ (123), textured structures can be obtained by a special thermal treatment. At the first stage 123 ceramic samples were prepared by a conventional solid-state-phase reaction method, the final thermal treatment was carried out in air at the temperature of 940 °C. The ceramic pellets obtained this way had critical currents j_c = 100-200 A/cm^2 at 77 K in zero magnetic field. After that the pellets were heated to the temperature 1030 °C at which the superconducting 123 phase was decomposed to the "green" phase Y_2BaCuO_5 (211) and $BaCuO_2$, the Ba and Cu oxides were also present in partially melted state. The process can be described by the reaction:

$$123 \rightarrow 211 + BaCuO_2 + L\ (Ba_{0.8}Cu_{2.2}O_y).$$

After the exposure to 1030 °C for about 10 min, the sample was cooled down to the temperature of 975 °C for about 10 min. The crystallites of the 211 phase have the form of needles 50-100 μm long, and the typical size in the cross-section about 5-10 μm. The samples are held at 975 °C for 5-6 h and during this time the superconducting phase 123 is restored in the processes:

$$211 + BaCuO_2 + L \rightarrow 123 + 211 + L.$$

$$211 + L \rightarrow 123.$$

It is important that the morphology of the needle-like grains is preserved after the restoration of 123 phase. After restoration of the oxygen content of the 123 phase by thermal treatment in air, the critical current values of the morphologically transformed non-oriented material is increased by an order of magnitude as compared with the initial ceramics, to the values $j_c \approx 1000$ A/cm^2 at 77 K. The further increase of the critical current is achieved by texturing in the thermal gradient, the method has been described in Sec.2. The material produced by this method consists of oriented crystallite blocks 50-100 μm long, with the thickness 5-10 μm, their *c*-axes are parallel. The material has the critical current $j_c \approx 10^4$ A/cm^2 at 77 K.

4. Textured HTSC coatings on metallic wires and substrates

The problem of producing long, uniform, flexible HTSC wires and tapes with high critical current densities is one of the most challenging problems on the way to real full-scale application of high-Tc superconductivity. As it was mentioned in Sec.2, the first stage of preparation of a high-Tc coating on a metallic substrate should be the deposition of a precursor amorphous layer with a good adhesion to the substrate. The precursor layer should also have the proper element content to form the stoichiometric superconducting compound after the thermal treatment. The procedure we have used included the various methods for the precursor preparation which are described below. We have also studied the processes of formation, growth and texturing of superconducting 2212 phase from the initial Bi-based amorphous layer on the substrate.

4.1. Methods for precursor preparation

4.1.1. Citrate sol-gel method

Precursors obtained by the sol-gel methods were viscous enough to be kept on the metallic surface as thick layers. It has been found that the metal surface should be treated by nitric acid and non-aqueous solutions of some surface-active substances (SAS) for good wetting of the surface by the solution precursor.

The preparation procedure consists of:

- preliminary treatment of the wire by nitric acid and SAS;
- adhesion of the solution precursor to the wire;
- heating of the wire in the furnace for fast evaporation of the solvent;
- treatment of the coating formed on the wire by SAS.

The procedure (except the first stage of the preparation) has been repeated 10-20 times before the 10-20 μm thick coating has been produced.

4.1.2. Carboxylate method

In this method the wire has been immersed in the non-aqueous carboxylate solution. The preparation procedure was similar to that described for the citrate method but the number of cycles was increased to 30-40 for preparing 20 μm thick coating.

4.1.3. Electrochemical deposition

We also used the procedure of the electrochemical deposition of stoichiometric alloy of HTSC components from aprotic solutions of corresponding nitrates or chlorides. This method included the following important steps:

- deep drying of reagents used and special surface prestarting procedure for the wire-cathode;
- empiric choice of the initial concentration of different salts at the given cathode potential;
- proper choice of the current density to maintain the smooth surface of the alloy deposited and to avoid the dendrite formation. In this method the thickness of the coating is determined by the current density and electrolysis time.

4.2. Annealing of coating

The 2212 superconducting phase is formed after annealing the coatings for 12-24 h at 830-870 °C in air. When the first two methods of the deposition were used, the preliminary step-by-step pyrolysis was applied, in which coated wires were placed in the furnace at the temperature 800-830 °C for 1-2 min (shock pyrolysis). This stage is needed to remove organics and to form fine-dispersed particles with high reaction ability.

4.3. The formation of the superconducting phase on wires

The kinetic of the superconducting phase formation was studied for the process starting with the rapid increase of the temperature, in which a partial melting of the coating occurred. The coating melting caused chemical reactions and growth of 2201 non-superconducting phase and calcium cuprate needle-like crystallites. The *c*-axes of 2201 crystals are normally oriented to the wire surface. During the annealing, the 2201 crystals grow larger thus forming long plates. We have estimated the role of elastic energy effects which become essential when the largest crystal dimension is of the order of the wire radius. The growth directions of long crystals are oriented along the wire axis. The effects of the substrate and elastic energy lead to natural texturing of the system. In the process of further thermal treatment, in the temperature range of the stability of 2212 phase, the long plate-like 2201 crystal regions transform into the superconducting ones. The final content of the HTSC 2212 phase in the coating reaches 90 %, the coatings have highly textured morphology, in which *c*-axes in all crystallites are perpendicular to the substrate surface and a good connection between thin plate 2212 crystallite is achieved. We also believe that the present procedure leads to the alignment of the crystallite axes in the *a*-*b* planes of the neighbouring crystallites.

The magnetic susceptibility measurements have shown the beginning of the superconducting transition at 116 K; this could be explained by the presence of the superconducting 2223 phase. The main transition was observed at 86 K.

The critical currents found from magnetization measurements of the coated wire samples are approximately $(3\text{-}4)\times10^3$ A/cm^2 at 77 K.

5. Conclusion

In this report we have considered only some aspects of preparing high-current HTSC materials. Beside the points which have been reviewed in the present report, we have not touched the problems connected with the pinning of Abrikosov's vortices inside superconducting grains and crystallites, as well as the very complicated question of the influence of chemical and physical properties of grain boundaries on the critical current densities of weak links. The methods of preparing textured structures starting from bulk ceramics or glass samples and obtaining textured coatings on metallic wires and tapes can be considered as the first but important step to produce materials with the morphology favourable enough to carry high superconducting currents.

RECENT TRENDS IN THE SYNTHESIS, OPTIMIZATION AND CHARACTERIZATION OF LAYERED LEAD AND BISMUTH CUPRATES.

B. RAVEAU, A. MAIGNAN, D. GROULT, C. MICHEL AND M. HERVIEU.
Laboratoire de Cristallographie et Sciences des Matériaux (CNRS URA 1318)
ISMRA - Boulevard du Maréchal Juin - 14050 Caen cedex - France

ABSTRACT.

Layered lead and bismuth cuprates form a large family of high T_c superconductors whose structural principles are well known. However the superconducting properties of these materials are closely related to complex problems of non-stoichiometry dealing not only with oxygen but also with lead, bismuth and strontium, complicated by the appearance of order-disorder phenomena. These issues are studied and discussed here for the "1212" and "0201" lead and for "2201" and "2212" bismuth cuprates. A particular attention is paid to the synthesis and to the study of superconducting properties of the pure phases $La_{2-x}Pb_xCuO_{4\pm\delta}$, $Bi_2Sr_2CaCu_2O_8$ and $Bi_{2.1}Sr_{1.9}CaCu_2O_{8+\delta}$.

1 - Introduction.

The recent studies of thallium cuprates have definitely shown that the redox reactions, involving the couples Cu(II) / Cu(III) and Tl(I) / Tl(III) in the same crystal, had a key-role for the optimization of the critical temperature of these materials (see for a review ref. 1). For instance one observes that the oxide $Tl_2Ba_2CuO_6$ synthesized under an oxygen pressure of ten bars is not a superconductor, but becomes a bulk superconductor with a critical temperature of 92K, by annealing at low temperature (270°C) under a reducing atmosphere. Moreover the HREM observations show that this dramatic variation of the superconducting properties cannot in any case be attributed to the formation of extended defects, since the regular stacking of the various layers are identical before and after annealing. Clearly, a very small deviation from oxygen stoichiometry is at the origin of the variation of Tc's. Such a phenomenom which corresponds to a drastic influence of the hole carrier density upon the superconducting properties of these materials should also be observed in other layered cuprates. In the present work we report on the recent results obtained for the lead and bismuth cuprates.

2 - Lead cuprates.

The control of the hole-carrier density in lead cuprates is a complex issue owing to the possible coexistence in the same crystal of the two redox couples Cu(II) / Cu(III) and Pb(II) / Pb(IV). Thus it appears as rather difficult to optimize the superconducting properties of these phases.

Three types of intergrowths have been isolated for the lead cuprates :

(i) The oxides $Pb_2Sr_2Ca_{1-x}Y_xCu_3O_8$[2-3] which consist of an intergrowth of single rock salt type layers with double pyramidal copper layers and single layers of Cu^IO_2 sticks (Fig. 1a).

(ii) The oxides $Pb_{1-x}Cu_xSr_2Ca_{1-y}Y_yCu_2O_{7-\delta}$[4-5] and $Pb_{0.5}Sr_{2.5}Ca_{0.5}Y_{0.5}Cu_2O_{7-\delta}$[6-8] which correspond to an intergrowth of single rock salt layers $[Pb_{0.5}Cu_{0.5}O]_\infty$ and $[Pb_{0.5}Sr_{0.5}O]_\infty$ respectively, with double pyramidal copper layers (Fig. 1b).

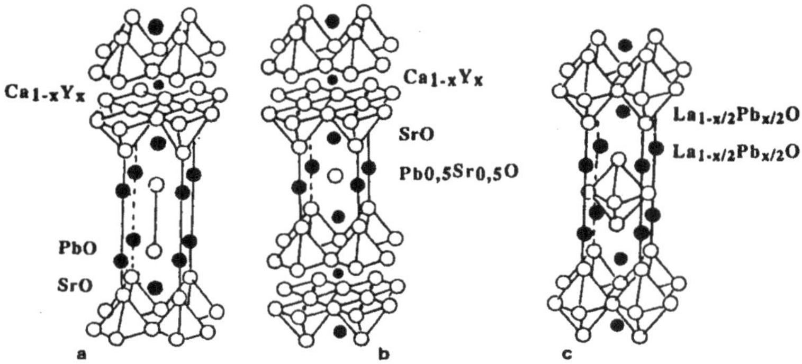

Figure 1 : Structure of $Pb_2Sr_2Ca_{1-x}Y_xCu_3O_8$ (a), $Pb_{0.5}Sr_{2.5}Ca_{1-x}Y_xCu_2O_7$ (b) and $La_{2-x}Pb_2CuO_{4\pm\delta}$ (c).

(iii) The pseudo-tetragonal oxide $La_{2-x}Pb_xCuO_{4\pm\delta}$[9], which exhibits a K_2NiF_4 type structure, i.e. corresponds to an intergrowth of single $[La_{1-x}Pb_xO]_\infty$ layers with single octahedral perovskite layers (Fig. 1c).

In all these compounds the appearance of superconductivity depends critically on the thermal treatment and especially on the oxygen pressure used for the synthesis.

In the cuprate $Pb_2Sr_2Ca_{1-x}Y_xCu_3O_8$[2] superconducting properties up to 68K were obtained by preparing this phase at a low oxygen pressure, i.e. in a nitrogen flow containing 1% oxygen. The critical influence of the oxygen is demonstrated by the disappearance of superconductivity by annealing this cuprate at higher oxygen pressure without destroying the structure, leading to a significant oxygen intercalation. The partial replacement of lead by bismuth[3] allowing zero resistance to be obtained at 79K supports also the idea that the oxygen pressure has to be carefully controlled in order to optimize superconductivity in this material.

The study of the "1212" lead cuprates illustrates the importance of oxygen stoichiometry on the optimization of the critical temperature of these compounds. The as-synthesized oxide, with nominal composition $Pb_{1-x}Cu_xSr_2Ca_{0.5}Y_{0.5}Cu_2O_7$, prepared in evacuated ampoule does not superconduct. Annealing of this phase at low temperature (400°C) and under different oxygen pressures ranging from 10 bars to 100 bars shows a spectacular appearance of superconductivity as soon as an oxygen pressure of ten bars is applied. The important feature deals with the evolution of the critical temperature of this phase versus the applied oxygen pressure (Fig. 2), which increases dramatically from 0 to 29K by increasing the pressure up to ten bars, then increases more smoothly from 29 to 40K until a pressure of 30 bars leading to a maximum of T_c of 42K for an oxygen pressure of 40 to 60 bars. Besides this "$Pb_{0.5}Cu_{0.5}$" cuprate, the isotypic oxide $Pb_{0.5}Sr_{0.5}Sr_2Ca_{0.5}Y_{0.5}Cu_2O_7$, which corresponds to the replacement of copper by strontium in the rock salt layers, exhibits a similar behaviour. The $\chi'(T)$ curves (Fig. 3) of the various samples prepared under different oxygen pressures using SrO_2 and PbO_2 as source of oxygen illustrate the same tendency. For low oxygen pressures the superconducting volume is very weak (lower than 1%) and then it increases up to 40% with a T_c (onset) of 90K for an oxygen pressure of 10 bars. In the same way, annealing of this phase at various temperatures and oxygen pressures allows Tc's to be increased. Thus both oxides, "$Pb_{1-x}Cu_x$" and "$Pb_{1-x}Sr_x$" exhibit very similar properties in agreement with their structure.

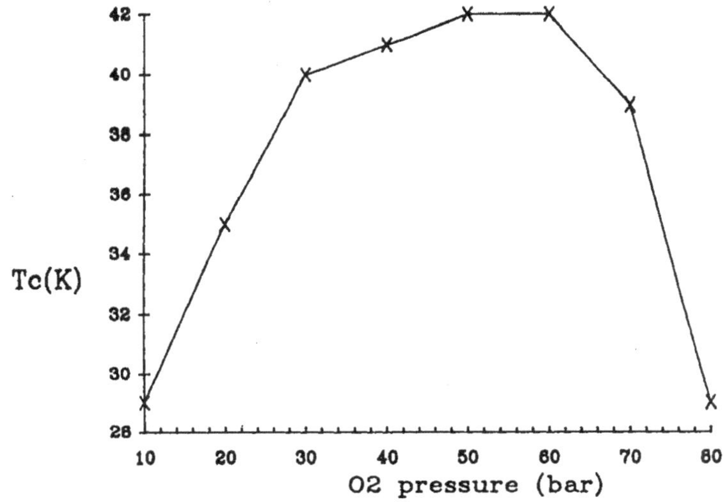

Figure 2 : Oxygen pressure dependence of Tc for $(Pb_{0.7}Cu_{0.3})Sr_2(Ca_{0.5}Y_{0.5})Cu_2O_7$ annealed at 400°C for 10 hours.

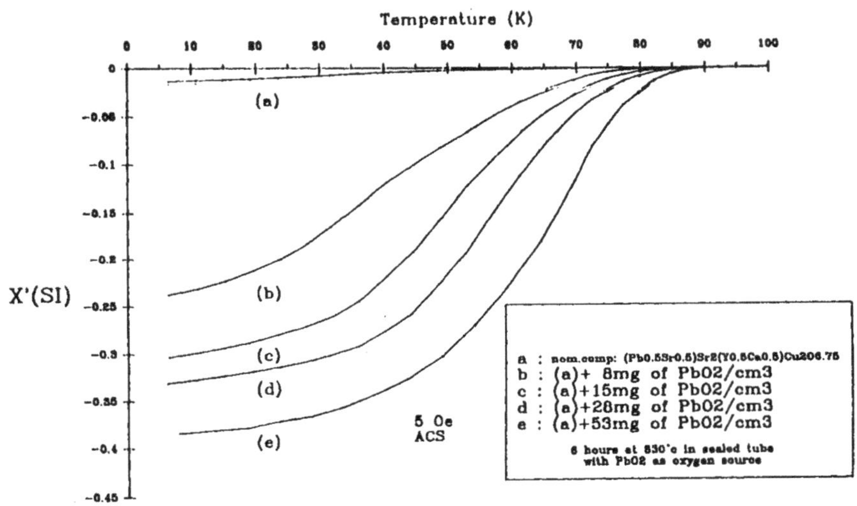

Figure 3 : AC susceptibility curves versus temperature for $Pb_{0.5}Sr_{2.5}Y_{0.5}Ca_{0.5}Cu_2O_{6.75}$ samples prepared in sealed tube at 830°C for 6 hours with PbO_2 as source of oxygen.

However, one observes that the maximum critical temperature of the "$Pb_{1-x}Sr_x$" cuprate is much higher than that of the "$Pb_{1-x}Cu_x$" cuprate. No conclusion can be drawn from this difference as long as we are not able to determine the oxygen stoichiometry of the two families of oxides, which certainly plays a key role for their critical temperature.

The recent discovery of the pseudo-tetragonal phase $La_{2-x}Pb_xCuO_{4\pm\delta}$[9] with a K_2NiF_4 structure shows also the ability of lead to be distributed with lanthanum in the rock-salt type layers. This oxide, synthesized in evacuated ampoule from nominal composition $La_{2-x}Pb_xCuO_{4+x/2}$, was obtained as a pure phase for $0.50 \leq x \leq 0.60$. The magnetic curves $\chi'(T)$ of the as-synthesized samples (Fig. 4) measured in a field of 7G show traces of superconductivity below 12K for x=0.50 and a significant superconducting volume for x=0.60 (4% at 4.5K). In order to discard the possibility of contribution of traces of La_2CuO_4 to this phenomenom, the two latter samples were annealed under different oxygen pressures. Fig. 5 shows the spectacular effect of oxygen pressure upon the superconducting properties of the "$La_{1.5}Pb_{0.5}$" phase. One indeed observes that the superconducting volume reaches 25% at 4.5K for an oxygen pressure of 40 bars and that Tc (onset) is of 15K. A similar effect is observed for "$La_{1.4}Pb_{0.6}$", for which it seems that there exists an optimum value of the oxygen pressure allowing maximum values of Tc's and of the superconducting volume to be reached. The XRD patterns of the annealed samples confirm that neither the cell parameters nor the peak intensities have been affected by the oxygen annealing, the phase remaining absolutely pure. These results demonstrate the existence of a new superconductor with a Tc (onset) of 15K and confirm the important role of oxygen content in these lead-based oxides for the optimization of superconductivity. Moreover it should be pointed out that higher Tc's should be obtained in this system, as shown for the sample with nominal composition $LaPbCuO_{4.5}$ (Fig. 6), for which a Tc of 18K is reached. Unfortunately this composition is biphasic, containing besides the tetragonal superconducting phase significants amounts of the pyrochlore $La_2Pb_2O_7$.

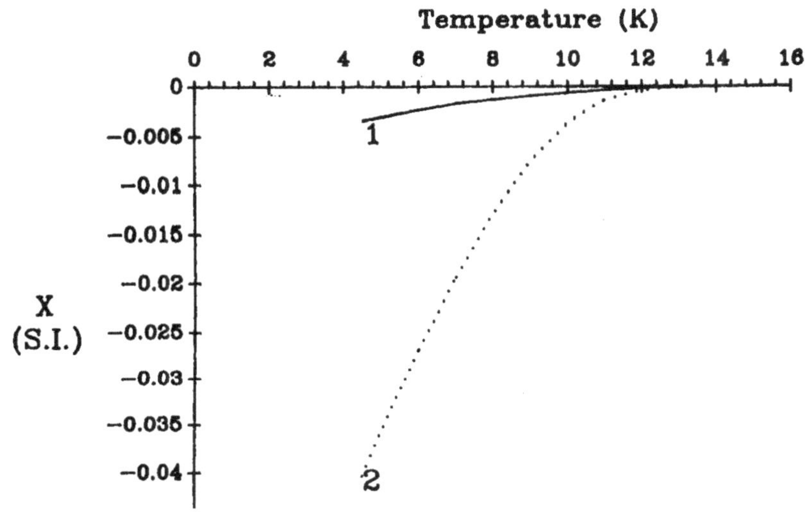

Figure 4 : Susceptibility χ versus T for the two nominal compositions $La_{1.5}Pb_{0.5}CuO_{4.15}$ (1) and $La_{1.4}Pb_{0.6}CuO_{4.25}$ (2).

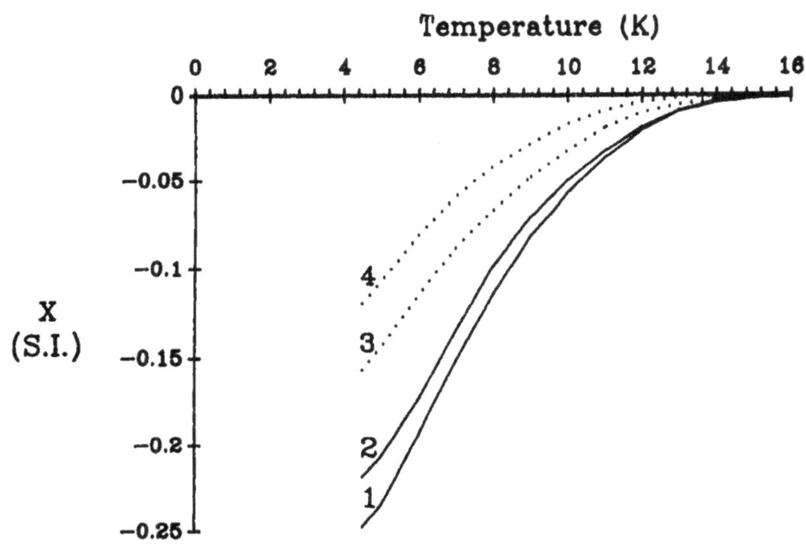

Figure 5 : DC susceptibility χ versus T for the two oxygen pressures 20 and 40 bars. $La_{1.5}Pb_{0.5}CuO_{4.25}$ samples : 20b (1) and 40b (2); $La_{1.4}Pb_{0.6}CuO_{4.3}$ samples : 20b (3) and 40b (4).

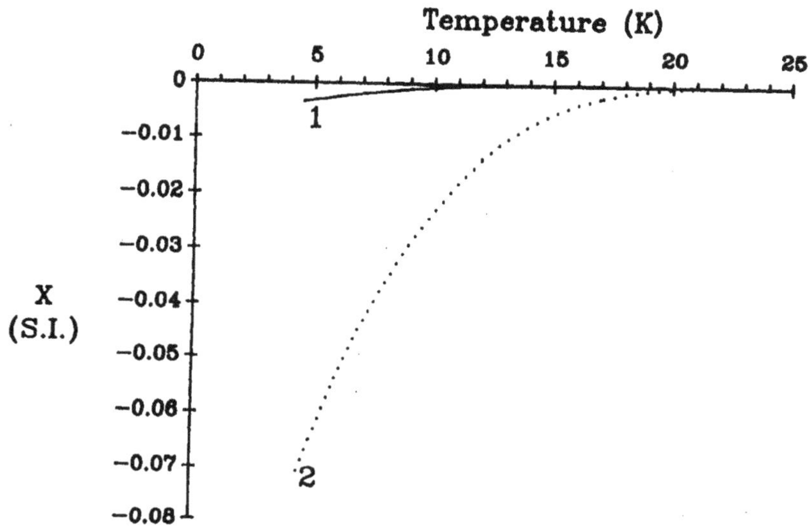

Figure 6 : DC susceptibility versus T for the two nominal compositions : $La_{1.5}Pb_{0.5}CuO_{4.25}$ (1) and $LaPbCuO_{4.5}$ (2).

There is no doubt that the superconducting properties of these lead cuprates are so far not optimized. The broad transition, shown by all the $\chi'(T)$ curves, suggests a dramatic inhomogeneity in the distribution of oxygen in the structure. The redox equilibrium Pb(II) + 2 Cu(III) $\rightleftarrows$ Pb(IV) + 2 Cu(II), will have to be taken into consideration in order to determine the role of the different species in the optimization of superconductivity.

3 - Bismuth cuprates.

A great deal of works has been devoted to the superconducting bismuth cuprates whose structural principles, based on the chemical formula $Bi_2Sr_2Ca_{m-1}Cu_mO_{2m+4}$, are theoritically very simple, i.e. corresponding to the intergrowths of multiple copper layers with triple rock salt type layers $[(BiO)_2SrO]_\infty$. The reality is much more complicated. Several single crystal X-ray diffraction investigations[10-21] and electron microscopy studies[12, 13, 22-36] have shown the existence of an excess oxygen and a dramatic modulation of the structure of these phases which hinders an accurate determination of the atomic positions. Moreover, the possibility of bismuth or strontium non stoichiometry, leading to order-disorder phenomena which may influence the superconducting properties, has to be considered. In this section, we present our very recent studies of the "2212" and "2201" bismuth cuprates which shed some light on the issues of non stoichiometry and modulation in these phases in connection with their superconducting properties.

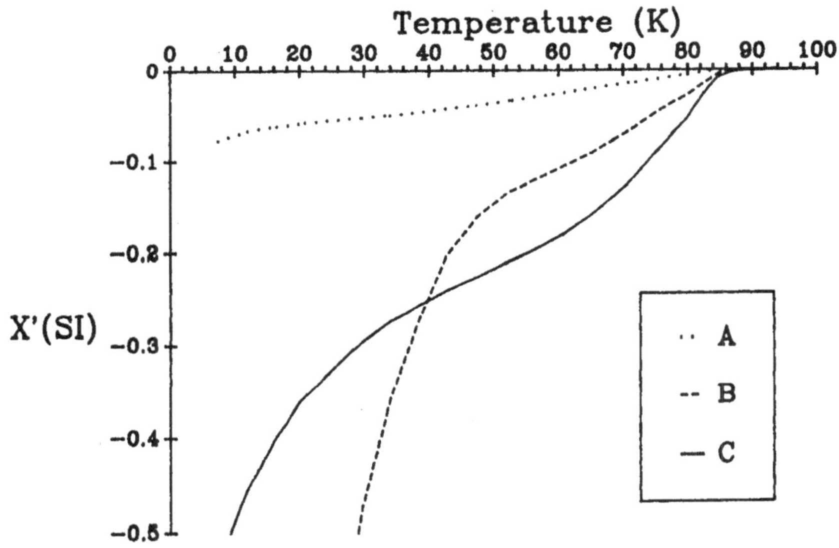

Figure 7 : AC susceptibility χ' (T) of the Ar prepared stoichiometric $Bi_2Sr_2CaCu_2O_8$. A, B, C correspond to the last step temperatures of 770, 775 and 780°C respectively.

The different studies of the "2212" bismuth cuprates have shown that the critical temperature of this phase depends on the oxygen non stoichiometry δ, according to the formulation $Bi_2Sr_2CaCu_2O_{8+\delta}$. However the different data are often divergent, so that it is not really possible to establish correlations between δ and Tc's[37, 38]. An important result has been obtained for this phase by Triscone et al[39] who have shown Tc's goes through a maximum of 94K for an optimum oxygen pressure of 10^{-5} bar. Among all these interesting results, an issue remains unanswered which deals with the exact compostion of the "2212" cuprate. Indeed, in all the synthesis carried out in air, one always observes an impurity besides the "2212" bismuth cuprate. The presence of such a secondary phase prevents any interpretation of the relationships between oxygen non-stoichiometry and superconductivity. The recent synthesis of the pure stoichiometric phase $Bi_2Sr_2CaCu_2O_8$ performed in an argon flow[40] shows definitely that an excess oxygen is not necessary to induce superconductivity in this oxide. Indeed the magnetic $\chi'(T)$ curves corresponding to three different single phase samples and which differ only by the last step temperature during synthesis - A (770°C), B (775°C) and C (780°C) - show the presence of bulk superconductivity with a Tc (onset) of 86K (Fig. 7). The fact that the superconducting volume increases with the temperature of the last step of synthesis is explained by a grain size effect; microstructure observations indeed show that the 770°C ceramic A-exhibits a grain size smaller than 1 μm, i.e. close to the penetration depth, whereas a size of about of several μm is observed for sample B (20% superconducting volume at 30K) and 10 μm for sample C (30% superconducting volume at 30K). The second transition observed at much lower temperature is due to a grain boundary effect as shown from the displacement of this transition toward higher temperature by decreasing the magnetic field (Fig. 8).

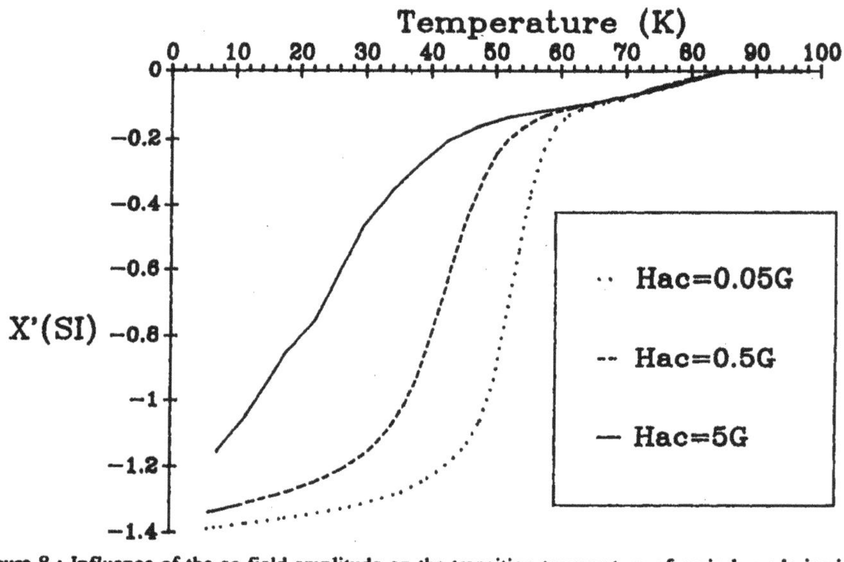

Figure 8 : Influence of the ac field amplitude on the transition temperature of grain boundaries in the single phase B sample $Bi_2Sr_2CaCu_2O_8$.

The transport measurements (Fig. 9) confirm the magnetic results, with a Tc (onset) of 95K and semi-metallic behaviour above this temperature. They also show that the broad character of the transition is not due to inhomogeneity but to the poor quality of the ceramic; indeed it is observed that increasing the pressure on the pellets before sintering from 1 $t.cm^{-2}$ to 10 $t.cm^{-2}$ allows Tc (R=0) to be increased from 66 to 72K, whereas a significant decrease of the normal resistance is obtained. Unfortunately, it was not possible to sinter this phase at higher temperature without any decomposition.

The annealing of one of these samples (C) in air at low temperature (400°C) for various times and under an oxygen pressure of 100 bars allows oxygen intercalation to be performed without any formation of a secondary phase. In the intercalated oxide $Bi_2Sr_2CaCu_2O_{8+\delta}$, δ can be increased from 0 to 0.10, whereas correlatively Tc decreases smoothly from 86K to 82K (table 1).

Table 1. : Modulation vector (q) and critical temperature (T_c) as a function of δ for biphasic $Bi_2Sr_2CaCu_2O_{8+\delta}$ (A), and monophasic $Bi_2Sr_2CaCu_2O_{8+\delta}$ (B) and $Bi_{2.1}Sr_{1.9}CaCu_2O_{8+\delta}$ (C).

δ	A		B		C	
	q	Tc(K)	q	Tc(K)	q	Tc(K)
-0.01			4.91	86		
+0.01			4.76	83		
+0.06	4.85	85	4.74	82		
+0.08			4.70	82		
+0.10	4.82	94				
+0.17					4.77	90
+0.20					4.71	76
+0.23					4.65	66
+0.32	4.68	70				

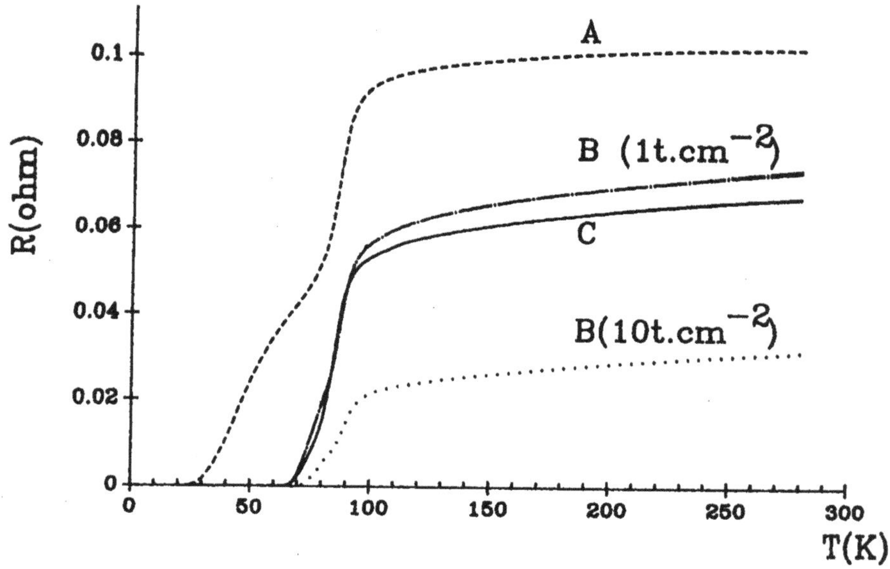

Figure 9 : Pure stoichiometric $Bi_2Sr_2CaCu_2O_8$ electrical resistance vs.T for the samples A, B, C. The effect of the pressure before sintering is shown for the B sample.

In order to understand the influence of the secondary phase upon Tc's in the air synthesized mixture $Bi_2Sr_2CaCu_2O_{8+\delta}$, attempts were made to synthesize a pure phase in air by changing the ratio Bi : Sr. A pure phase $Bi_{2.1}Sr_{1.9}CaCu_2O_{8+\delta}$ was synthesized in air, for δ=0.20[41], whose purity was checked by X-ray diffraction and electron diffraction on about 100 crystals. The critical temperature of this phase, 76K, is significantly smaller than that of $Bi_2Sr_2CaCu_2O_8$. Annealing this phase under argon at low temperature (400°C), allows oxygen desintercalation to be performed without any destruction of the structure, δ decreasing from 0.20 to 0.17. In the same way, annealing the phase under 100 bars oxygen pressure at 400°C allows the oxygen content to be increased from 0.20 to 0.23. But the most striking feature deals with the evolution of Tc's versus δ (table 1). One indeed observes like for $Bi_2Sr_2CaCu_2O_8$, that Tc's decreases as δ increases from 90K for δ=0.17 to 66K for δ=0.23. However contrary to $Bi_2Sr_2CaCu_2O_8$, a small variation of the oxygen stoichiometry of $Bi_{2.1}Sr_{1.9}CaCu_2O_{8+\delta}$, involves a dramatic variation of the critical temperature.

On the opposite, we could confirm in agreement with previous authors that the air synthesized composition $Bi_2Sr_2CaCu_2O_{8+\delta}$ is always a mixture, the "2212-type" phase being accompanied by a secondary phase identified as $Ba_9Sr_{11}Ca_5O_{30-y}$; moreover the amount of the latter varies with the oxygen content δ. For these biphasic compositions, it is observed (table 1) that Tc's goes through a maximum value of 94K for an optimum δ value of 0.10, in agreement with the results previously established by Triscone et al[39].

These results demonstrate that there is not only one "2212-phase" with a fixed ratio Bi : Sr equal to 1, but that besides the pure phase $Bi_2Sr_2CaCu_2O_{8+\delta}$, there exists a large possibility of non-stoichiometry on the cationic sites, such as the solid solution $Bi_{2+x}Sr_{2-x}CaCu_2O_{8+\delta}$ or the non-stoichiometric oxides $Bi_{2\pm x}Sr_{2\pm y}CaCu_2O_{8+\delta}$. Consequently, the evolution of Tc's observed for the air synthesized mixture "$Bi_2Sr_2CaCu_2O_{8+\delta}$" can only be interpreted as the result of the existence of distinct "2212"-type phases, which differ not only by their oxygen content but also by their cationic distribution.

Another important issue of the "2212" bismuth cuprates deals with the modulation of the structure and its relationships with oxygen non-stoichiometry and superconductivity. The origin of this phenomenon, which was evidenced by several authors, is always subject to controversy since some of them think that it results from an excess oxygen in the structure[20, 21, 39] whereas others suggest that the $6s^2$ lone pair of Bi(III) is an important factor[43]. The systematic measurements of the q vector on enlarged electron diffraction patterns of the two series of pure phases, $Bi_2Sr_2CaCu_2O_{8+\delta}$ and $Bi_{2.1}Sr_{1.9}CaCu_2O_{8+\delta}$, allow further conclusions to be drawn. The first important point deals with the fact that the pure stoichiometric phase $Bi_2Sr_2CaCu_2O_8$ exhibits, like other oxides, incommensurate satellites (Fig. 10) with a q vector close to 4.91; this shows definitely that an excess oxygen is not necessary for the appearance of the modulation, and supports more strongly the effect of the $6s^2$ lone pair of Bi(III). The second feature concerns the evolution of the q vector (table 1); one observes that in each series of oxides $Bi_2Sr_2CaCu_2O_{8+\delta}$ and $Bi_{2.1}Sr_{1.9}CaCu_2O_{8+\delta}$, the q vector decreases significantly as δ increases, from q=4.91 for δ=0 to q=4.70 for δ=0.8 for the first series and from q=4.77 (δ=0.17) to q=4.65 (δ=0.23) for the second series. However the comparison of the two series, shows that the oxygen stoichiometry is not the only factor which affects the modulation, but that the cationic composition, i.e. likely the ordering-disordering phenomena of bismuth and strontium in the distorted rock-salt layers have a contribution to the modulation. These observations show that oxygen intercalation is not at the origin of the modulation and reinforce the hypothesis that Bi(III), due to its lone pair, induces the buckling and corrugation of the layers. On return, the modulation can be modified by oxygen intercalation or desintercalation and by the distribution of bismuth and strontium on the cationic sites. In any case, no correlation can be established between the modulation and superconductivity.

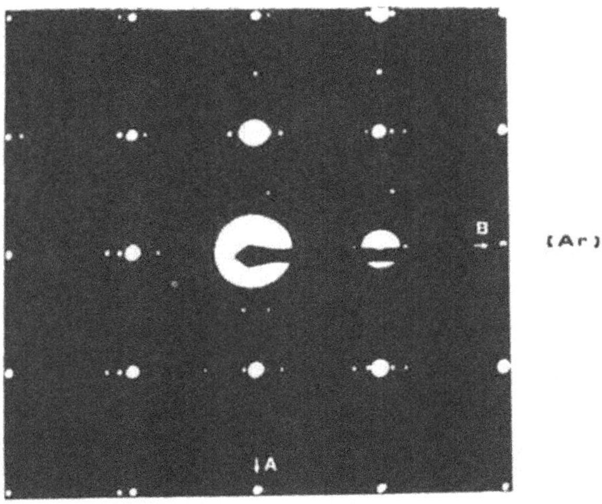

Figure 10 : [001] electron diffraction pattern for pure stoichiometric $Bi_2Sr_2CaCu_2O_8$.

The "2201" bismuth cuprate, with ideal composition $Bi_2Sr_2CuO_6$, although it superconducts at lower temperature[22], is an interesting material for the understanding of the relationships between structure, non-stoichiometry and superconductivity owing to its more simple composition. Like the "2212-type" structure, this phase exhibits incommensurate satellites. Several single crystal X-ray investigations have been performed using the main reflections[15, 18, 44] which allowed the structural model of the mean structure to be proposed. Other studies taking into account satellites reflections[45-47] allowed conclusions to be drawn about the modulation of the distribution and displacements of heavy atoms and especially of bismuth in the different layers. However, no complete single crystal structure determination had been performed on this phase owing to the poor quality of the crystals. For this reason a complete study of the modulated structure of a "2201" crystal with nominal composition $Bi_2Sr_2CuO_6$ was undertaken[48], using all the informations provided by first order and second order satellites corresponding to the vectors $q_1^* \approx 0.20$ (along a*) and $q_3^* \approx 0.50$ (along c*). The first important result deals with the composition of the crystal deduced from the structure refinements and elementary analysis. One observes indeed a Bi/Sr molar ratio significantly greater than 1, which leads to a cationic non-stoichiometry according to the formula $Bi_{2.08}Sr_{1.84}CuO_{6-\delta}$. Moreover it is demonstrated that a significant part of the Sr sites are occupied by bismuth in the mean structure (up to about 8%), whereas the part of the bismuth sites are vacant (less than 4%). These features are of great interest since they confirm that the cationic non-stoichiometry and order-disorder phenomena must be considered as a general property of these bismuth cuprates, as discussed above for the "2212" structure.

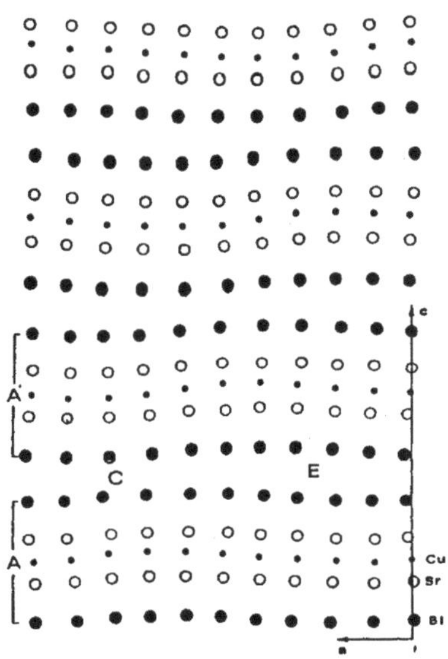

Figure 11 : Waving of the Bi, Sr and Cu sheets in the actual crystal $Bi_{2.08}Sr_{1.84}CuO_{6-\delta}$, projected on the (010) plane. The atomic displacements from the basic positions are not magnified.

The displacive modulation of Bi, Sr and Cu is almost harmonic along $\vec{c}$ (direction perpendicular to the layers) whereas it deviates significantly from harmonicity along $\vec{a}$; the anharmonicity effect is enhanced for the oxygen atoms. Fig. 11 gives an image of this modulated structure by considering a restricted number of unit cells, five along $\vec{a}$ (q_1 being close to 1/5) and two along $\vec{c}$ (q_2 being close to 1/2). It can be seen that the structure consists of blocks of five corrugated layers "Bi-Sr-Cu-Sr-Bi" (labelled A). Within each A block the five layers exhibit approximately the same curvature, i.e. the different cations of these layers display rather similar displacements along $\vec{c}$. Two successive blocks A and A' are inversed with respect to each other. Correlatively the two bismuth layers which form the junction of two adjacent blocks do not exhibit the same waving. The distance along $\vec{c}$ between these two Bi layers varies continuously, forming expanded regions (labelled E) where the interlayer distances are maximum and compressed regions (labelled C) where the interlayer distances are the smallest. An unexpected result deals with the fact that the oxygen apical sites of the CuO_6 octahedra (labelled O and O') are only partly occupied, with a mean occupancy factor of about 75%. The occupancy probability function P' of these sites and those of Sr sites versus the t parameter which characterizes the $\vec{P}$ unit cells of the modulated structure are drawn on Fig. 12. It can be seen that the occupancy probabilities of the O and O' sites and Sr sites are correlated in a spectacular way. In the parts of the crystal where the occupancy probabilities of the Sr sites by strontium are minimum, i.e. where the Sr sites are occupied by bismuth, the occupancy probabilities of O and O' sites are about 1. On the opposite, the apical O and O' sites exhibit low occupancy probabilities when the Sr(1) sites are likely occupied by strontium only. The last important feature of this very accurate structure determination concerns the oxygen content. Indeed no excess of oxygen could be detected with respect to the ideal formula $Bi_2Sr_2CuO_6$, but on the opposite the maximum oxygen content deduced from this X-ray study leads to the formula $Bi_{2.08}Sr_{1.84}CuO_{5.5}$. This suggests the existence of extra oxygen diluted in different sites of the structure, not detectable by X-ray diffraction. A careful analysis of the modulated structure, taking into account the steric effects between oxygen atoms in the different cells with variable cationic occupancies, leads to the conclusion that extra oxygen atoms are probably located both within and between the Bi-O layers giving rise to a complex coordination of bismuth. But in any case, this additional oxygen, cannot overpass the amount of 0.4 to 0.5 mole per formula in agreement with the electrical neutrality involving Bi(III) and Cu(II). These results which show cationic disorders and oxygen vacancies in the "2201" structure explain the difficulty of optimization and even of appearance of superconductivity in this material.

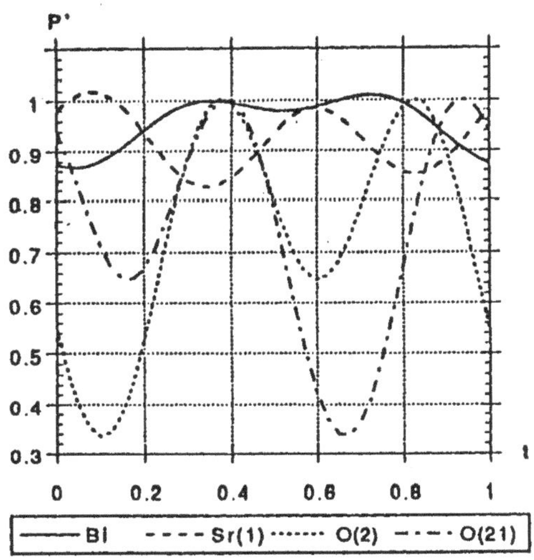

Figure 12 : Occupancy probabilities P' of the Bi, Sr, O and O' apical sites in $Bi_{2.08}Sr_{1.84}CuO_{6-\delta}$.

4 - Concluding remarks.

The structural principles which govern the solid state chemistry of lead and bismuth cuprates are now well established. However these recent results show definitely that non-stoichiometry in these oxides is much more complicated than expected, dealing not only with oxygen but also with cations such as Pb, Sr, Bi and even Ca and order-disorder phenomena. All these issues will have now to be studied in detail in order to be able to synthesize pure phases and to optimize their superconducting properties.

5 - References.

1 - B. Raveau, A. Maignan, C. Michel, C. Martin, M. Hervieu, V. Caignaert and J. Provost, *Proceedings Internat. Conf. High Tc Superconductivity and Localization phenomena*, Moscow 11-17 May 1991.

2 - R.J. Cava, E. Battlogg, J.J. Krajewski, L.W. Rupp, L.F. Schneemeyer, T. Siegrist, R.B. Van Dover, P. Marsh, W.F. Peck, P.K. Gallagher, S.H. Glarum, J.H. Marshall, R.C. Farrow, J.V. Waszczak, R. Hull and P. Trevor, *Nature*, 336, (1988) 211.

3 - R. Retoux, C. Michel, M. Hervieu and B. Raveau, *Modern Physics Lett.* 3, (1989) 591.

4 - M.A. Subramanian, J. Gopalakrishnan, C.C. Torardi, P.L. Gai, E.D. Boyes, T.R. Askew, R.B. Flippen, W.E. Farneth and A.W. Sleight, *Physica* C 157, (1989) 124.

5 - J.Y. Lee, J.S. Swinnea and H. Steinfink, *J. Mat. Res.* 4, (1989) 763.

6 - T. Rouillon, J. Provost, M. Hervieu, D. Groult, C. Michel and B. Raveau, *Physica C*, 159, (1989) 201.

7 - T. Rouillon, A. Maignan, M. Hervieu, D. Groult, C. Michel and B. Raveau, *Physica C* 171, (1990), 7.

8 - A. Maignan, T. Rouillon, D. Groult, J. Provost, M. Hervieu, C. Michel and B. raveau, *Physica C* (1991), in press.

9 - A. Maignan, C. Michel, M. Hervieu, J. Provost, D. Groult and B. Raveau, *Physica C* (1991), in press.

10 - J.M. Tarascon, Y. Lepage, P. Barboux, B.G. Bagley, L.M. Greene, W.R. Mc Kinnon, G.W. Hull, M.G. Giroud, B.W. Hwang, *Phys. Rev.* B37, (1988) 9382.

11 - M.A. Subramanian, C.C. Torardi, J.C. Calabrese, J. Gopalakrishnan, K.J. Morrissey, T.R. Askew, R.B. Flippen, U. Chowdry, A.N. Sleight, *Science* 239, (1988) 1015.

12 - M. Hervieu, B. Domenges, C. Michel, B. Raveau, *Mod. Phys. Lett.* 2, (1988) 835.

13 - M. Hervieu, C. Michel, B. Domenges, Y. Laligant, A. Lebail, G. Ferey, B. Raveau, *Mod. Phys. Lett.* 2, (1988) 491.

14 - P. Bordet, J.J. Capponi, C. Chaillout, J. Chenavas, A.W. Hewat, E.A. Hewat, J.L. Hodeau, M. Marezio, J. Tholence, D. Tranqui, *Physica C* , 153-155, (1988) 623.

15 - H.G. Von Schnering, L. Walz, M. Schwarz, W. Becker, M. Hartweg, T. Popp, B. Hettich, P. Muller, G. Kampf, *Angew. Chem.* Int'l Ed. Engl. 27, (1988) 574.

16 - T. Kajitani, K. Kusuba, M. Kikuchi, N. Kobayashi, Y. Syono, T. Williams, M. Irabayashi, *Jpn. J. Appl. Phys.* 27, (1988) L587.

17 - S.A. Sunshine, T. Siegrist, L.F. Schneemeyer, D.W. Murphy, R.J. Cava, B. Batlogg, R.B. Van Dover, R.M. Fleming, S.H. Glarum, S. Nakahara, R. Farrow, J.J. Krajewski, S.M. Zahurak, J.V. Waszczak, J.H. Marshal, P. March, L.W. Rupp, W.F. Peck, *Phys. Rev.* B38, (1988) 893.

18 - C.C. Torardi, M.A. Subramanian, J.C. Calabrese, J. Gopalakrishnan, E.M. Mc Carron, K.J. Morrissey, T.R. Askew, R.B. Flippen, U. Chowdry, A.W. Sleight, *Phys. Rev.* B38, (1988) 225.

19 - V. Petricek, Y. Gao, P. Lee, Ph. Coppens, *Phys. Rev.* B, 42, (1990) 387.

20 - E.A. Hewat, J.J. Capponi and M. Marezio, *Physica C* 157, (1989) 502.

21 - Y. Le Page, W.R. Mc Kinnon, J.M. Tarascon and P. Barboux, *Phys. Rev.* B, 40, (1989) 6810.

22 - C. Michel, M. Hervieu, M.M. Borel, A. Grandin, F. Deslandes, J. Provost, B. Raveau, *Z. Phys.* B68, (1987) 421.

23 - G. Van Tendeloo, H.W. Zandbergen, S.A. Amelinckx, *Solid State Commun.* 66, (1988) 927.

24 - G. Van Tendeloo, H.W. Zandbergen, J. Van Lenduy, S.A. Amelinckx, *Appl. Phys. A* 46, (1988) 233.

25 - E. Kakayama, E. Muromachi, Y. Uchida, A. Ono, F. Izumi, M. Onoda, *Jpn. J. Appl. Phys.* 27, (1989) L365.

26 - T. Kijima, J. Tanaka, Y. Bando, M. Onoda, F. Izumi, *Jpn. J. Appl. Phys.* 27, (1988) L369.

27 - P.L. Gai, P. Day, *Physica C* 152, (1988) 335.

28 - B. Raveau, C. Michel, M. Hervieu, J. Provost, *Physica C* 153-155, (1988) 3.

29 - E.A. Hewat, P. Bordet, J.J. Capponi, C. Chaillout, J.L. Hodeau, M. Marezio, *Physica C* 153-155, (1988) 619.

30 - E. Hewat, *Physica C* 152, (1988) 413.

31 - M. Hervieu, C. Michel, N. Nguyen, R. Retoux, B. Raveau, *Eur. J. Solid State Inorg. Chem.* 25, (1988) 1.

32 - Y. Matsui, H. Maeda, Y. Tanaka, S. Horiuchi, *Jpn. J. Appl. Phys.* 27, (1988) L372.

33 - J. Darriet, A. Le Lirzin, E. Marquestaut, B. Lepine, B. Chevalier, J. Etourneau, *Solid State Commun.* 69, (1989) 739.

34 - A.K. Ganguli, R. Nagarajan, K.S. Nanjrindaswamy, C.N.R. Rao, *Mater. Res. Bull.* 24, (1989) 103.

35 - J.M. Tarascon, P. Barboux, G.W. Hull, R. Ramesh, L.H. Greene, M. Giroud, M.S. Hedge, W.R. Mc Kinnon, *Phys. Rev.* B 39, (1989) 4316.

36 - S. Olivier, W.A. Groen, C. Van Der Beek, H.W. Zandbergen, *Physica C* 157, (1989) 531.

37 - C. Namgung, J.T.S. Irvine, J.H. Binks, E.E. Lachowski and A.R. West, *Supercond. Sci. Technol.* 2, (1989) 181.

38 - W.A. Groen, D.M. de Leeuw and L.F. Feiner, *Physica C* 165, (1990) 55.

39 - G. Triscone, J.Y. Genoud, T. Graf, A. Junod and J. Müller, *Physica C*, 176, (1991) 247.

40 - A.Q. Pham, A. Maignan, M. Hervieu, C. Michel, J. Provost and B. Raveau, *Physica C*, accepted (1991).

41 - A.Q. Pham, M. Hervieu, A. Maignan, C. Michel, J. Provost and B. Raveau, to be submitted.

42 - H.W. Zandbergen, W.A. Groen, F.C. Mijlhoff, G. Van Tendeloo and S. Amelinckx, *Physica C* 156, (1988) 325.

43 - C. Michel, C. Martin, M. Hervieu, D. Groult, D. Bourgault, J. Provost and B. Raveau, *Proc. Intern. Symp. on Superconductivity, Hsinshu*, Taïwan (1989).

44 - K. Imai, I. Nakai, T. Kawashima, S. Sueno, A. Ono, *Jpn. J. Appl. Phys.* 27, (1988) 1661.

45 - Y. Matsui, S. Takekawa, S. Horiuchi and A. Umezono, *Jpn. J. Appl. Phys.* 27, (1988) 1873-1876.

46 - M. Onoda and M. Sato, *Solid State Commun.* 67, (1988) 799-804.

47 - Y. Gao, P. Lee, J. Ye, P. Bush, V. Petricek and P. Coppens, *Physica C* 160, (1989) 431-438.

48 - H. Leligny, S. Durcok, P. Labbé, M. Ledesert and B. Raveau, *Acta. Cryst.* submitted.

THE COMPREHENSIVE PHASE DIAGRAM $BaO-La_2O_3-CuO-O$

F.ABBATTISTA, D.MAZZA and M.VALLINO
Dipartimento di Scienza dei Materiali e Ingegneria Chimica,
Politecnico di Torino, C.so Duca degli Abruzzi 24, 10129 Torino(Italy)

ABSTRACT

The equilibrium relationships were established among the different binary and ternary phases in the titled system, at 950°C and in CO_2-free air.

The different ternary phases are characterized by atomic ratios (La+Ba)/Cu corresponding to 3,2,1.5,1.33 and 1. Excluding the ratio value 3, to which corresponds the mixed crystals series $La_{2-x}Ba_{1+x}CuO_{5-x/2+\delta}$ ($0.1 \le x \le 0.8$), all other values correspond to terms of the series $AO\cdot(ABO_3)_n$ (A=La,Ba;B=Cu) with simple ($n=\infty$) or layered (n=1,2,3) perovskite structure. For the ratio value 2 two solids solutions are stable: $Ba_{2-x}La_xCuO_{3+x/2+\delta}$ ($0 \le x \le 0.667$) and $La_{2-x}Ba_xCuO_{4-x/2+\delta}$ ($0 \le x \le 0.2$); for the ratio value 1.5 mixed crystal series $La_{2-x}Ba_{1+x}Cu_2O_{6-x/2+\delta}$ originates ($0.25 \le x \le 0.8$); for the ratio 1.33 we evidenced for the first time $La_{1.33}Ba_{2.67}Cu_3O_{7.67+\delta}$; finally for the ratio of 1 results stable the phase defined by the stoichiometry $La_4BaCu_5O_{12+\delta}$ and the solid solution $La_{3-x}Ba_{3+x}Cu_6O_{13.5-x/2+\delta}$ ($0.2 \le x \le 0.8$).

INTRODUCTION

In three previous notes concerning the investigations about limited zones of the $BaO-La_2O_3-CuO-O$ system, the chemical-structural characteristics of the mixed crystals series $Ba_{2-x}La_xCuO_{3+\delta}$ [1], $La_{2-x}Ba_{1+x}CuO_{5-x/2+\delta}$ [2] and $La_{2-x}Ba_{1+x}Cu_2O_{6-x/2+\delta}$ [3] were described by us.

In the same note [3] the existence was also signaled, for the first time, of the phase $Ba_{2.67}La_{1.33}Cu_3O_{7.67+\delta}$, having a layered perovskite structure $Sr_4Ti_3O_{10}$-type.

With the aim to conclude our reserches about the quoted system with the comprehensive phase diagram, we examined many other compositions suitable to verify the relations of bi- and tri-phasic equilibria between the different phases in the whole system.

The results of this last research, completed with those already reported in the three preceding communications, are here discussed and compared with the most recent data in the literature about this subject.

EXPERIMENTAL PROCEDURE

The samples were prepared according to a technique already described by us in the previous works, starting from $Ba(NO_3)_2$, La_2O_3 and CuO. The heat treatments were performed at 950°C in gold vessels and in a stream of CO_2-free air, for times never shorter than 50 hours.

The samples were analyzed by X-ray powder diffraction, using CuK_α radiation, graphite monochromator on the diffracted beam and sample spinner. The average oxidation number of copper was determined by measuring the oxygen evolution of the sample after dissolution in 1N perchloric acid.

RESULTS AND DISCUSSION

The results we obtained are reported in the diagram of Fig.1. The various ternary phases, some of which originated from binary compounds by mutual substitution of barium and lanthanum, are characterized by atomic ratios (La+Ba)/Cu corresponding to the values 3, 2, 1.5, 1.33 and 1 respectively.

If we exclude the ratio (La+Ba)/Cu = 3, corresponding to the series of mixed crystals $La_{2-x}Ba_{1+x}CuO_{5-x/2+\delta}$, all other values correspond to terms of the series $AO.(ABO_3)_n$ (A = La, Ba; B = Cu) with layered perovskite type structures (n = 1, 2, 3) or simple perovskite type (for n = ∞).

As already shown in the note [2] the structural characteristics of the terms of the mixed crystals series $La_{2-x}Ba_{1+x}CuO_{5-x/2+\delta}$ $(0.1 \le x \le 0.8)$ close agree with the model proposed by Michel [4] for the phase with ideal composition "La_2BaCuO_5" (S.G. = P4/mbm). The parameters of

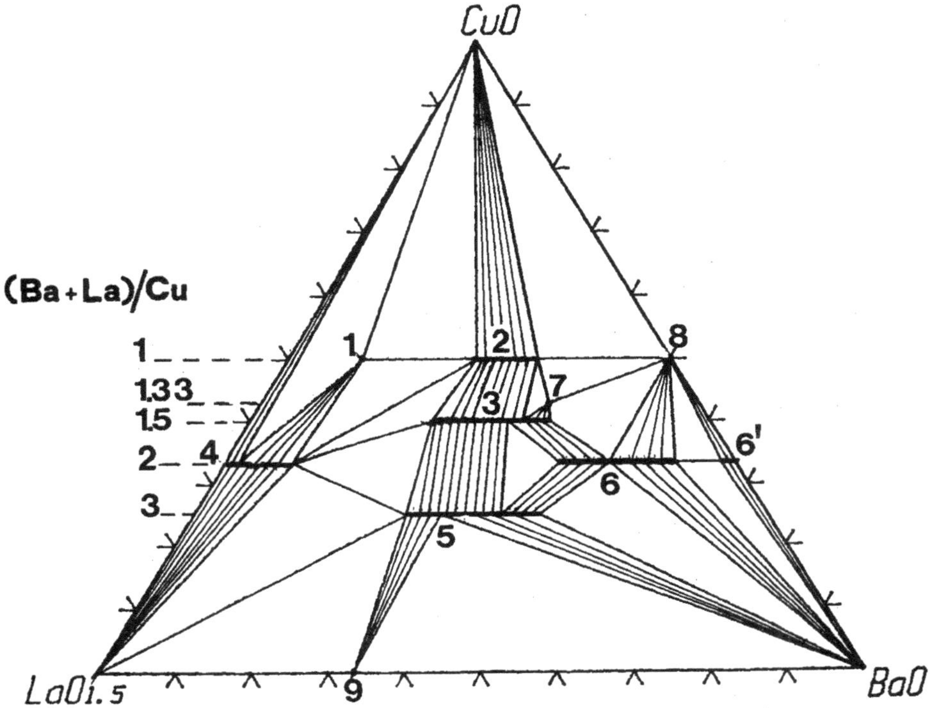

Fig.1 The phase diagram BaO - La_2O_3 - CuO - O in CO_2-free air at 950°C.
1) $BaLa_4Cu_5O_{12+\delta}$ 2) $Ba_{3+x}La_{3-x}Cu_6O_{13.5-x/2+\delta}$ 3) $Ba_{1+x}La_{2-x}Cu_2O_{6-x/2+\delta}$
4) $Ba_xLa_{2-x}CuO_{4-x/2+\delta}$ 5) $Ba_{1+x}La_{2-x}CuO_{5-x/x+\delta}$ 6) $Ba_{2-x}La_xCuO_{3+x/2+\delta}$
($0.25 \leq x \leq 0.667$) 6') $Ba_{2-x}La_xCuO_{3+x/2+\delta}$ ($0 \leq x \leq 0.05$)
7) $Ba_{2.67}La_{1.33}Cu_3O_{7.67+\delta}$ 8) $BaCuO_2$ 9) La_2BaO_4

TABLE I. Unit cell parameters and oxygen content for some terms of the solid solution $La_{2-x}Ba_{1+x}CuO_{5-x/2+\delta}$.

x	=	0.1	0.2	0.4	0.6	0.8	
a_o	=	6.847(4)	6.877(3)	6.898(4)	6.936(3)	6.946(4)	(A)
c_o	=	5.876(4)	5.877(3)	5.871(3)	5.877(3)	5.871(3)	(A)
δ	=	0.03	0.08	0.17	0.23	0.28	

the tetragonal unit cell of the different terms of this series together with the correspondent δ values are repoted in table 1.

This same phase was reported in the literature by other authors, however with incorrect stoichiometry: the range of the mixed crystals field of this phase were reported as $La_{4-2x}Ba_{2+2x}Cu_{2-x}O_{10-2x}$ ($0.1 \le x \le 0.25$) [4] [5] and $La_6Ba_4Cu_3O_{15}$ [6] and $La_{6-x}Ba_{4+x}Cu_3O_{15-x/2}$ ($1 \le x \le 1.7$) [7].

For the ratio (La+Ba)/Cu = 2 two restricted series of mixed crystals have been reported, $Ba_{2-x}La_xCuO_{3+x/2+\delta}$ and $La_{2-x}Ba_xCuO_{4-x/2+\delta}$, that originate from the binary compounds $Ba_2CuO_{3+\delta}$ and La_2CuO_4 respectively, by mutual substitution between barium and lanthanum. The phase $Ba_2CuO_{3+\delta}$ of the system BaO-CuO-O shows tetragonal symmetry above 740°C and orthorhombic at lower temperatures.

Our researches [1] about solids belonging to the series $Ba_{2-x}La_xCuO_{3+x/2+\delta}$ ($0 \le x \le 0.667$), quenched in liquid N_2 from 950°C, allowed us to ascertain the presence of two distinct monophasic fields for $0 \le x \le 0.05$ and $0.25 \le x \le 0.667$, characterized by mixed crystals with different structures. The terms of the first field show tetragonal symmetry and K_2NiF_4-type structure, analogous to that of the high-temperature (T>750°C) form of the dimorphous phase $Ba_2CuO_{3+\delta}$. The terms of the second monophasic field show, on the contrary, orthorhombic symmetry and Sr_2CuO_3-type structure as for the low-temperature modification of the phase $Ba_2CuO_{3+\delta}$ itself.

A single orthorhombic monophasic field with undefined width was instead reported in the note [7]. According to the authors, it can be pointed out in the examined samples only if they are slowly cooled in furnace.

No mention about the above mixed crystals series appears on the contrary in the note [5], in which moreover the authors rise doubs about the existence of the binary compound $Ba_2CuO_{3+\delta}$ itself.

The literature data concerning the mixed crystals series $La_{2-x}Ba_xCuO_{4-x/2+\delta}$ ($0 \le x \le 0.2$) [8][9][10][11], confirmed also by our experiences, agree on the contrary enough each other. All terms of this solid solution have a K_2NiF_4-type structure: the terms poorest in barium, including the compound La_2CuO_4 itself, show orthorhombic symmetry

(S.G. = Fmmm n°69) with $a_o \approx b_o \approx \sqrt{2} \cdot a(K_2NiF_4)$ and $c_o \approx c(K_2NiF_4)$, caused by the Jahn-Teller deformation of the octahedra CuO_6; the terms richest in barium (x>0.1) result on the contrary tetragonal, isomorphous with K_2NiF_4.

For the ratio (La+Ba)/Cu = 1.5 the mixed crystals series $La_{2-x}Ba_{1+x}Cu_2O_{6-x/2+\delta}$ takes rise, whose structure derives from that of the $Sr_3Ti_2O_7$ titanate and is analogous to that of the phases $La_{2-x}Sr_{1+x}Cu_2O_{6-x/2+\delta}$ ($0 \leq x \leq 0.14$) and $La_{2-x}Ca_{1+x}Cu_2O_{6-x/2+\delta}$ (x=0.1) already described by Nguyen and co-workers [12].

The data obtained by us [3] show that, at 950°C, the field of the examined mixed crystals extends within the limits ($0.25 \leq x \leq 0.8$) which well agree with the late recent results of De Leeuw [7]. Wholly different result, on the contrary, the data of Klibanow [5] that reports a compositional range ($0 \leq x \leq 0.2$) completely out of the field defined by us.

All terms of the series have tetragonal symmetry except the limit one (x=0.8) whose XRD patterns show clear signs of a transition from tetragonal to orthorhombic symmetry.

The extent of the above mixed crystals field appreciably increases with the temperature and at 1000°C the limit term takes the composition $LaBa_2Cu_2O_{5.5+\delta}$. In the terms very rich in barium ($0.8 \leq x \leq 1$) sintered at 1000°C, a progressive increase of the orthorhombic modification of the lattice has been remarked.

In Table 2 are reported the cell parameters and the Cu^{3+} content, for the various terms of the above series, including the term with x=1 obtained at 1000°C .

TABLE II. Cell parameters and Cu^{3+} content of some terms of the series $La_{2-x}Ba_{1+x}Cu_2O_{6-x/2+\delta}$.

x =	0.25	0.50	0.75	1.00	
a_o =	3.869	3.878	3.892	3.894	(A)
b_o =	-	-	-	3.943	(A)
c_o =	20.813	20.865	20.891	20.899	(A)
δ =	0.09	0.10	0.12	0.12	

For the ratio (La+Ba)/Cu = 1.33 the existence has been for the first time remarked by us [3] of a new phase with composition $La_{1.33}Ba_{2.67}Cu_3O_{7.67+\delta}$. Its XRD spectrogram is indexable according to a lattice with orthorhombic symmetry (S.G. = I/mmm) and structure derived from that of the compound $Sr_4Ti_3O_{10}$. The cell parameters (a_o = 3.956; b_o = 3.903 and c_o = 28.744A) agree with the structural model of the quoted titanate, characterized by the superposition of triple perovskite layers $[(Ba,La)CuO_{3-\delta}]_3$ to NaCl-type layers (Ba,La)O. Other data about the formation kinetics and the stability field of this phase are reported in the already cited note [3]. This phase originates only by mean of thermal treatment in the temperature range 960-1000°C; the same phase however may be brought to room temperature by slow cooling in furnace.

Finally, for an atomic ratio (La+Ba)/Cu = 1, Michel [13] and Er Rakho [14] identified for the first time the two new phases $La_4BaCu_5O_x$ and $La_3Ba_3Cu_6O_y$, both with perovskite like structure ordered in the A sites.

Further researches followed later [15] and proposed that the $La_3Ba_3Cu_6O_y$ phase should be considered as limit term (x=0) of the series of mixed crystals $La_{3-x}Ba_{3+x}Cu_6O_y$ ($0 \le x \le 1$) and the other term (x=1) should display superconductive properties, with $T_c \approx 90K$, comparable with those of the YBCO itself [16][17].

Until now, indeed, the opinions of the various researchers about the extension of this mixed crystals series fails to agree. At 950°C the data reported by Klibanow and co-workers ($0 \le x \le 0.8$)[6] result actually significantly different from those ($-0.4 \le x \le 0.8$) by Wong-Ng [16] and ($0 \le x \le 0.75$) by Segre[18] at 975°C.

The results of our experiments showed that the field width of the examined mixed crystals series is sensibly affected by the temperature. The XRD analysis of samples quenched in air from 950°C furnished the limits ($-0.2 \le x \le 0.8$), both with tetragonal symmetry and cell parameters a_o = 3.910; c_o = 11.730A for x=-0.2 and a_o = 3.932 ; c_o = 11.796A for x=0.8 .

The solids heat treated at 1000°C furnished, on the contrary, the limits ($-0.4 \le x \le 1$). The XR powder diffraction patterns allowed us to verify that the term x=-0.4 still display tetragonal symmetry (a_0 = 3.901 ; c_0 = 11.703A), while the term very rich in barium (x=1) shows orthorhombic symmetry, with a_0 = 3.913 b_0 = 3.952 ; c_0 = 11.739A.

REFERENCES

1) F. Abbattista, C. Brisi, S. Delmastro, M. Lucco-Borlera, D. Mazza and M. Vallino,Mater. Chem. Phys., 24 (1989) 147.
2) F. Abbattista. M. Vallino and D. Mazza, Mater. Chem. Phys.,24 (1990) 363.
3) F. Abbattista, D. Mazza and M. Vallino, Mater. Chem. Phys., 25 (1990) 251.
4) C, Michel, L. Er-Rakho and B. Raveau, J. Solid State Chem., 39 (1981) 161.
5) D. Klibanow, K. Sujata and T. O. Mason, J. Am. Ceram. Soc., 71 (1988) C267.
6) J. B. Torrance, Y. Tokura and A. Nazzal. Chemtronics,2 (1987) 120.
7) D. M. De Leeuw, J. of Less-Common Metals, 150 (1989) 95.
8) C. Michel and B. Raveau, Revue de Chimie Minerale, 21 (1984) 4.
9) S. Uchida, H. Takagi, K. Kitazawa and S. Tanaka, Jpn. J. Appl. Phys., 26 (1987) L1.
10) H. Takagi, S. Uchida, K. Kitazawa and S. Tanaka, Jpn. J. Appl. Phys., 26 (1987) L123.
11) S.A. Sunshina, L.F. Schneemeyer, J.V. Waszczak, D.W. Murphy, S. Miraglia, A. Santoro and F. Beech, J. Cryst. Growth, 85 (1987) 632.
12) N. Nguyen, L. Er Rakho, C. Michel, J. Choisnet and B.Raveau, Mater. Res. Bull., 15, (1980) 891.
13) C. Michel, L. Er Rakho and B. Raveau, Meter. Res. Bull., 20 (1985) 667.

14) L. Er Rakho, C. Michel, J. Provost and B. Raveau, J. Solid State Chem., 37 (1981) 151.
15) A. Maeda, T.Yabe, K.Uchinokura and S. Tanaka, Jpn. J. Appl. Phys., 26 (1987) L 1368.
16) W. Wong-Ng, B. Paretzkin and E.R. Fuller jr., J. Solid State Chem. 85 (1990) 117.
17) C. Michel, F. Deslandes, J.Provost, P. Lejay, R. Tournier, M. Hervieu and B. Raveau, C. R. Acad. Sci. Paris, 304 II (1987) 1169.
18) C.U. Segre, B. Dabrowski, D.G. Hinks, K. Zhang, J.D. Jorgensen, M.A. Beno and I.K. Schuller, Nature, 329 (1987) 227.

THE $(Bi,Pb)_2Sr_2(Y,Ca)Cu_2O_z$ PHASE: THE VANISHING OF STRUCTURAL MODULATION IN A 2212 PHASE AND ITS EFFECT ON THE SUPERCONDUCTING PROPERTIES

G. CALESTANI, M.G. FRANCESCONI, G. SALSI, G. D. ANDREETTI
Istituto di Strutturistica Chimica, Universita' di Parma and Centro di Studio per la Strutturistica Diffrattometrica del C.N.R., Viale delle Scienze, 43100 Parma, Italy

and

A. MIGLIORI
C.N.R.- Istituto LAMEL, Via Castagnoli 1, Bologna, Italy

ABSTRACT

The high-T_c superconductors based on bismuth cuprates (BSCCO) are characterized by a unidimensional structural modulation, quasi-commensurate to the fundamental cell, with a wave vector close to $1/5a^*$. The $(Bi,Pb)_2Sr_2(Ca,Y)Cu_2O_z$ system was investigated and the existence of a modulation free 2212 phase with a wide range of composition ($Bi_{2-x}Pb_xSr_2Y_{1-y}Ca_yCu_2O_z$, $0\leq y\leq 0.8$ and $x=(1-y/2)\pm 0.2$) was pointed out by electron diffraction. In comparison with $Bi_2Sr_2CaCu_2O_8$ the new phase exibits a larger orthorhombic distorsion. The crystal structure of the unmodulated phase was determined by single crystal X-ray diffraction. The structure differs from the fundamental one of $Bi_2Sr_2CaCu_2O_8$: the unit cell is primitive and a different relative arrangement of the ribbons in adjacent BiO layers was pointed out. The non modulated $Bi_{2-x}Pb_xSr_2Y_{1-y}Ca_yCu_2O_z$ phase still shows superconductivity in the Ca-rich region with T_c decreasing with increasing the Y content, supplying a further example of superconductivity not related to structural modulation in BSCCO.

1. Introduction

The high-T_c superconductors based on bismuth cuprates (BSCCO) consist of three phases of general formula $Bi_2Sr_2Ca_{n-1}Cu_nO_{2n+4}$ (with n=1,2 and 3) denoted as 2201, 2212 and 2223 from the stoichiometric ratio of the cations in the chemical formula. The layered structures of these phases, deriving from intercalations of double BiO layers and perovskite blocks of variable thickness depending on n, are clearly related and their fundamental cells differ in the *c* axis (the stacking axis) lenght. The structures are characterized by a unidimensional structural modulation, quasi-commensurate to the fundamental cell, with a wave vector ***k*** close to $1/5a^*$. In spite of the numerous structural analyses, performed in particular on the 2201 and 2212 phases, the features of the structural modulation are still matter of discussion. It is now commonly accepted that the modulation arises from the structural mismatch between the BiO layers and the perovskite blocks produced by the stereochemical requirements of bismuth (due to the presence of a $6s^2$ lone pair), which in the idealized structure should have too long bond distances with the oxygen atoms.

The presence of extra oxygen within the BiO layers, for a partial adjustment of the structural mismatch, was first proposed by Zandbergen et al.[1] on the basis of high resolution electron microscopy (HREM) images. Several attempts, mostly based on single crystal X-ray diffraction data, to determine the real structure of the 2212 BSCCO phase were performed, utilizing both conventional three-dimensional approachs as well as superspace symmetry and formalism. The presence of extra oxygen was reported by some authors even if disagreement exists on the position and occupation of the extra oxygen atoms: in particular for the Y containing 2212 phase Torardi et al.[2] found extra oxygen in the BiO layer between adjacent BiO ribbons whereas Tarascon et al.[3] found for $Bi_{10}Sr_{15}Fe_{10}O_{46}$, a compound with a fundamental structure similar to the 2212 BSCCO but with a different modulation symmetry, extra oxygen in the BiO ribbon (one for each modulation period) suggesting the presence of extra oxygen as origin of the modulation. Petricek et al. [4] analized the 2212 BSCCO structure on the basis of the four-dimensional formalism and symmetry with results compatible with the existence of extra oxygen in a site similar to that found by Tarascon et al.[3]. Wether the extra oxygen or the structural mismatch is the origin of the modulation remains however for the authors a subject for discussion. Other authors[5,6] found similar features of the displacive modulation but no clear evidence of extra oxygen in the BiO layers and attribute the origin of the modulation to the structural mismatch between the BiO layers and the perovskite blocks.

When Bi is partially replaced by Pb the 2212 structure is retained (the orthorhombic distorsion becomes larger) but the structural modulation is strongly influenced: the modulation period increases ($p \sim 7$), the symmetry of the modulation changes completely and the intensity of the satellite reflections decreases considerably indicating that the deviations of the local structure from the fundamental structure become smaller. In this optics the modulation is expected to vanish for larger Pb contents but unfortunately this cannot be proved since a limit in the substitution range (from 0.4 to 0.6 in the formula unit depending on authors) exists. The knowledge of the structural variations induced by the substitution in these phases would be very interesting also for a better understanding of the origin of the structural modulation in BSCCO; up to now however the real structure of the Pb-containig (BPSCCO) phase was not determined principally because of the weak intensities of the satellite reflections.

If the charge difference (Pb^{2+} substituted for Bi^{3+}) is responsible for the limit in the substitution range in BPSCCO, the substitution of Ca^{2+} with a trivalent cation is expected to increase the possible Pb content of the 2212 phase. Following this idea modulation-free BSCCO- type phases were recently produced. Tarascon et al.[7] reported the preparation of a series of compounds $Bi_{2-x}Pb_xSr_{2-y}La_yCuO_z$ that do not show modulation for $x \sim y+0.2$ but still show superconductivity. The vanishing of the structural modulation was ascribed to the lack of extra oxygen within the Bi-O layers. In a similar way a modulation-free 2212 phase[8-9] was produced by simultaneous replacement of Bi by Pb and Ca by Y, but superconductivity was reported to be destroyed before the modulation. The structural properties of the non modulated semiconductor $BiPbSr_2YCu_2O_8$ were investigated by electron diffraction by Zandbergen et al.[9], which also investigate the variation of the structural modulation in the $Bi_{2-x}Pb_xSr_2YCu_2O_8$ samples.

We have investigated the Bi-Pb-Sr-Y-Ca-Cu-O system in order to establish the existence range of the 2212 phase and in particular to determine the variations produced by the simultaneous substitutions (Pb for Bi and Y for Ca) on the 2212 structure.

2. Experimental

The samples were prepared by solid state reactions starting from Bi_2O_3, PbO, Y_2O_3, CuO, $SrCO_3$ and $CaCO_3$ in the appropriate ratios required for an overall 2212 stoichiometry. The intimate mixtures were prereacted at 780-820°C depending on the composition (higher temperatures for higher Y contents) for 8 hours and then annealed (with intermediate grinding) at 860-920°C (depending on the composition) for periods ranging from 24 to 120 hours until the formation of a single 2212 phase or the coexistence of different phases was evidenced by X-ray powder diffraction. The highest temperatures and the longest heating periods were required by the Pb-free Y-containig samples owing to the inhomogeneity (Y and Ca rich regions) produced in the initial stage of the reaction. The annealing was prolonged in these cases until the linewidth of the $00l$ diffraction peaks (typically broadened for inhomogeneous samples) became comparable with that of the other reflections at similar diffraction angles. The samples utilized for the electricl measurements were prepared by pressing the powder into pellets (ϕ= 12mm, thickness 0.8-1.0mm), which were sintered for further 8 hours at the same temperature utilized for the solid state reaction. All the thermal tratments were performed in air. Crystals suitable for single crystal X-ray diffraction were isolated from solidified melts slow cooled belove the solidification temperature. The crystal were grown in alumina crucibles from 5g load; the mixture was heated at 1000°C for 30 minutes and then cooled at 800°C with a cooling rate in the range 2-5 °C h^{-1}. Stoichiometric compositions were utilized since it has been shown[10] that, in spite of its incongruent melting, the 2212 phase can be grown by partial melting almost free from 2201 impurities if appropriate cooling rates are utilized.

The samples were characterized by X-ray powder diffraction by using the CuKα radiation on an automatized Philips PW1050 powder diffractometer. The digitized spectra were recorded with 2θ step of 0.05° by counting 2 sec/step. Single crystal data were collected on a Philips PW1100 four circle diffractometer utilizing graphite monocromatized MoKα radiation. Selected area electron diffraction (SAD) was performed by using a Philps CM30 TEM operating at 300 KV. Resistivity measurements were performed by a standard DC four probe method.

3. Results

By keeping constant the overall 2212 stoichiometry the whole range of substitutions Pb→Bi and Y→Ca was investigated in a pseudoquaternary phase diagam (Fig. 1) involving $Bi_2Sr_2CaCu_2O_x$,$Bi_2Sr_2YCu_2O_x$ and the two corresponding hypothetical phases, $Pb_2Sr_2CaCu_2O_x$ and $Pb_2Sr_2YCu_2O_x$, resulting from a complete replacement of Bi with Pb. It was found by powder X-ray diffraction that a single 2212 phase exists in a wide region of the phase diagram characterized by high Bi contents. The Pb-rich region results on the contrary to be polyphasic at least in the temperature range under study: $Pb_2Sr_2(Y,Ca)Cu_3O_8$[11] and a mixed oxide, whose composition is yet not well identified, represent the dominant phases at the Y- and Ca-rich corners respectively.

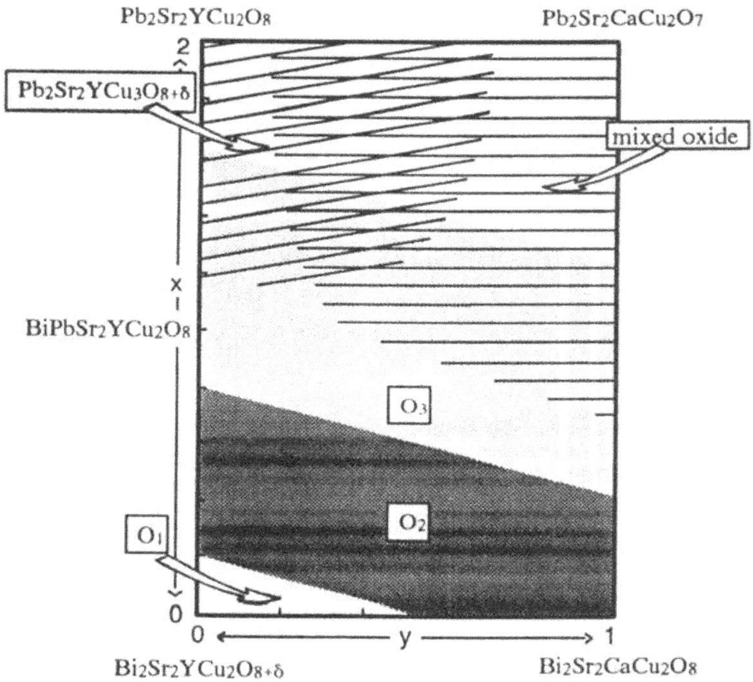

Figure 1: Phase diagram for $Bi_{2-x}Pb_xSr_2Y_{1-y}Ca_yCu_2O_z$ in the xy plane.

The variation of the lattice parameters of the 2212 phase with the composition (Fig. 2 and 3) makes however possible to divide its existence range in three structural regions caracterized by ***a/b*** ratios lower than, close to and grather than one (O_1, O_2 and O_3 respectively). The O_1 structure is typical of $Bi_2Sr_2YCu_2O_x$ and characterizes the Y-rich region for low Pb content. O_3 is on the contrary observed for the higher possible Pb content of 2212 phase independently from the Y/Ca ratio. The pseudotetragonal O_2 cell, similar to those of $Bi_2Sr_2CaCu_2O_x$, characterizes the central region of the 2212 existence range in the phase diagram. In reality this cell is orthorhombic but the difference between a and b is too small to be resolved by powder x-ray diffraction. Contrary to $Bi_{2-x}Pb_xSr_{2-y}La_yCuO_z$, for which the structural transition between the pseudotetragonal and the O_3 ($a<b$) phases seems to be discontinuous in a relative wide range of x (≈ 0.3)[7], in this case all the structural transitions involved seems to be continuous or at least to shows a narrow discontinuity range which

could also be related to some sample inhomogeneity The *c* lattice parameter decreases, as expected, with the Y content. However, for a fixed Y/Ca, ratio it remains almost independent from the Pb content with the exception of the pure Y compounds for which *c* shows an initial progressive increase by substituting Bi with Pb up to ~30.35Å and than remains constant up to the upper limits of the Pb substitution.

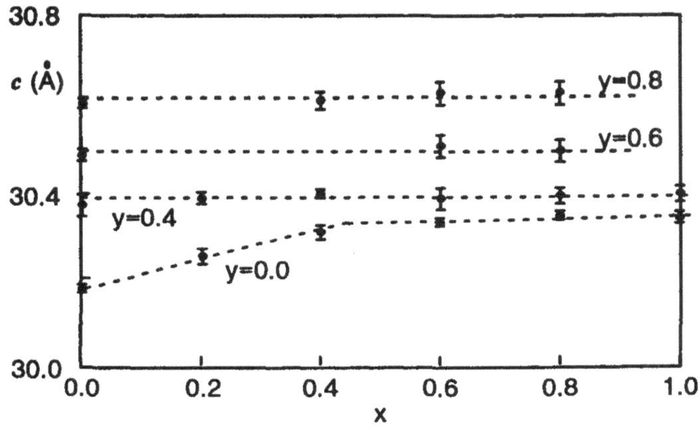

Figure 2. Variation of the *c* lattice parameter with the Pb content in $Bi_{2-x}Pb_xSr_2Y_{1-y}Ca_yCu_2O_z$.

In order to verify if the vatiation of the lattice parameters with the composition is related to structural modifications, in particular for what concernes the $O_2 \rightarrow O_3$ transition, electron diffraction was performed on a serie of compounds of the three structural domains. SAD patterns taken along the [001] direction are shown in Figs. 4-7. Fig. 4 shows the variation of the structural modulation along the $Bi_{2-x}Pb_xSr_2YCu_2O_z$ serie for x=0.2, 0.4, 0.6, 0.8 and 1.0 which agrees with the results of Zandbergen et al.[9]. At x=0.2 (near the upper limit of the O_1 phase) the structure shows the complex modulation typical of $Bi_2Sr_2YCu_2O_z$ (Y-type) consisting of two modulation wave vectors k_a~(1/4)***a*** and k_b~(1/8.5)***b***. The Bi-type modulation satellites show here an increased (k_a~(1/4.5)***a*** instead of (1/4)***a***) modulation period with respect to $Bi_2Sr_2YCu_2O_z$ and the associated additional satellites are reduced to strikes indicating the increasing of their modulation period. By increasing x to 0.4, the modulation along ***b*** disappeares and only the Bi-type modulation, with a further increase of wave vector k_a~(1/5)***a***, is observed. For x=0.6 the the modulation become complicated and requires two modulation wave vectors k_{a1}~(1/5)***a*** and k_{a2}~(1/7.5)***a*** which are close to those of the Bi- and Pb-type of modulation. For x>0.8 and up to the 2212 stability limits no satellites are visible in the diffraction patterns. Fig. 5 shows the variation of the structural modulation along the $Bi_{2-x}Pb_xSr_2Y_{0.8}Ca_{0.2}Cu_2O_z$ serie for x=0.2, 0.4, 0.6, 0.8 and 1.0. In this case the Y-type modulation in not more visible at x=0.2 where it is substituted by a Bi-type modulation with a wave vector k_a~(1/4.7)***a***. By increasing x an evolution of the structural modulation, similar to that previously described, is observed.

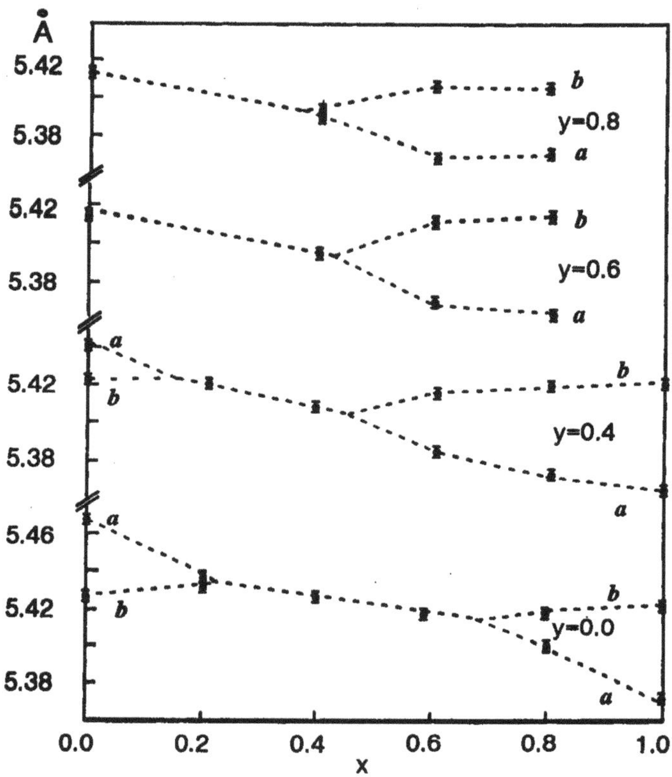

Figure 3: Variation of the *a* and *b* lattice parameters with the composition in $Bi_{2-x}Pb_xSr_2Y_{1-y}Ca_yCu_2O_z$.

On the basis of these diffraction analyses it seems that the Y-type modulation characterizes the O_1 region while in the O_2 region the Bi-type modulation is progressively substituted by the complex ($k_{a1} \approx (1/5)a$ and $k_{a2} \approx (1/7.5)a$) modulation reported by Zandbergen et al.[9]. These results have been confirmed by further investigation of other series in the O_2 region. The O_3 region seems on the contrary to be free from structural modulation. Figs. 6 and 7 show the results of SAD analysis performed in the O_3 region for compositions $Bi_{1.2}Pb_{0.8}Sr_2Y_{1-y}Ca_yCu_2O_z$ (y=0.0, 0.2, 0.4, 0.6 and 0.8) and $Bi_{1.4}Pb_{0.6}Sr_2Y_{1-y}Ca_yCu_2O_z$ (y=0.4, 0.6, 0.7, 0.8, 0.9 and 1.0) respectively. In the first case the structure remains non modulated in all the investigated range whereas for the lower Pb contents the typical Pb-type modulation appears for y≥0.9. In all cases the electron diffraction pattern are consistent with a primitive lattice indicating that the non modulated phase shows a decrease of symmetry with respect to the fundamental cell (A-centered) of the modulated one.

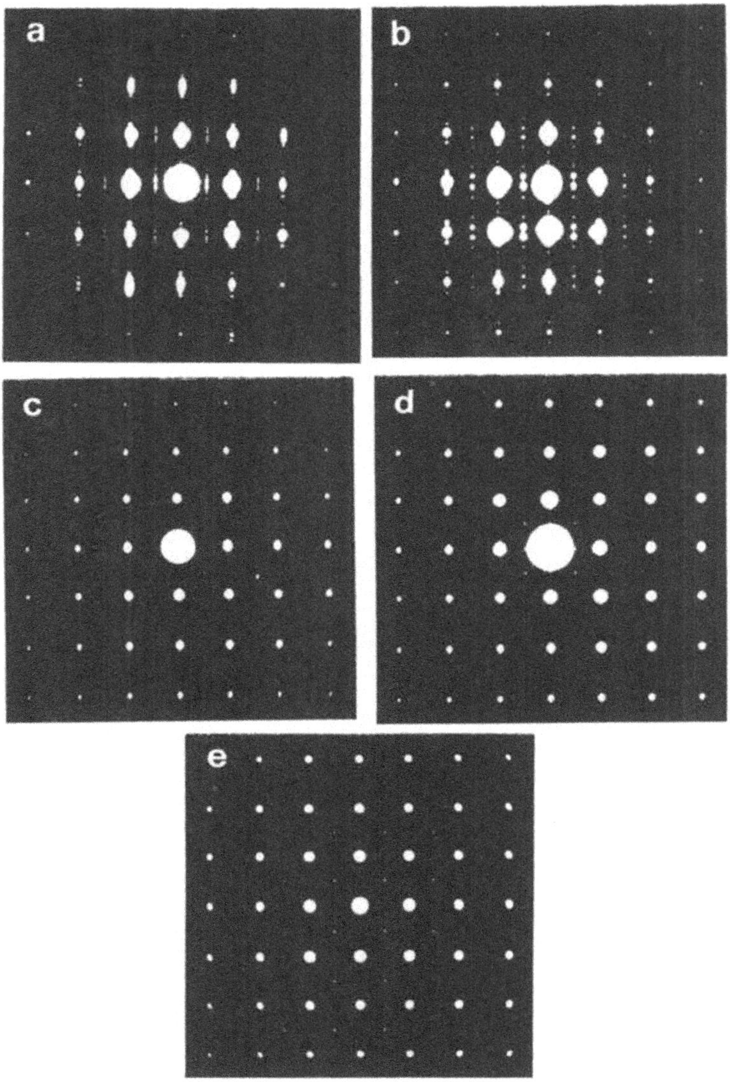

Figure 4. Selected-area diffraction patterns along the 001 direction for $Bi_{2-x}Pb_xSr_2YCu_2O_z$: a) x=0.2; b) x=0.4: c) x=0.6; d) x=0.8; e) x=1.0.

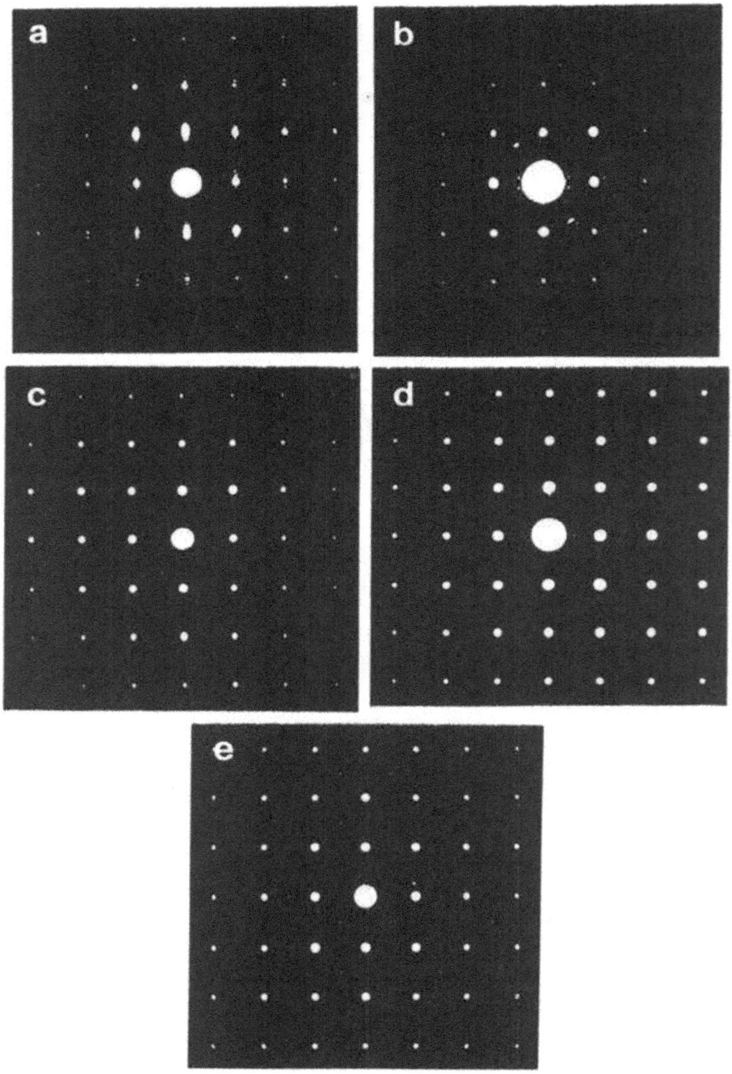

Figure 5. Selected-area diffraction patterns along the 001 direction for $Bi_{2-x}Pb_xSr_2Y_{0.6}Ca_{0.4}Cu_2O_z$: a) x=0.2; b) x=0.4: c) x=0.6; d) x=0.8; e) x=1.0.

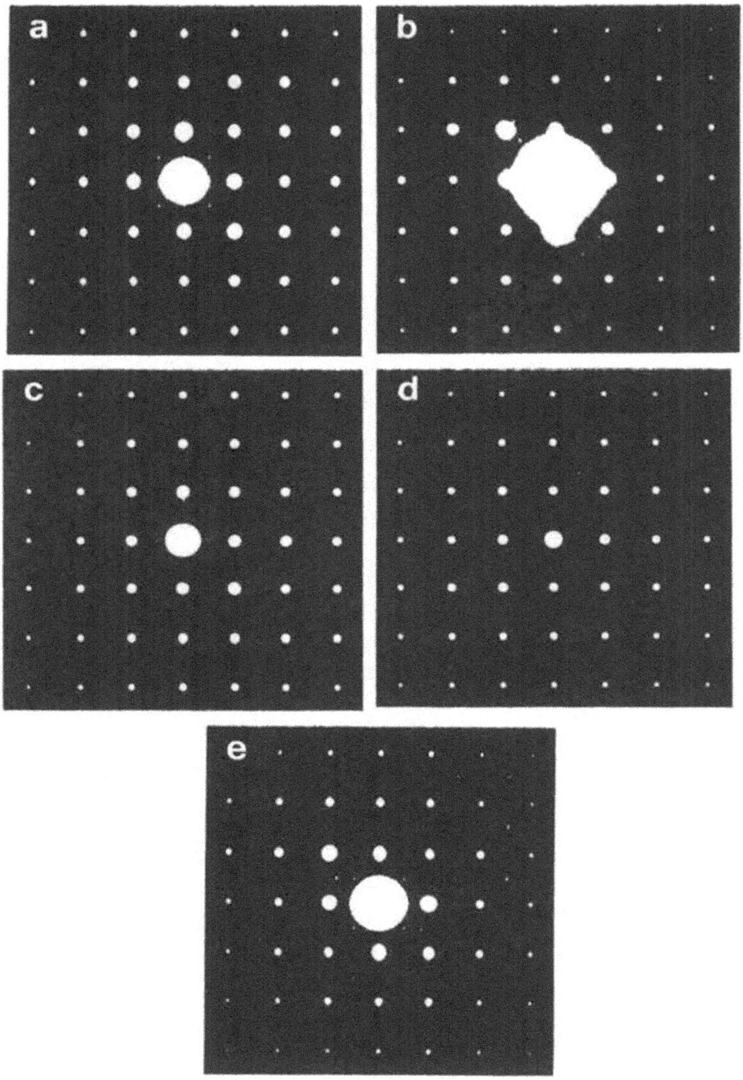

Figure 6. Selected-area diffraction patterns along the 001 direction for $Bi_{1.2}Pb_{0.8}Sr_2Y_{1-y}Ca_yCu_2O_z$: a) y=0.0; b) y=0.2: c) y=0.4; d) y=0.6; e) y=0.8.

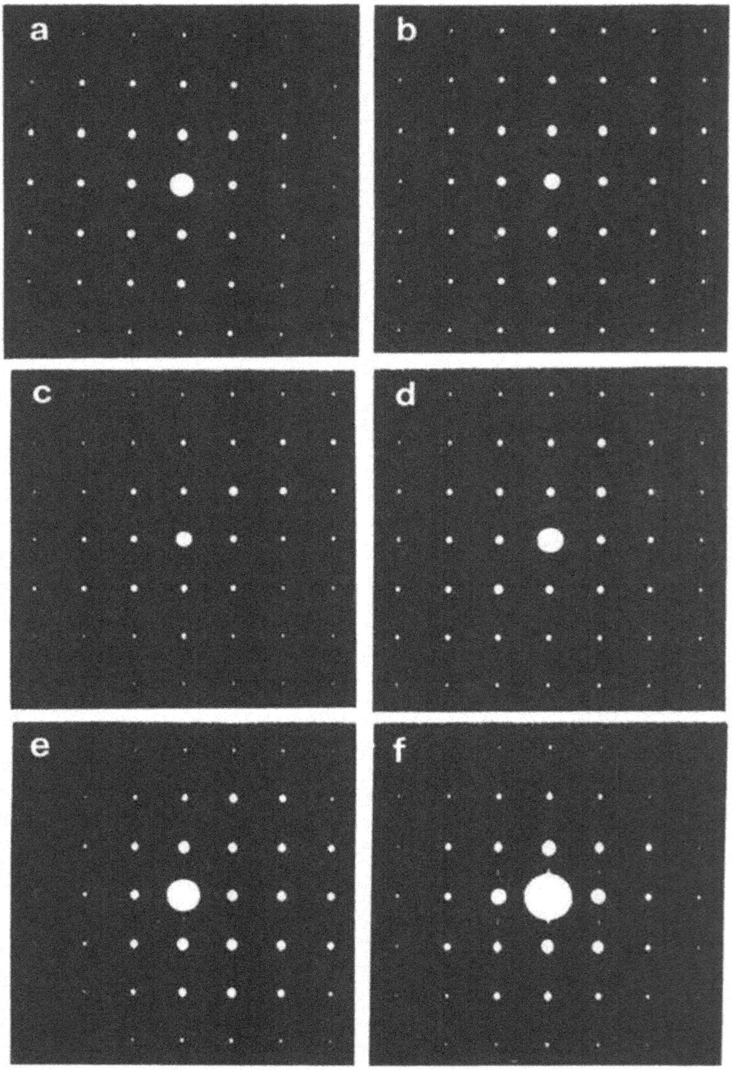

Figure 7. Selected-area diffraction patterns along the 001 direction for $Bi_{1.4}Pb_{0.6}Sr_2Y_{1-y}Ca_yCu_2O_z$: a) y=0.4; b) y=0.6: c) y=0.7; d) y=0.8; e) y=0.9; f) y=1.0.

In fact an accurate analysys of the diffraction patterns shows that *hk0* reflections with $h+k\neq 2n$ are present, even if very weak if compared to the reflections with all-even indices. The intensity of the *hk0* reflections with not all-even indices seems to depend from the composition since some classes of these weak reflections show a different relative intensity in different compounds. This suggests, as possible origin of these weak reflections, the ordering of Pb and Bi and (for y~0.5) of Y and Ca in the structure but also a variation in the oxygen stoichiometry could be taken into account.

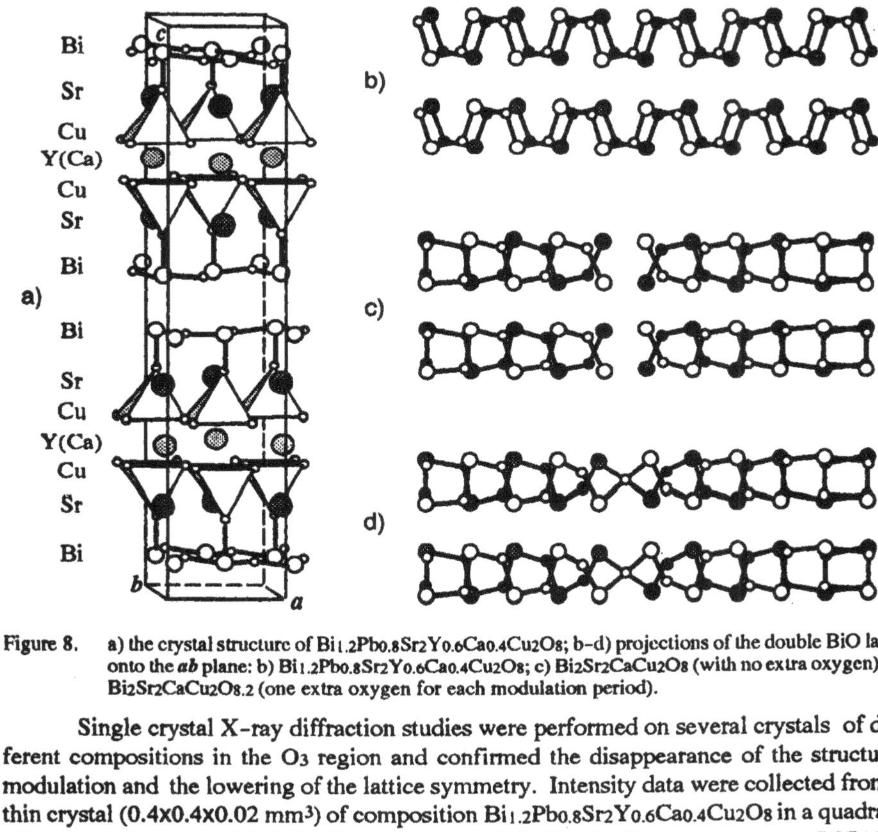

Figure 8. a) the crystal structure of $Bi_{1.2}Pb_{0.8}Sr_2Y_{0.6}Ca_{0.4}Cu_2O_8$; b-d) projections of the double BiO layer onto the *ab* plane: b) $Bi_{1.2}Pb_{0.8}Sr_2Y_{0.6}Ca_{0.4}Cu_2O_8$; c) $Bi_2Sr_2CaCu_2O_8$ (with no extra oxygen); d) $Bi_2Sr_2CaCu_2O_{8.2}$ (one extra oxygen for each modulation period).

Single crystal X-ray diffraction studies were performed on several crystals of different compositions in the O_3 region and confirmed the disappearance of the structural modulation and the lowering of the lattice symmetry. Intensity data were collected from a thin crystal (0.4X0.4X0.02 mm^3) of composition $Bi_{1.2}Pb_{0.8}Sr_2Y_{0.6}Ca_{0.4}Cu_2O_8$ in a quadrant of reciprocal space ($\pm h,k,l$) in the θ range 3-30°. The lattice parameters a=5.374(1), b=5.421(1) and c=30.397(3)Å, in agreement with those obtained on ceramic samples with the same nominal composition, were refined by least squares fitting of the setting angles of 25 selected reflections with $\theta>15°$. 417 independent reflections with $I>2.5\sigma(I)$ were considered observed and, corrected for Lorentz, polarization and absorption effects, used in the structure analysis. The space group Pnan was chosen for the structure refinement on the basis of the systematic absences. In reality, if some weak reflections are taken into account, the absences were consistent with the non isomorphic subgroup $P222_1$, but the small number (only few of those observed by electron diffraction) and the very weak intensities of the reflections suggesting the non centrosymmetric space group do not allow the doubling of the number of the parameter to be refined introducing moreover strong correlation among them. The refinement, performed with isotropic thermal parameters for all the atoms, converged to R=0.053 with unit weights. The occupancy refinement of the (Y,Ca) site confirmed the presence of both the cations in a ratio close to the theorical one.

Table I. Fractional atomic coordinates and isotropic thermal parameters for $Bi_{1.2}Pb_{0.8}Sr_2Y_{0.6}Ca_{0.4}Cu_2O_8$.

Atom	site	x	y	z	B_{iso}
Bi(Pb)	8e	-0.0100(6)	0.7665(4)	0.0492(1)	0.92(4)
Sr	8e	0.4715(12)	0.7492(11)	0.1371(1)	0.81(8)
Y(Ca)	4d	0.4764(19)	3/4	1/4	0.41(5)
Cu	8e	-0.0257(14)	0.7498(14)	0.1958(2)	0.52(9)
O1	8e	0.224(20)	-0.019(17)	0.199(2)	0.92(58)
O2	8e	0.228(21)	0.516(20)	0.202(2)	0.95(56)
O3	8e	0.458(8)	0.275(8)	0.117(1)	0.83(63)
O4	8e	0.116(11)	0.152(10)	0.053(2)	2.4(11)

Table II. Bond distances (Å) for $Bi_{1.2}Pb_{0.8}Sr_2Y_{0.6}Ca_{0.4}Cu_2O_8$ with e.s.d's in parentheses.

Bi(Pb) - O3	2.09(4)	Sr - O1	2.61(8)
Bi(Pb) - O4	2.20(6)	Sr - O1	2.73(8)
Bi(Pb) - O4	2.06(6)	Sr - O2	2.68(8)
Bi(Pb)....O4	3.19(6)	Sr - O2	2.80(9)
Bi(Pb)....O4	3.40(6)	Sr - O3	2.64(4)
Bi(Pb)....O4	3.40(6)	Sr - O3	2.69(4)
		Sr - O3	2.83(4)
		Sr - O3	2.91(4)
		Sr - O4	2.72(6)
Cu - O1	1.83(9)		
Cu - O1	1.99(9)	Y(Ca) - O1	2.42(9) x 2
Cu - O2	1.87(9)	Y(Ca) - O1	2.52(9) x 2
Cu - O2	1.96(9)	Y(Ca) - O2	2.35(9) x 2
Cu - O3	2.39(4)	Y(Ca) - O2	2.46(9) x 2

The atomic coordinates and thermal parameters are reported in table I and relevant bond distances and angles in table II. The structure (Fig. 8a) is essentially that found for

$Bi_2Sr_2CaCu_2O_8$: in both structure the presence of the lone pair on Bi(Pb) atoms give rise to an arrangement of short and long Bi-O bonds, which distorts the ideal stacking and forms ribbons along ***a***. The relative arrangement of the ribbons in adjacent BiO layers of the two compounds changes however completely: the skeletons of the -Bi-O- chains are in this case almost superimposed (with a small shift due to the different polarity of the ribbons in adjacent layers) whereas in BSCCO they are mutually shifted by *a*/2 (Fig. 8b). The arrangement of the BiO ribbons is, on the other hand, very similar to that found in the corresponding modulation-free 2201 phase[7].

The disappearance of the structural modulation is not related, as in the corresponding 2201 case, to a disappearance of the superconductivity. Fig. 9 shows the resistivity vs. temperature measuremets performed on a series of $Bi_{1.2}Pb_{0.8}Sr_2Y_{1-y}Ca_yCu_2O_z$ samples ($0.2<y<0.8$). The T_c shows a maximum at the highest Ca content and decrease with decreasing it, similarly to the corresponding lead-free samples.

4. Discussion

By substituting Pb for Bi and Y for Ca in $Bi_2Sr_2CaCu_2O_8$ it is possible to prepare a $Bi_{2-x}Pb_xSr_2Y_{1-y}Ca_yCu_2O_z$ phase with a 2212-type structure that does not show structural modulation in a wide region (roughly defined by $0\leq y\leq 0.8$ and $x=(1-y/2)\pm 0.2$) of the phase diagram. Except for the highest Ca contents ($y>0.8$), for which the structure shows the typical Pb-type modulation, the existence range of the modulation-free phase is almost coincident with the O_3 domain of the 2212 structure. The vanishing of the structural modulation involves a change of the symmetry of the double BiO layers just revealed in the corresponding 2201 phase ($Bi_{2-x}Pb_xSr_{2-y}La_yCuO_6$).

In $Bi_{2-x}Pb_xSr_{2-y}La_yCuO_z$ (the corresponding 2201 phase) the absence of modulation seems to be related to an 0.2 excess in the Pb substitution ($x=y+0.2$) so that the vanishing of the modulation has been attributed to the disappearance of the extra oxygen in the BiO layers[7]. If a similar straight line (in our case it would be $x=(1-y)+0.2$) is considered in the present phase diagram , it will cross regions with different structural properties, going from pseudotetragonal to orthorhombic cells, from modulated to non modulated structures. Therefore the situation in $Bi_{2-x}Pb_xSr_2Y_{1-y}Ca_yCu_2O_z$ is more complicated than in $Bi_{2-x}Pb_xSr_{2-y}La_yCuO_z$: the existence range of the modulation-free structure seems to be still related (with a $x \approx 0.4$ amplitude) to a straight line but its slope in the xy plane can not be related in an easy way to the oxygen content as deduced to the formal valence of the cations. If we assume that the valences of bismuth and lead in these compounds are Bi(III) and Pb(II) respectively, the oxygen stoichiometry of the modulation-free phase is expected to vary remarkably.

The non modulated $Bi_{2-x}Pb_xSr_2Y_{1-y}Ca_yCu_2O_z$ still shows superconductivity in the Ca-rich region; the T_c is somewhat lower than in the corresponding Pb-free samples and decreases in a similar way with increasing the Y content. However the decrease of T_c with the increase of the lead content is much less dramatic as expected by supposing that all the charge, deriving from the $Pb^{2+}\rightarrow Bi^{3+}$ substitution, would be transferred into the CuO_2 planes. In particular $Bi_{1.4}Pb_{0.6}Sr_2Y_{0.2}Ca_{0.8}Cu_2O_z$ and $Bi_{1.2}Pb_{0.8}Sr_2Y_{0.2}Ca_{0.8}Cu_2O_z$ show

T_c values comparable with $Bi_2Sr_2CaCu_2O_8$. This could be explained by hypothesizing that the double (Bi, Pb)O layer play a competitive role (with respect to the CuO_2 plane) in accomodating the extra charge deriving from the Pb→Bi substitution. The oxygen stoichiometry of these compounds could be therefore close to 8, the excess of oxygen being compensate from a partial oxydation of Pb^{2+} to Pb^{4+} as observed in lead cuprates. We have not performed determination of the oxygen content by chemical titration methods since they are based on assumptions on the valence of the elemets: in our case is not clear wether the oxygen excess, the variation of the valence of one or more elements or a mixing of the two phenomena are measured.

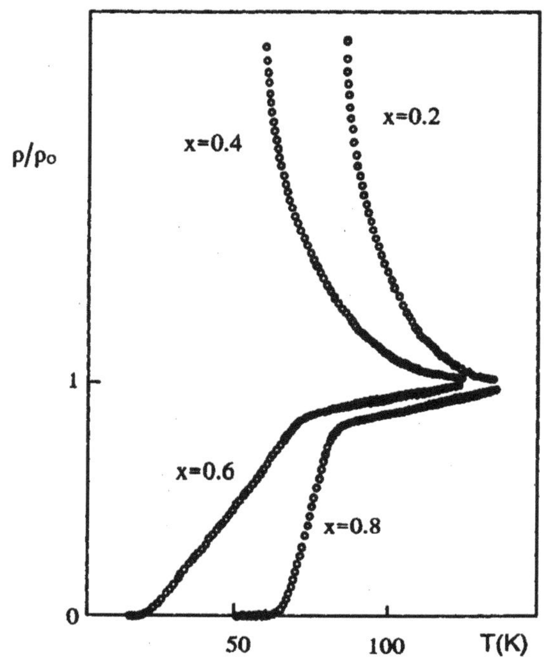

Figure 9 Normalized resistivity vs. temperature for a series of $Bi_{1.2}Pb0_{0.8}Sr_2Y_{1-y}Ca_yCu_2O_z$ samples ($0.2<y<0.8$) showing modulation-free 2212 structure.

The ability of the BiO layer in accomodating charge is revealed by the magnetic phase diagram of $Bi_2Sr_2Ca_{1-y}Y_yCu_2O_z$, which shows a large plateau of T_c by increasing the Y content and the disappearing of superconductivity only at y=0.6. The presence of partially reduced bismuth in the BiO layers of BSCCO superconductors was first proposed by Calestani et al.[5] on the basis of the structural analysis of $Bi_2Sr_2CaCu_2O_8$. The hypothesis was confirmed by XAS measurements at the BiL_{III}-edge[12] showing a significant shift of the curves toward lower energy with respect to Bi_2O_3 (or the non superconducting

$Bi_2Sr_2CaFe_2O_9$). Such a feature suggests that Bi exhibits in the superconducting cuprates a formal oxydation state intermediate between that of Bi(III) and metallic bismuth, interpreted as a semimetallic character of the BiO layers, which would result from a narrow band built from overlapping of hybrydized *s–p* bismuth orbitals with 2*p* oxygen orbitals. In this model electrons would be transferred from CuO_2 layers in this band leading to the mixed valence of copper in these layers. All these phenomena must be taken into account when the mechanism leading to the appearing of superconductivity in the bismuth cuprate is considered. The presence of extra oxygen in the BiO layers as origin of the superconductivity represents an over-simplified model which moreover seems to be in contrast with the increase of T_c produced in $Bi_2Sr_2CaCu_2O_8$ by annealing in reducent atmosphere (argon or nitrogen). These annealing do not produces appreciable structural variation and experiments performed by single crystal X-ray diffraction[13] showed that also the modulations is not affected by similar treatments. This can be considered as an indication of the structural mismatch as origin of the modulation in BSCCO since, if a variation of the oxygen content is produced by the annealing, no variation of the modulation is observed. An additional indication comes from the fact that the slope, which determines in the xy plane the modulation-free domain, seems to be a constant of the phase diagram; in fact it determines, beside the $O_2 \rightarrow O_3$ phase transition, the $O_1 \rightarrow O_2$ transition as well as the change, in period and symmetry, of the modulation in the O_2 region. The increase of x, required for producing similar structural effects by increasing the Y content, could be related to the expansion of the basal plane of the structure produced by the yttrium insertion.

A deeper characterization of $Bi_{2-x}Pb_xSr_2Y_{1-y}Ca_yCu_2O_z$, including the study of the influence of the annealing atmosphere on structural and superconductive properties, seems to be necessary for a better definition of the mechanism controlling the superconductivity in this system.

5. Acknowledgments

Acknowledgments are due to Dr. M. Sparpaglione and Dr. L. Bonoldi for the useful discussions and to Mrs. R. Magnani for the help in sample preparation. This work is partially supported by the CNR (National Council of Research) in the *Progetto Finalizzato Materiali per Tecnologie Avanzate*.

6. References

1. H. W. Zandbergen, W. A. Groen, F. C. Mijlhoff, G. van Tendeloo and S. Amelinckx, *Physica C* **156** (1988) 325.
2 C. C. Torardi, J.B. Parise, M. A. Subramanian, J. Gopalakrishnan and aA. W. Sleight, *Physica C* **157** (1989) 115.
3 J. M. Tarascon et al., *Phys. Rev. B* **39** (1989) 11587
4 V. Petricek, T. Gao, P. Lee and P. Coppens, *Phys. Rev. B* **42** (1990) 387
5 G. Calestani, C. Rizzoli, M.G. Francesconi and G. D. Andreetti, *Physica C* **161** (1989) 598

6 X. Kan, *PHD thesis*, University of Huston (1990)
7 J. M. Tarascon, W. R. McKinnon, Y. Le Page, K. Remschnig, R. Ramesh, R. Jones, G. Pleizier and G. W. Hull, *Physica C* **172** (1990) 13
8 Y. Gao, P. Lee, H. Graafsma, J. Yeh, P. Bush, V. Petricek and P. Coppens, *Chem. Mater.* (1990) 323.
9 H. W. Zandbergen and W. A. Groen, *Physica C* **166** (1990) 282
10 L. Dimesso, M.G. Francesconi, G. Calestani, E. Silva, R. Fastampa, R. Marcon, D. Fiorani, E. Agostinelli,and A. M. Testa, *Physica C* **176** (1990) 216
11 R. J. Cava et al., *Nature* **336** (1988) 211
12 R. Retoux, F. Studer, C. Michel, B. Raveau,, A. Fontaine and E. Dartyge, *Phys. Rev. B* **41** (1990) 193
13 G. Calestani et al., to be published.

A STUDY ON THE DOUBLE SUBSTITUTED $Bi_{1.6}Pb_{0.4}Sr_{2-x}Pr_xCuO_{6+\delta}$ SUPERCONDUCTORS

F. C. MATACOTTA[1], N. LEYAROVSKA[2], P. ŠŤASTNÝ[3], Q. JIANG, D. PAL,
IITM/ICTP HTS Laboratory, P.O.Box 586, Trieste, Italy

S. BERNIK
IJS, Jamova 39, Ljubljana, Slovenia

C. SCHMID
Institute for Applied Chemistry, University of Trieste, Trieste, Italy

ABSTRACT

The final properties of Pb- and Pr-substituted 2201 superconductors depend on several processing parameters like annealing temperature, reaction time and atmosphere and cooling procedure. As the Pr-content is reduced, their effect becomes more important and less reproducible. For samples having the composition $Bi_{1.6}Pb_{0.4}Sr_{1.65}Pr_{0.35}CuO_{6+\delta}$ the T_c value varies from less than 10 K up to 30 K. Moreover only part of the samples show regular dependencies of their properties on preparation conditions. We found that during formation there is a sizable loss of lead so that the superconducting phase grows while cation disproportioning takes place. The irreproducibility found in this compound is tentatively explained in terms of different cation redistribution processes which are kinetically in competition.

1. Introduction

The $Bi_2Sr_2CuO_6$ compound (2201) [1] is the n=1 term of the large family of bismuth based superconducting cuprates, represented by the formula $Bi_2Sr_2Ca_{n-1}Cu_nO_{2n+4}$ [2, 3].

The nominal 2201 composition is not found either in single crystal nor ceramics. Several authors showed that there are a few phases with similar structure near the formal composition, among them a monoclinic insulating phase (usually called collapsed phase) and an tetragonal Sr-deficient $Bi_{2+x}Sr_{2-x}CuO_{6+\delta}$ [4, 5] with T_c depending on x . The structure of the latter compound exhibits an incommensurate modulation with commensurate repetition of $5b$ [5, 6].

The double substitution of Pb on Bi-site and rare earth on Sr-site in the 2201 system enhances the phase stability, increases T_c values [7] and depresses the modulation [8]. Several studies on $(Bi,Pb)_2(Sr,RE)_2CuO_{6+\delta}$, with RE=La,Nd,Pr,Sm have been published so far [7, 8, 9, 10, 11, 12] but the reported results concerning structural and superconducting properties show some inconsistencies reflecting the complexity of the phase diagram of the system.

In this paper we present the results on structural and superconducting properties of $Bi_{1.6}Pb_{0.4}Sr_{2-x}Pr_xCuO_{6+\delta}$. The effect of several parameters related to the preparation conditions, like annealing and quenching temperature, gas flow rate and atmosphere composition, resulted to be much stronger than the variation of nominal composition especially for low values of x at which the structure of the coumpound becomes markedly orthorhombic. The parameters interplay in a combined way so that their final effect is not simply explainable. We think that the Pb volatility at the temperatures at which these compounds are formed, is one key to interpret the observed results.

2.Experimental

The samples were prepared by mixing proper amounts of high-purity Bi_2O_3, PbO, $Sr(NO_3)_2$ Pr_6O_{11} and CuO powders. The mixtures were ground thoroughly and fired twice in air at 860°C and 900°C, respectively. For the samples with x=0.35 and x=0.65 the prereacted powders were pelletized and sintered at various temperatures in the interval 850°C–940°C for 24–240 hours either in flowing oxygen or nitrogen. The pellets were quenched from the annealing temperature (T^{an}) or slowly cooled to intermediate temperatures (T_q) and then quenched. The pellets were weighed before and after the final annealing to find out the total mass loss during heat treatment. For the other compositions, the annealing temperature was 920°C and the duration 96 hours, followed by quenching.

Powder X-ray diffraction was performed using a Siemens D 500 diffractometer with monochromatic Cu Kα radiation. Lattice constants and other structural parameters were calculated using the Rietveld profile refinement [13] of the diffraction patterns. For all concentrations the *Pnan* space group symmetry and the crystal data reported by Tarascon et. al. [7] were used. SEM EDAX analysis were carried out using a JEOL JXA-840A electron probe microanalyser. Standards of all involved elements (their oxides and $SrTiO_3$) and ZAF correction were used in order to have a resolution better than 0.05 atomic per cent on the molar composition of the observed regions. A few DTA and thermogravimetry (TG) measurements were performed in order to find decomposition and melting temperatures in the investigated system and rates and temperature dependencies of mass loss. TG analysis was repeated with samples with no lead substitution in order to determine the nature of the weight loss. The total oxygen content was determined by wet iodometric titration. Resistivity was measured down to 10 K using standard four probe method. As the $T_c(\rho = 0)$ values are influenced by granularity effects due to non perfect homogeneity of samples, we report the transition temperature as the temperature of maximum derivative of the $\rho(T)$ curve in order to get a more reliable parameter to evaluate the intrinsic behavior of the material. A high sensitivity mutual inductance bridge was used to detect the a.c.susceptibility of sintered bars and powders down to 4.2 K. The χ' signal of the powdered samples at 6 K was compared to the a.c. susceptibility of vanadium bar (with equal volume and shape), to determine the screening factor.

3.Results

The lattice constants and critical temperature values are plotted as functions of praseodymium content x in Fig.1.

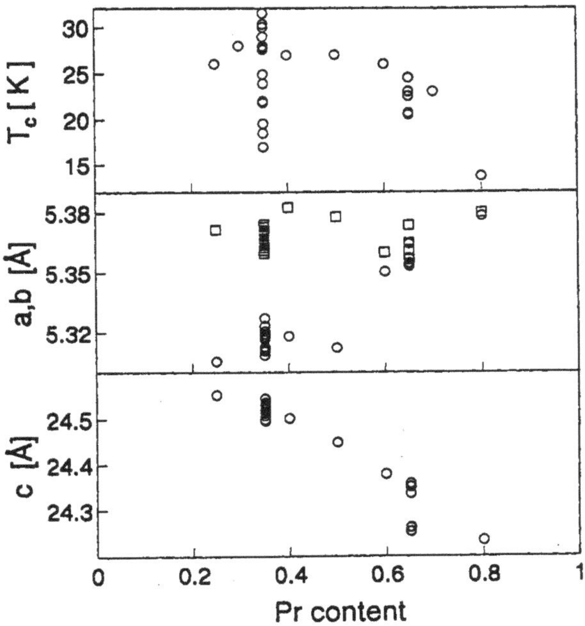

Fig.1: The critical temperature and lattice constants b,a and c (from top to bottom, respectively) as a function of Pr-content. For x=0.35 and x=0.65 the process parameters have been varied, causing a large scatter in the data which can be of the order of the total variation induced by compositional effects.

It is evident that the effect of varying the process parameters (for the two pref-erence compositions x=0.35 and x=0.65) exceeds the apparent effect of Pr-substitution. Before attending any discussions on compositional variations, careful study of the phenomena occurring during the preparation of these compounds has to be carried out.

3.1 Composition and structure

X-ray and SEM results show that there is a very narrow interval of T^{an} which yields single phase samples, the best T^{an} in oxygen being close to 920°C. Samples annealed at

$T^{an} < 900°C$ are not fully reacted whereas the samples treated at $T^{an} > 930°C$ are partially melted. The decomposition temperature is decreasing as the partial pressure of oxygen decreases. We were not able to prepare unmelted pellets in nitrogen at temperatures exceeding 870°C. In Fig.2 a,b, the SEM images of polished surfaces of samples (treated in oxygen at $T^{an} \simeq 920°C$ and then quenched) with x=0.65 and x=0.35, are shown.

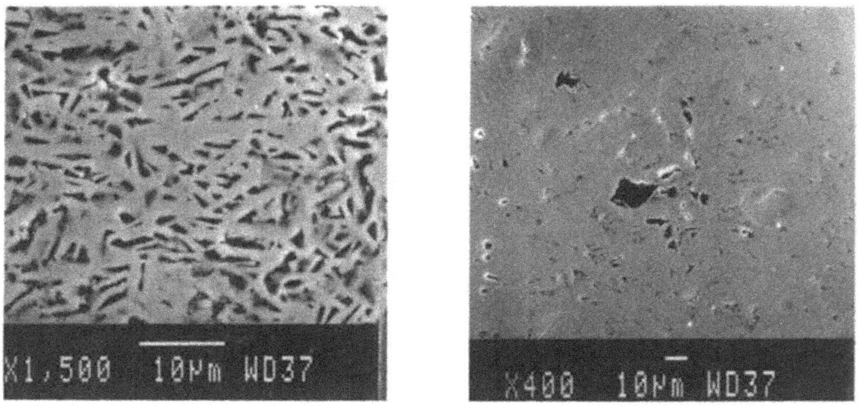

Fig.2: SEM pictures of polished surface of samples with Pr-content (a) x=0.65 and (b) x=0.35.

No traces of secondary phases are observed in sample with x=0.65 while a small amount of impurity phase occurs in sample with x=0.35 (darker spots in Fig.1b). The quantitative EDAX analysis, collected and averaged from five different places for each phase, indicates for both samples almost nominal composition for the main phase; in the sample with Pr-content x=0.35, however, small grains of strontium-rich bismuthates were detected. Small additional peaks in XRD pattern were indentified comparing to standard powder diffraction file as Sr_2CuO_3.

In several XRD patterns, first- and possibly second-order satellites of the $00l$ peaks appear. The satellite peaks have been assigned to the modulation along b-axis, as reported in [5]. As it will be discussed in detail later, the presence of impurity and modulation peaks is strongly influenced by preparation conditions and lead volatility. No trace of the so called collapsed phase [4] or other phases with $c > 15$ Å, was observed.

3.2 Weight loss

Prolonged heat treatment at temperature higher than 900°C always resulted in remarkable total weight losses (from 1% up to 5%, depending on starting lead content, temperature, anneling time, gas flow rate etc.). At the same time at the cold end of the quartz tubes, where the annealing were taking place, yellow deposits were observed, indicating that some metal had been released from reacting samples. The EDAX analysis of this deposit showed that it is pure lead oxide.

In Fig.3, the dependence of mass loss on temperature during heating cycle (heating rate 10 K/min) in static oxygen (curves a and c) and nitrogen (curve b) atmosphere for samples with x=0.35 prefired at 900°C are shown. At lower temperatures, not shown in Fig.3, the system loses weight due to the desorption of water ($T < 250°C$) and the decomposition of unreacted precursors ($T \approx 700°C$). Comparing curves a and c, the effect of the Pb-substitution is clear as these curves refer to the samples with Pb-content 0.4 and no Pb-substitution, respectively.

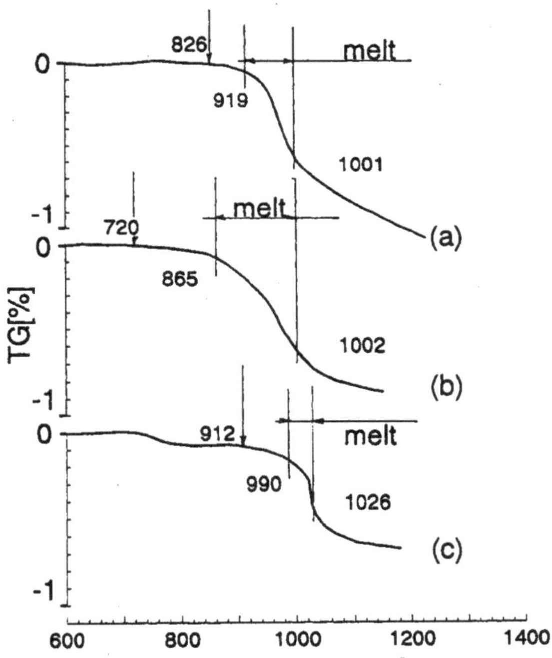

Fig.3: The weight loss of x=0.35 samples during heating cycle in static oxygen, curves a and c, and nitrogen curve b, atmosphere. (a),(b): $Bi_{1.6}Pb_{0.4}Sr_{1.65}Pr_{0.35}CuO_{6+\delta}$; (c): $Bi_2Sr_{1.65}Pr_{0.35}CuO_{6+\delta}$

The temperatures at which a steady loss of material starts, are 826°C and 912°C for samples with and without lead, respectively; in nitrogen atmosphere the sample containing lead starts to lose weight at 720°C. The corresponding DTA curves show that decomposition and melting temperatures depend on both Pb-content and reaction atmosphere and the resulting melting regions are indicated in Fig.3. The static isothermal TG measurements at various temperatures for analogous samples are shown in Fig.4. These curves proof that in oxygen at $T^{an} \approx 920°C$ the only metal evaporating from the compound is lead. As the preparation of fully reacted superconducting compounds requires long annealing time and high T^{an} (in general, the T^{an} for the cuprates increases as the number of Cu-O_2 layers decreases), the Pb-loss can not be avoided.

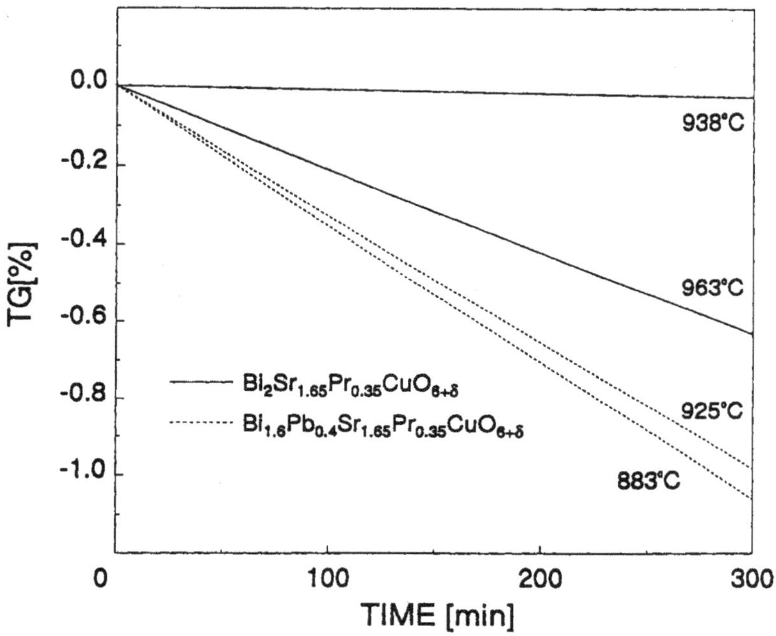

Fig.4: Static isothermal TG measurements. Solid lines refer to $Bi_2Sr_{1.65}Pr_{0.35}CuO_{6+\delta}$, dashed lines to $Bi_{1.6}Pb_{0.4}Sr_{1.65}Pr_{0.35}CuO_{6+\delta}$. (The sample heated at 925°C was not in powder form).

Combining the total weight loss measured after annealing with TG results, we estimate that the total Pb loss during a typical 96 hour treatment at 920°C is of the order of 0.07 formula unit.

The cooling cycles in static atmosphere always showed a weight increase due to partial incorporation of lead into the sample. This process obviously depends on the partial pressure of lead containing gas.

3.3 $(Bi_{1.6}Pb_{0.4})(Sr_{1.35}Pr_{0.65})CuO_{6+\delta}$

Samples with higher Pr-content show very weak dependence on preparation conditions (different T^{an}, T_q and O_2/N_2 atmosphere). For all the samples, the XRD profile refinement gave the cell parameters a=5.383±0.002, b=5.374±0.002 and c=24.351±0.005, the orthorhombicity factor, $ort. = [2*(a-b)/(a+b)] \approx 0.0016$, being negligible. The overall oxygen content variation is within the error and its value is 6.27±0.02 which is what should be expected by substituing the given amount of Sr^{2+} by Pr^{3+}. All samples were metallic with very similar values of resistivity. The T_c values were found in the range 19–22 K. A.c. susceptibility measurements on bulk bars showed quite sharp transitions and not very pronounced field dependence. However, similar measurements of powdered samples did not show a.c. magnetic screening. This is probably related to the fact, that the thickness of grains (see Fig.2a) is comparable with the penetration depth of this material.

3.4 $(Bi_{1.6}Pb_{0.4})(Sr_{1.65}Pr_{0.35})CuO_{6+\delta}$

The properties of samples with x=0.35 exhibit the highest T_c values and much stronger dependence on preparation conditions.

The resistivity at 300 K varies over one order of magnitude from 0.5 mΩcm up to 8 mΩcm.

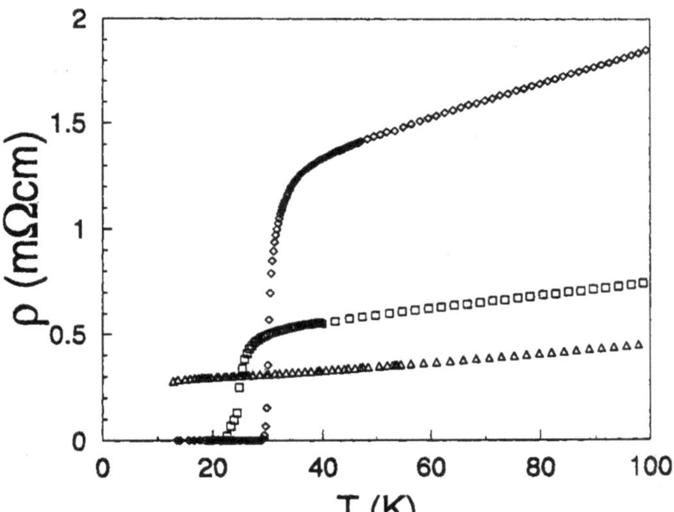

Fig.5: The influence of different preparation conditions on resistivity and T_c of three x=0.35 samples annealed for 4 days in oxygen at: (△) 900°C then slowly cooled down to room temperature, (□) 920°C slowly cooled down to 820°C and then quenched, (◇) 920°C then quenched

The slope of the $\rho(T)$ curve above the SC transition changes from positive values to zero or slightly negative ones. The $\rho = 0$ temperature ranges from below 10 K up to 29.5 K (see Fig.5). Many samples, however, show a two step transition consisting in a quite sharp drop at T-onset and a tail slowly approaching $\rho = 0$. As the shape of the high temperature part of the transition (involving always more than 50% of the total ρ variation) is the same for all samples, irrespective of the presence and slope of the tail, we took the value of T_c at the maximum transition slope.

Varying the annealing time at the T^{an}=920°C, no change in T_c was observed, but the metallic character of normal state resistivity is maximal after 80-100 hours (Fig.6). Correspondingly, the impurity and modulation peaks in XRD spectra reach minimum intensity for that time. The weight loss is increasing monotonically with increasing annealing time as could be expected from TG measurements.

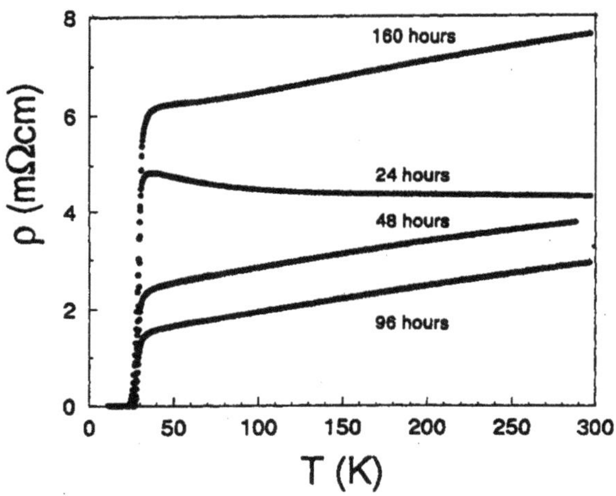

Fig.6: The resistivity curves for samples with x=0.35 annealed in oxygen at 920°C for (a) 24, (b) 48, (c) 96 and (d) 160 hours, and then quenched.

The a.c. susceptibility measurements on powdered samples resulted always in high values of the screening factor (20% to 38% with respect to the vanadium reference). The susceptibility curves obtained from bulk bars show different dependencies on the magnetic field: some samples exhibit a strong field dependent dissipation at temperatures much lower than T_c, indicating the presence of normal barriers to a very microscopic levels. As this occurrence is not always related to reduced normal state electrical conductivity, we suppose that some of these impurities have a metallic character.

The relationship between T_c and orthorhombicity for all the x=0.35 samples is shown in Fig.7a. The data are grouped into two linear parallel dependencies (branches (1) and (2)). The samples belonging to branch (1) show a linear decrease of the total oxygen content as a function of orthorhombicity (Fig.7b), while the corresponding values of the samples in branch (2) are scattered, but always lying on or below the line set by the other branch. However, it is to mention that the average oxygen content can be affected by an error due to the Pb-loss which results in a shift from the nominal composition stoichiometry and may give rise to the formation of small quantities of spurious phases with possible different oxidation state. Under these conditions the total oxygen content is not an intrinsic property of the superconducting phase but a parameter reflecting different cation distributions.

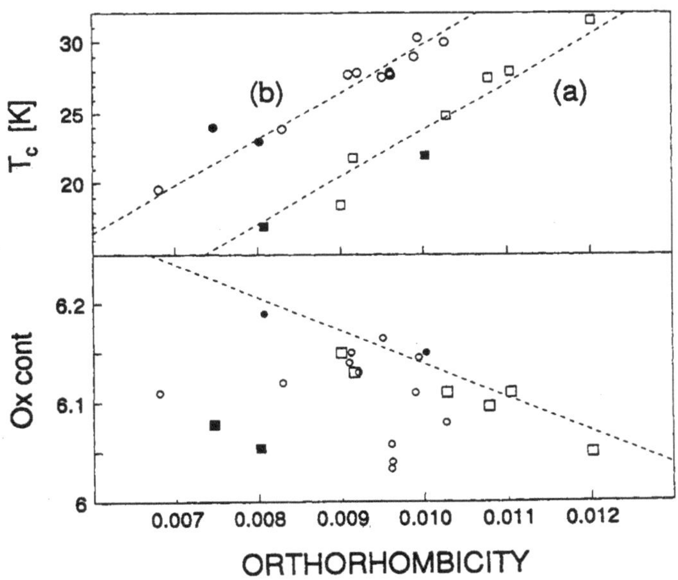

Fig.7: The relationship between (a) T_c and (b) total oxygen content versus orthorhombicity for x=0.35 samples. Empty markers - oxygen treated samples, filled markers nitrogen treated samples. Fig.7a defines two families of samples: squares belong to branch 1 and circles to branch 2. In following figures the symbols keep the same meaning.

4.Discussion

Our results show that at formation temperature several effects influence the stoichiometry and structure of the 2201 system. The lead doped compounds have to be reacted at temperatures where lead oxide is volatile and this has important consequences. As the matter of fact, during reaction the cation ratio is changed, consequences. As the matter of fact, during reaction the cation ratio is changed, giving rise to different processes which, in our opinion, are kinetically in competition. Under these conditions we should expect both stoichiometric shifts in the SC phase and formation of second phases. The changes in composition should account for changes in structural parameters and T_c, while the occurrence of second phases affect parameters like total oxygen content, normal state resistivity and field dependence of a.c. susceptibility.

According to EDAX analysis in the main phase the Bi:Pb ratio does not vary so much to account for the total mass loss. Furthermore, XRD and EDAX data revealed the presence of two main secondary phases in the x=0.35 samples, namely Sr_2CuO_3 and a strontium rich bismuthate.

Upon lead loss the stoichiometry unbalance could simply give rise to the formation of strontium cuprate and to a change in Bi:Pb ratio. As the latter remains practically constant, another mechanism must be present, probably Bi-substitution on Sr-site. Indeed, the Rietveld refinement of the powder XRD profiles turned out to be very sensitive to the occupation number of Bi at the Sr-site. Although the method can not distinguish between Pb and Bi , from stoichiometric reasons we believe that Pb is not available for such a substitution. Remarkably this occupation discriminates the two branches appearing in Fig.7a. Samples in branch (2) resulted to have a higher bismuth occupation on Sr-site (0.07-0.5 formula unit) while refinements of samples in branch (1) converge with this parameter close to zero.

Moreover, restricting our analysis to samples annealed for 4 days and plotting lattice parameters, T_c and total oxygen content as functions of this new parameter (Fig.8,9), we find that samples from branch (1) show well defined dependencies whereas the samples from branch (2) have more scattered values and less regular behavior.

The refinements of samples with higher Pr-content do not show such a occupation number variation of Sr-site probably due to the fact that higher content of Pr^{3+} cations prevents Bi to occupy this site. In our picture it is consistent with the lesser variability of other intrinsic parameters.

It has to be pointed out that the refinements of XRD data can not be an absolute technique to determine the distribution of cations within the unit cell but considering also the the different scattering factors of Bi- and Sr-atoms, this result is significant at least in qualitative way.

As the matter of fact, whatever is the meaning of the Bi on Sr-site value, it seems that samples lying on the right side in Fig.8 and Fig.9 and are of different quality even though XRD and EDAX results do not show macroscopic differences.

Very subtle mechanisms related to the thermal history of the samples, and probably to its early steps, determine the kind of intrinsic cation rearrangement induced by reaction in an increasingly Pb-poor environment. According to our results, in most of the samples from branch (2) Bi is allowed to substitute at the Sr-site, producing a much higher content of Sr-rich impurities and resulting also in a less homogeneous superconducting compound.

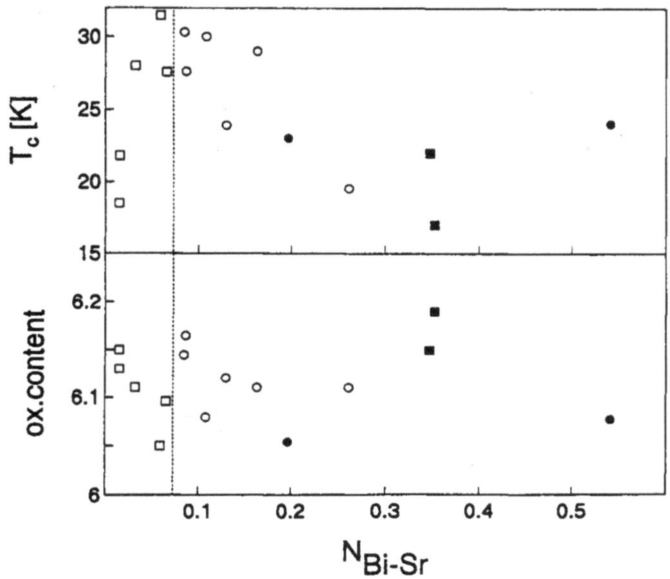

Fig.8: The relationship between the occupation number of Bi on Sr site and T_c (a), total oxygen content (b) of x=0.35 samples. The assignement of markers is the same as in Fig.7a,b.

In these samples T_c can still exceed 30 K but the intrinsic structural and physical properties are masked by the presence of impurities so that the results are quite scattered and no simple relation can be extracted.

We can not explain, however, why at a given orthorhombicity, T_c is higher for branch (2) as it is known that in the case of unsubstituted 2201 samples, T_c decreases monotonically as Bi substitutes on Sr-site [14, 15].

Contradicting some recent papers [11, 4] we could not observe any positive effect on T_c reducing the partial oxygen pressure. Actually, the best nitrogen annealed samples had T_c=24 K; as the kinetics of the system in nitrogen is different, the optimum process to get high T_c could not be the same.

5. Conclusion

Structural and superconducting properties of the $Bi_{1.6}Pb_{0.4}Sr_{1.35}Pr_{0.65}CuO_{6+\delta}$ and $Bi_{1.6}Pb_{0.4}Sr_{1.65}Pr_{0.35}CuO_{6+\delta}$ compounds have been investigated as functions of process parameters.

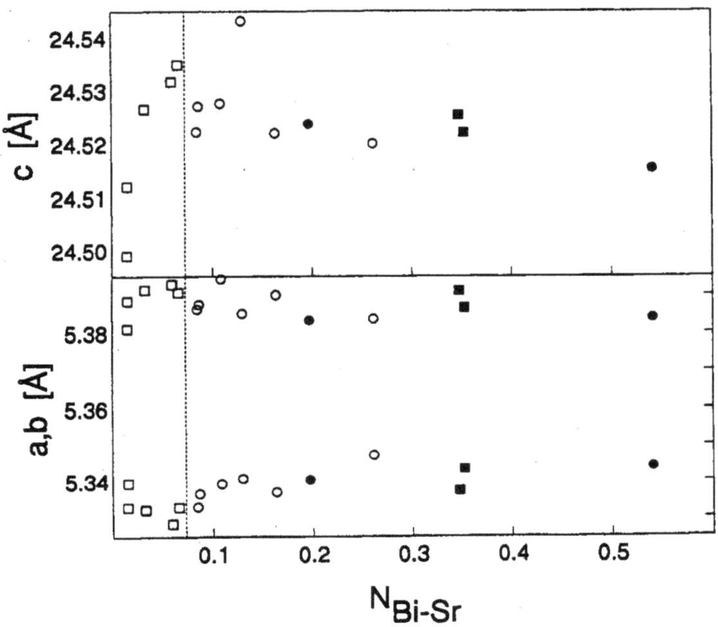

Fig.9: The relationship between the occupation number of Bi on Sr site and lattice parameters of x=0.35 samples. The assignement of markers is the same as in Fig.7a,b.

At the temperatures requested for synthesis the lead evaporates at such a rate that a total weight loss of the order of 0.07 formula unit is found after an optimum annealing process. Samples with higher Pr-content are not affected by changes in preparation but the x=0.35 samples showed to be extremely sensitive to that. The best properties are found in samples annealed at and quenched from temperatures as close as possible to the melting temperature. As the lead evaporates, several processes are taking place and are kinetically in competition so that the final results depend on the detailed thermal history of a given sample.

Qualitative results indicate that the occurrence of the substitution of bismuth at the Sr-site (one of the possible consequences of lead loss) is one cause of the extreme variation of the intrinsic properties of this compound.

No positive effect of nitrogen atmosphere has been detected but as the weight loss rate and decomposition temperature depend very strongly upon atmosphere composition, more data are needed to clarify this point.

Permanent addresses:
[1] CNR-ITM, Via Bassini 15, Milan, Italy
[2] Institute of Solid State Physics, 1784 Sofia, Bulgaria
[3] Dept. of Semicon. Physics, Charles University, Prague, Czechoslovakia

References

[1] C. Michel, M. Hervieu, M.M. Borel, A. Grandin, F. Deslandes, J. Provost, B. Raveau, *Z. Phys. B* **68** (1987) 421

[2] For a review see Sleight A.W. *MRS Bulltin* **1989** (1989) 45

[3] J.M. Tarascon, W.R. McKinnon, P. Barboux, D.M. Hwang, B.G. Bagley, L.H. Greene, G.W. Hull, Y. LePage, N. Stofel, M. Giroud *Phys. Rev. B* **38**, (1991) 8885

[4] B.C. Chakoumakos, P.S. Ebey, B.C. Sales, E. Sonder *J. Mater. Res.* **4** (1989) 767

[5] Z. Hiroi, Y. Ikeda, M. Takano, Y. Bando, *J. Mater. Res.* **6** (1991) 435

[6] H.W. Zandbergen, W.A. Groen, F.C. Miljhoff, G. Van Tedeloo, S. Amelinckx, *Physica C* **156** (1988) 325

[7] J.M. Tarascon, W.R. McKinnon, Y. LePage, K. Remschnig, R. Remesh, R. Jones, G. Pleizier, G.W. Hull, *Physica C* **172** (1990) 13

[8] V. Manivannan, J. Gopalakrishnan, C. N. R. Rao, *Phys. Rev. B* **43** (1991) 8686

[9] S. Kambe, M. Kawai, T. Kawai, *Solid State Comm.* **75** (1990) 435

[10] B.C. Sales, B.C. Chakoumakos, *Phys. Rev. B* **43** (1991) 12994

[11] Y. Murakosi, S. Kambe, M. Kawai, *Physica C* **178** (1991) 71

[12] K. Zhang, G. Sedler, B.H. Ma and C.U. Segre *Physica C* **179** (1991) 405

[13] D.B. Wiles and R.A. Young, *J.Appl.Cryst.* **14** (1981) 149

[14] R.M. Fleming, S.A. Sunshine, L.F. Schneemeyer, R.B. Van Dover, R.J. Cava, P.M. Marsh, J.V. Waszczak, S.H. Glarum, and S.M. Zahurak, *Physica C* **173** (1991) 37

[15] J. Akujieze, Ph. D. Thesis, unpublished

Part 2

Critical currents, flux motion and flux pinning enhancement. Magnetic shielding

FLUX LATTICE RELAXATION IN YBCO AND BSCCO SINGLE CRYSTALS IN AN EXTENDED FREQUENCY AND TEMPERATURE RANGE

V.Calzona, M.R.Cimberle, C.Ferdeghini, G.Grasso, A.Iorio, F.Licci*, M.Putti, C.Rizzuto, W.P.Seminozhenko**, A.S.Siri.

C.N.R. I.N.F.M. Dipartimento di Fisica via Dodecaneso 33 16146 Genova
*MASPEC Institute of C.N.R. via Chiavari 18/a 43100 Parma
**Institute for single crystal Kharkov USSR

ABSTRACT

The flux-line lattice dynamics has been studied in YBCO and BSCCO single crystals with different experimental techniques: measurements of frequency shift of a.c. susceptibility, measurements of magnetization decay and measurements of hysteresis loops at various field sweep rates will be presented, so spanning the frequency interval 10^{-3} - 10^{4}Hz.

It will be shown that the simplified Thermally Activated Flux Creep (T.A.F.C.) and Thermally Assisted Flux Flow (T.A.F.F.) models are not able to explain satisfactory the experimental results on the whole temperature range.

By observing the ρ vs.J characteristic it comes out that, differently from the traditional superconductors, an experiment can involve an extended range of J values between 0 and J_c and, correspondingly, an effective potential barriers between U_o and zero.

It is proposed to solve numerically, without approximation, the flux lines diffusion equation with the aim to represent various physical experiments.

Some simple examples for the magnetization decay will be presented.

1. Beyond the critical state approximation.

In this work we assume the classical schema of thermal activation for flux lines motion[1], disregarding any theory of glass superconducting state or melting transition[1,2].

In a type II superconductor in mixed state a microscopic relationship connect the resistivity to the current flowing in the sample; the effective current flowing in the sample depends on this relationship and on the particular experiment performed.

In a simple experiment in which the magnetization of an ideal tubular thin sample is measured the situation can be sketched as in fig 1.

In this figure the lnρ versus J curve named 'a' represents the microscopic relationship for a superconductor; ρ is the resistivity due to the

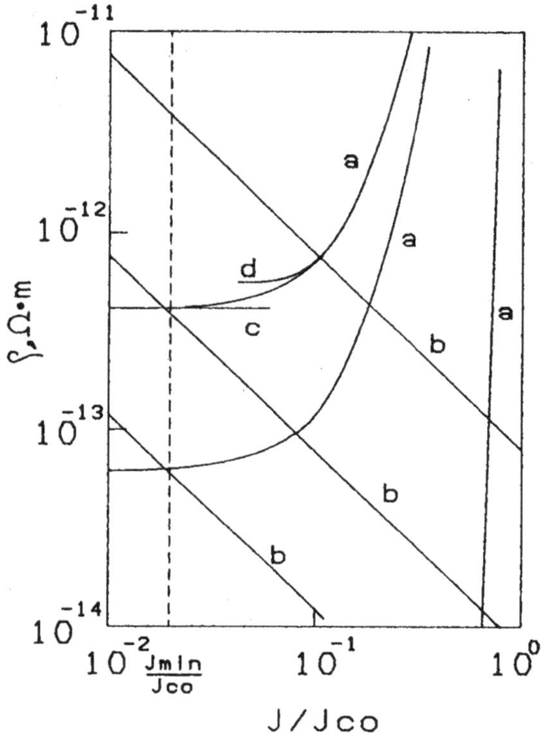

FIG. 1 a) ln(ρ) vs. ln(J/J_{co}) as deduced by microscopic properties (see formula (4)), with $\omega_o x_o$=500 m/sec, $J_{co}=10^8 A/m^2$, U_o/kT=20, U_o/kT=22 and U_o/kT=60 (the last one well represents a typical behaviour of a conventional low Tc superconductor).
b) ln(ρ) vs. ln(J/J_{co}) related to the experimental conditions (see formula (7)), with $l=10^{-3}$m, $J_{co}=10^8 Am^{-2}$ and dB/dt=16 mT/sec, dB/dt=1.6 mT/sec, dB/dt=0.25 mT/sec.
c) TAFF approximations of curve a), (see formula (5)).
d) TAFC approximations of curve a) (see formula (6)).

dissipative motion of the fluxons .

It results $\rho = E/J = Bv/J$ where B is the magnetic field, v the velocity of fluxons and J the current.

The fluxons velocity is given by the maximum velocity for the fluxoids at the critical state $x_o\omega_o$ time the difference between the probability of direct and reverse hopping (x_o is the hopping distance and ω_o is the attempt frequency).

The two effective energy barriers, U_d and U_r, respectively in the case of direct and reverse hopping, differ in the work of the Lorentz force. The expression depends on the kind and shape of the potential well. We can write:

$$v = x_o \omega_o \left[\exp\left(-\frac{U_d}{kT}\right) - \exp\left(-\frac{U_r}{kT}\right)\right] \tag{1}$$

and so we obtain:

$$\rho = \frac{B}{J} x_o \omega_o \left[\exp\left(-\frac{U_d}{kT}\right) - \exp\left(-\frac{U_r}{kT}\right)\right] \tag{2}$$

In the case of triangular potential wells the well known dependence of U from J is valid[4]:

$$\begin{cases} U_r = U_o + U_L = U\left(1 + \frac{J}{J_{co}}\right) \\ U_d = U_o - U_L = U_o\left(1 - \frac{J}{J_{co}}\right) \end{cases} \tag{3}$$

where J_{co} is the critical current density value before the creep mechanism takes place that means $U_L=U_o$.

Using this approximations we obtain:

$$\rho = 2\frac{B}{J} x_o \omega_o \exp\left(-\frac{U_o}{kT}\right) \sinh\left(\frac{J}{J_{co}}\frac{U_o}{kT}\right) \tag{4}$$

In fig 1 two different "a" curves are reported for two different temperatures.

Curves 'c' and 'd', in the same figure, represent the approximations of the curve 'a'in the two well known asymptotic situations[5,6]:

$\frac{J}{J_{co}}\frac{U_o}{kT} \ll 1$ Thermally Assisted Flux Flow model (TAFF)

$\begin{cases} \frac{J}{J_{co}}\frac{U_o}{kT} > 1 \\ U_o - U_L \gg kT \end{cases}$ Thermally Activated Flux Creep model (TAFC)

In these cases we obtain respectively:

$$\rho = \rho_{TAFF} = 2\, B\, x_o \omega_o \exp -\left(\frac{U_o}{kT}\right)\left(\frac{1}{J_{co}}\frac{U_o}{kT}\right) \tag{5}$$

$$\rho = \frac{B}{J} x_o \omega_o \exp\left(-\frac{U_o}{kT}\right) \exp\left(\frac{J}{J_{co}}\frac{U_o}{kT}\right) \tag{6}$$

We point out that TAFF solution is an analytical solution for the general diffusion equation only when the driving force is small, compared to kT, so that the resistance and the related diffusion coefficient is constant; TAFC is also an analytical solution when the reverse hopping for the fluxoids may be neglected, that means a strong driving force compared with a small thermal activation energy.

The curve 'b' is related to the experimental conditions; we have calculated it in the case of constant field sweep rate and when steady state is reached.

In such a case the variation of the internal field ΔB with time, can be considered uniform inside the sample and equal to the external field variation in the same interval of time and independent from the particular shape of the internal profile.

We can write the continuity equation for the flux of magnetic field in the sample considering a slab of thickness l and width d as:

$$\frac{d\Phi}{dt} - [B\, v\, d - B\,(-v)\, d] = 0 \qquad (7)$$

where v, in this case, is the velocity of the fluxons at the surface of the sample and B is the field at the surface. So we obtain

$$\rho = \frac{B\, v}{J} = \frac{dB}{dt}\,\frac{l}{2J} \qquad (8)$$

This equation is represented by the curve b in fig.1 (varying, for example the field sweep rate the curve b changes).

The intercept between curves a and b gives directly the current flowing in the sample: it's easy to see that lowering the temperature, maintaining fixed the curve 'b', it's possible to change the flux motion from TAFF to TAFC regime; we want to point out that it should be possible change in the same way the flux motion regime fixing curve 'a' and increasing the external field sweep rate or the sample dimension.

This aspect is not usually taken into account dealing with flux motion problem.

This feature is typical for the HTSC: in the low temperature superconductors, essentially because of a larger value of U_o/kT, (curve 'e' in fig.1), the intercept remain very near J_c where the well known critical state model (TAFC) works for any realistic experimental condition.In the case of HTSC on the contrary, varying the experimental parameter, a wider range of J/J_{co} is covered; in particular also intermediate regions between TAFC and TAFF may be reached.

For a real experiment, as for example a.c. susceptibility or hysteresis loops or magnetization decay measurements, in which the spatial dimension of the sample becomes important it is no more possible to disregard the spatial and time convolution and it is necessary to solve the complete diffusion equation for the flux lines motion without any simplification. But this schema, although exact only in an ideal condition, as already pointed out, is anyway a very useful guide to individuate the effective critical current density flowing in the sample and its variation changing the experimental conditions or the microscopic characteristics.

In the literature on HTSC flux line lattice dynamics some anomalies are reported. The two main features are: an increase of the effective pinning energy with the temperature for $T < T_c/2$ from magnetization decay measurements[7,8], and a magnetization decay often observed to be not logarithmic in time. Two specific models ave been proposed to explain these experimental results:

a) a U_o distribution function[10]

b) a not linear U vs J relationship[8].

Both the hypothesis lead to an effective U_o increasing with the temperature and a not logarithmic magnetization decay.

We want to point out that these two reasonable items, introduced in TAFC approximation are a priori, not necessary.We think that it is necessary before to solve the diffusion equation without approximation in a numerical way.

2. Experimental

We use for a.c. susceptibility measurements a home made mutual inductance susceptometer. The sample is introduced in a pick up coil (of 1000 turns) opposite to another identical pick up coil 1 cm. apart. The system is made by non-metallic materials and is placed inside a bore of a copper coil able to produce an a.c. magnetic field ranging from 10^{-6} to 10^{-2} T. This coil is inside a superconducting 8 magnet. The frequency range is $10 \div 2.5 \cdot 10^4$ Hz. With this apparatus we measure single crystals of $Y_1Ba_2Cu_3O_7$ produced in the Maspec laboratories.

The magnetization loops and the magnetization decay measurements are performed in an experimental set up which is described in details elsewhere[11].

For our purpose is sufficient to see that is possible to change the field sweep rate from 0.002 to 0.1 T/sec. The temperature range is 4.2 K $\div$ 90 K and the field range is up to 5 T.

The $Bi_2Sr_2Ca_1Cu_2O_8$ single crystal (whose dimensions are $6 \times 4 \times 2$ mm^3) was supplied by the Institute for Single Crystal in Kharkov.

3. Experimental results and discussion.

3.1. Ybco sample

As well known, in the a.c. susceptibility measurements in high magnetic field on a YBCO sample the χ'' peak shifts at higher temperature when the

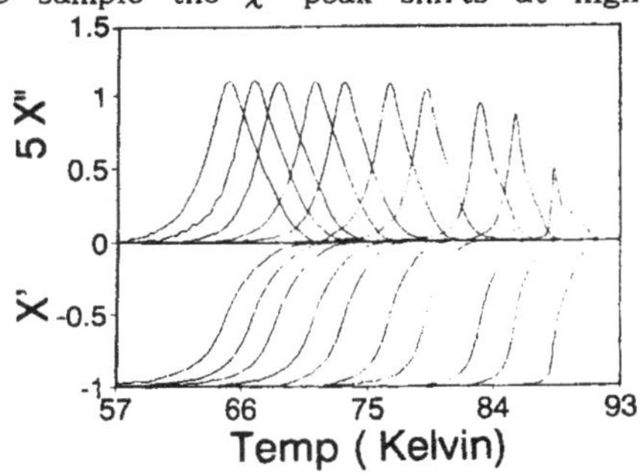

FIG.2 χ' and χ'' vs T at fixed frequency (f = 500 Hz) and fixed $Bac=10^{-4}$T, with Bdc ranging from zero to 8 T.

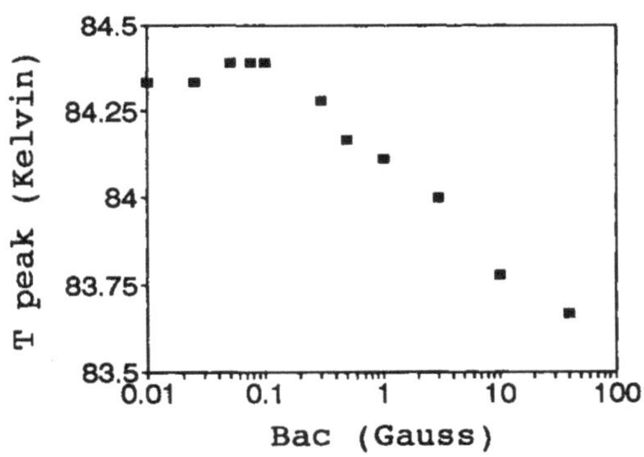

FIG.3 T_{peak} vs $\ln(B_{ac})$ at B_{dc} = 1 T, f = 1 Khz.

frequency is increased. This effect, relatively small at low field, increases with the field. In fig.2 the shift is observed varying the external applied field at fixed frequency. The shift reach a maximum of about 25 K at 8 T.

Our data show a frequencies dependence similar to the ones usually presented in literature. For example if, following the approach of Palstra et al.[12], we plot the logarithm of frequency versus $1/T_{peak}$ we found a linear dependence.

But this is valid only if Bac is large enough. In fig.3 we show the T_{peak} versus B_{ac} behavior. In the low B_{ac} region T_{peak} remain constant indicating a sample magnetization linear with the B_{ac} value. This behaviour indicate clearly that the TAFF regime is reached[5].

To determine experimentally the irreversibility line we define it operatively as the line in the T,B plane where the χ'' values have the minimum measurable value in the limit of our experimental set-up: $\chi_{min} \simeq 2\cdot10^{-3}$; which corresponds approximately to a sample surface current J_{min} = 30 mA/cm^2.

In fig.4 the irreversibility line for an YBCO sample, so defined, is reported with the frequency as parameter.

We observe that, when the system is on irreversibility line, the value of

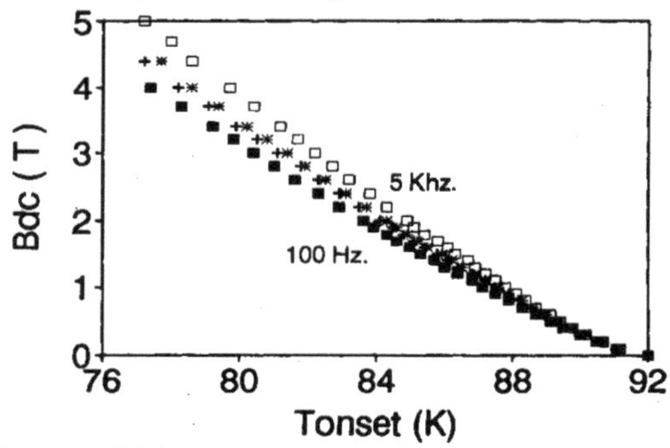

FIG.4 Irreversibility line as defined in the text at f = 0.1, 0.5, 1, 5 KHz.

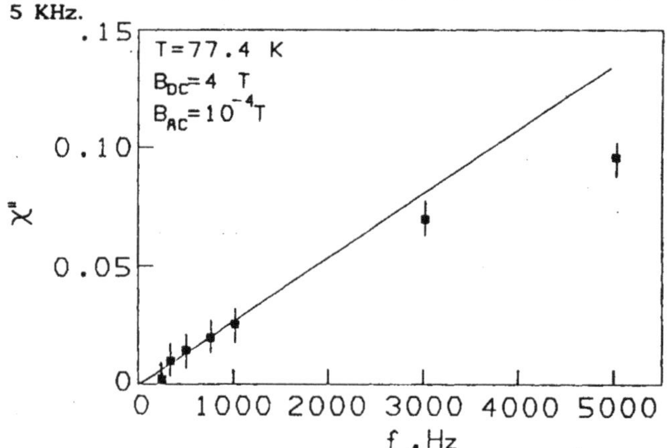

FIG.5 χ'' vs f at T=77.4 K, B_{dc}=4 and $B_{ac}=10^{-4}$ T: + experimental data; full line calculated in TAFF approximation.

J is J_{min} (vertical line in fig.1), and the working point falls in TAFF region.

Now we can calculate χ'' in TAFF regime as proposed by Kes[5]. To compare it with our experimental data we must cut the χ'' curves at fixed temperature: we choose T = 77.4 K, B_{dc} = 4 T.

The results are presented in fig 5. The agreement between experimental data and theoretical curve is substantially good, not in disagreement with a TAFF regime hypothesis, considering that the fit is performed without any adjustable parameters. The deviation at high frequencies may be due to the deviations from a pure TAFF regime.

3.2. Bscco sample

Fig.6 shows some hysteresis loops measured at T = 10 K and various dB/dt in the range 0.1-0.002 T/sec.

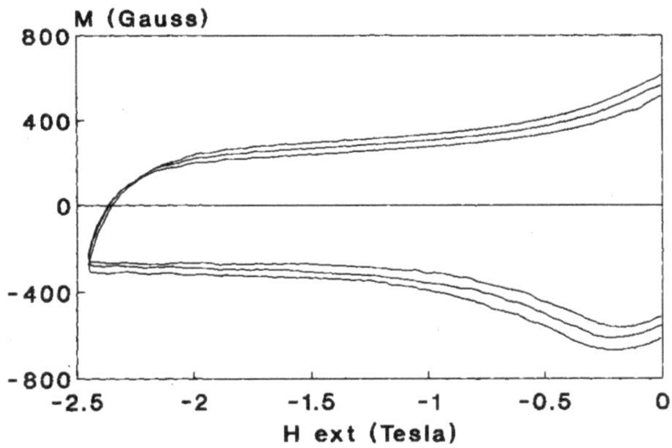

FIG.6 Hysteresis loops at T=10 K and dB/dt in the range 0.1-0.002 T/sec in BSCCO single crystal.

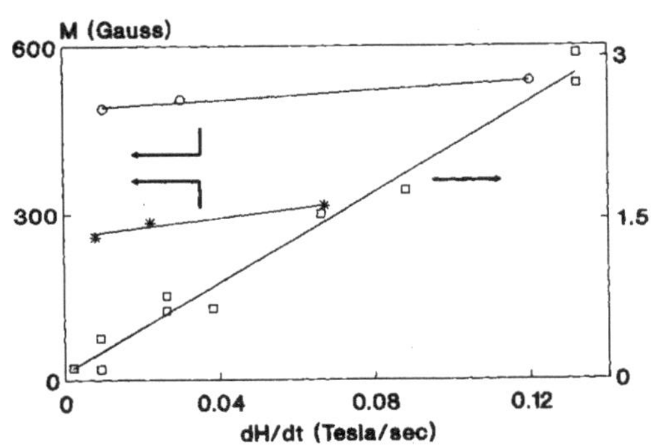

FIG.7 M vs dB/dt at: 1) ⊙ T=4.6 K and B=2.7 T; 2) * T=15 K and B=1.5 T; 3) ◻ T=35 K and B=0.25 T.

The hysteresis loop amplitude increases significantly as the sweep rate increases. Similar results were obtained in the temperature range 4.6–40 K for the same dB/dt values.

Considering the magnetization value at a constant external field, we can plot the amplitude of the magnetization versus the sweep rate dB/dt. The results are shown in fig. 7 for T = 4.6 K and B = 2.7 T, T = 15 K and B = 1.5 T, T = 35 K and B = 0.25 T.

At T = 35 K we observe a zero crossing linear behaviour while, decreasing the temperature the linear interpolation of the data crosses the y axis at higher magnetization values.

We interpretate this feature as an evidence at T = 35 K of the TAFF regime and, for lower temperatures, we note a drift toward a TAFC like behaviour.

At T = 35 K it is possible to calculate the TAFF resistivity through the relationship[13]:

$$\rho_{TAFF} \cong (\frac{dH}{dT}\frac{l^2}{3}\mu_o)/M_{irr} \cong 6.8\cdot10^{-10}\ \Omega m \qquad (9)$$

that is of the same order of magnitude of the copper resistivity at the same temperature.

In fig.8 we present some magnetization decays observed at fixed values of the external B. For each temperature value we choose B > $2B_p$ (where B_p is the full penetration field).

The first feature is that a large relaxation is observed and the decay is far to be logarithmic in t at least at higher temperatures.

It can be observed that for T = 20 K the time in which the magnetization decays to zero is very short (100 sec) and a continuous slope variation is observed.

It is possible using a not linear U vs. J relationship and hypotizing a U_o distribution function to fit this experimental data.The detailed procedure is reported elsewhere[13]. A U_o value of 20 meV was obtained in agreement with the current literature.

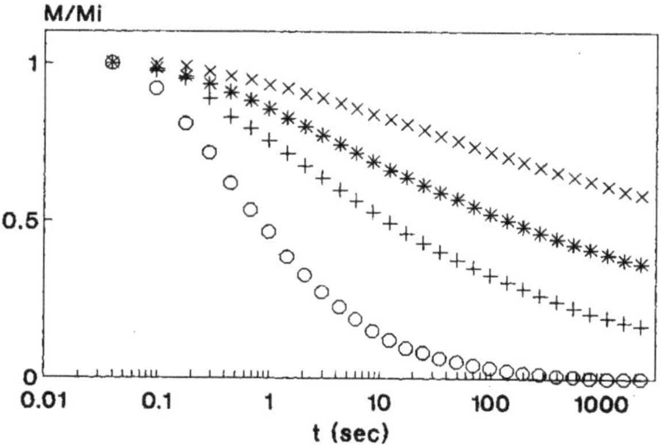

FIG.8 M/Mi vs. ln(t) at: x T=4.6 K and B=2.7 T; * T=10 K and B=1.5 T; + T=15 K and B=1.5 T;o T=20 K and B=1.5 T. Mi is the first experimental value at t=40 msec after the external field is stopped.

But we think that a more general approach is needed. So we try to solve the diffusion equation for the F.L.L. and we will show that the general behaviour can be described.

4. Numerical solution of the diffusion equation.

Following Beasley, Labush and Webb[6] we start from the diffusion equation for the vortex lines:

$$\frac{\partial B}{\partial t} = \frac{\partial}{\partial x}\left(D \frac{\partial B}{\partial x}\right) \qquad (10)$$

were the diffusion coefficient D is given by:

$$D = \frac{\rho}{\mu_o} = \frac{B\, x_o\, \omega_o}{J\, \mu_o} \left[\exp(-\frac{U_d}{kT}) - \exp(-\frac{U_r}{kT})\right] \qquad (11)$$

The effective energy for direct and the reverse hopping are taken both into account and by inserting the equations (3) in the relation (10) we obtain the well known expression :

$$D = \frac{B\, x_o\, \omega_o}{J\, \mu_o} \exp(-\frac{U_o}{kT}) \sinh\left(\frac{J}{J_{co}} \frac{U_o}{kT}\right) \qquad (12)$$

which implies the following approximations:

a) triangular wells and

b) single effective depth of the well.

By the relation (12) we can solve numerically the diffusion equation, but the strong D vs J dependence may causes serious numerical problems. At the moment only the magnetization decay is simulated, while technical problems prevent us to solve the equation in the case of hysteresis loop and a.c. susceptibility.

We tested our program comparing the numerical results with the analytic solution TAFF and TAFC. This comparison ensure us on the goodness of the numerical results.

In particular it's important to study the magnetization decay in an intermediate region where

$$\frac{J}{J_{co}} \approx \frac{kT}{U_o} \qquad \text{or} \qquad \frac{U_o}{kT} \approx \ln\left(\frac{\omega_o\, x_o\, B_{DC}}{B_{AC}\, l\, f}\right)$$

This first relationship came from imposing near to unity the argument of sinh in equation (12) ; the second from imposing sinh near the unity in the equation $\rho_a = \rho_b$ (being ρ_a and ρ_b the "microscopical" and "experimental" curve defined above).

We consider the magnetization decay in a slab of thickness d and we fix the boundary conditions:

$$\frac{dB}{dt} = \text{const} \qquad 0 < t < t_o$$

$$\frac{dB}{dt} = 0 \qquad t > t_o$$

In this simple case the diffusion equation has been solved and some results are plotted in fig. 9.

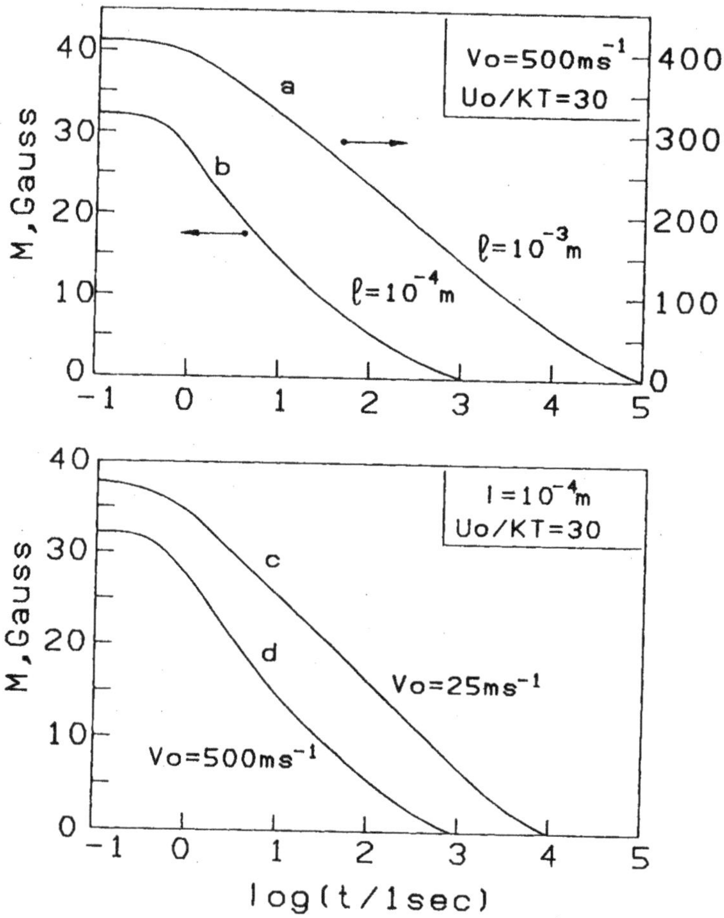

FIG.9 Upper part: Magnetization decay as obtained by numerical solution of eq.(9), without the approximations, varying only the sample dimension. $U_o/kT=30$, $v_o=\omega_o x_o=500$ m/sec, $J_{co}=4\ 10^8$ A/m^2 and: a) $l=10^{-3}$m b) $l=10^{-4}$ m.

Lower part: the same as in upper part varying only the v_o value. $U_o/kT=30$, $l=10^{-4}$m, $J_{co}=4\cdot10^8 A/m^2$ and c) $v_o=$ 25 m/sec d) v_o=500 m/sec.

The parameter involved are both microscopic both fixed by the experiment and the results depends in a complicated way on these values.

The preliminary results in fig.9 are been chosen to study the passage between the logarithmic and the non-logarithmic magnetization decay as experimentally observed.

The relaxation curves have been obtained with $U_o/kT = 30$ and $J_{co} = 4 \cdot 10^8$ A/m^2. In fig. 8(a) we fix $\omega_o x_o = 500$ m/sec and we change the thickness of the slab ($d = 10^{-3}$ m higher curve, $d = 10^{-4}$ m lower curve). The decrease of the sample dimensions obviously produces lower magnetization values, but what it is not obvious it is the change from logarithmic to not logarithmic decay. If we calculate U_o/kT from the slope of the curve we obtain correctly 30 in the large sample, but for the small sample any choice of the slope gives $U_o/kT \approx 23 - 25$ not in agreement with the imposed value. We note that the dimension of the sample, generally, is not take into account .

In fig. 9b we compare the magnetization decay in the smaller sample ($d = 10^{-4}$m) when we change $\omega_o x_o$ (500 m/sec lower curve, 25 m/sec higher curve). Decreasing the fluxon velocity $\omega_o x_o$ the logarithmic behaviour is still observed.

These results are indicative of the fact that, in some conditions, can be incorrect to evaluate U_o from relaxation measurements. In particular it is possible to pass from a logarithmic to a not logarithmic behaviour changing either microscopic or experimental parameter.

The work is in progress to fit the experimental data.

This work is partially supported by INFM and by CNR under the Progetto Finalizzato "Superconductive and Cryogenic Technologies".

References

1. P.W.Anderson, *Phys. Rev. Lett.* 9 (1962) 309
2. Mattew P.A.Fisher, *Phys. Rev. Lett.* **62** (1989) 1415
3. P.L.Grammel, L.F.Scneemeyer, J.V.Waszczak and D.J.Bishop, *Phys. Rev. Lett.* **61** (1988) 1666
4. R.Griessen, *Phisica C* **172** (1991) 441
5. P.H.Kes, J.Aorts, J. van der Berg, C.J. van der Beck and J.A.Mydosh, *Supercond. Sci. Technol.* 1 (1989) 242
6. M.R.Beasley, R.Labush and W.Webb *Phys. Rev.* **161** (1969) 682
7. Y.Xu, M.Suenaga, A.R.Moodenbaugh and D.O.Welch, *Phys. Rev. B* **40** (1990) 10882
8. I.A.Campbell, L.Fructer and R.Cabanel, *Phys. Rev. Lett.* **64** (1990) 1561
9. M.P.Maley, J.O.Willis, H.Lessure and M.E.McHenry, *Phys. Rev. B* **42** (1990) 2639
10. C.W.Hagen and R.Griessen *Phys. Rev. Lett.* **62** (1989) 2857
11. M.R.Cimberle, C.Ferdeghini, A.S.Siri, *Cryogenics* **29** (1989) 69
12. T.T.M.Palstra, B.Battlog, R.B. van Dover, L.F.Schneemeyer, J.V.Waszczak *Phys. Rev. B* **41** (1990) 6621
13. J. van der Berg, C.J. van der Beck P.Koorevaar,P.H.Kes, J.A.Mydosh, M.J.V.Menken and A.A.Menovsky, *Physica C* **162-164**, (1989) 1189
14. V.Calzona, M.R.Cimberle, C.Ferdeghini, A.Iorio, M.Putti and C.Rizzuto, in press on *Supercond: Sci. Technol.*

Dissipative phenomena in Bi(Pb)-Sr-Ca-Cu-O

D.Fiorani

and

A.M.Testa

ITSE, CNR, Area della Ricerca di Roma, P.B. 10-00016 Monterotondo Stazione- ITALY

ABSTRACT

Pinning and dissipative effects have been investigated in Bi-Pb-Sr-Ca- CuO samples (sintered and single crystals) by measurements of :
a) hysteresis cycles ;
b) the time decay of the zero field cooled magnetization (M_{zfc});
c) the (H,T) irreversibility line .
The temperature dependence of the reduced pinning force, the crossover from a logarithmic to a non logarithmic time decay of M_{zfc} (for T>50K) and the observed power law for the (H,T) line ($0.30<n<0.40$), are discussed in relation to the flux creep, to its extensions and to the collective pinning theory .

1. Introduction

Since the discovery of high temperature superconductors, extensive studies of dissipative effects have been carried out by several groups [1-5]. The fast decay of the magnetization, firstly reported by Muller et al.[6] on policrystalline materials and also observed in single crystals [4], was initially explained in terms of refined versions [4] (the so called "giant " flux creep) of the phenomenological model proposed by Anderson[7]. In addition to magnetic relaxation, also the broadening of the resistive transition with increasing magnetic fields and the thermally activated behaviour of the resistivity in the low temperature region, were attributed the thermally activated flux motion[8], with rather low activation energies as compared to conventional superconductors.

However, activation energies derived from magnetic and transport analyses have been found to vary widely, even for a given compound: i.e. in YBCO for instance, 0.1÷1 eV by resistive transition [9,10]; 10 ÷ 100 meV by magnetic relaxation [11]. Moreover, the temperature and field dependence of the pinning potential (U_o) deduced from experiments has been found to vary from a Ginzburg-Landau like form,$U_o \,\alpha\, (1-T/T_c)^n/H$, to an anomalous increase with temperature at low fields . The high degree of mobility of the magnetic flux lines is also reflected in the existence of an irreversibility line[6] in the (H,T) phase diagram, well below the upper critical field. As a consequence of these results, an extensive theoretical work on the flux motion has been stimulated and various models have been proposed.

M.P.A.Fisher[12,13] suggested that in a type II superconductor with impurity induced disorder, at low temperature the crystalline Abrikosov flux lattice may be replaced by a thermodynamically distinct phase, "vortex glass", with a true phase transition line separating from a high temperature "vortex liquid". Nelson et al.[14,15] proposed the existence of an entangled flux line phase between the Abrikosov phase and the mean field $H_{c2}(T)$, and pointed out the role of thermally induced collisions in a dense vortex liquid. The idea of an "intrinsic " pinning provided by the CuO planes (with a substantial effectiveness when the field is parallel to the layers), as a consequence of the weak interplanar Josephson coupling and small coherence length, has been discussed by some authors [16,17]. Finally, the theory of collective pinning of Larkin et al.[18] was extended by Feigelman et al. [19-21] to describe flux creep in high T_c superconductors. Unlike the conventional flux creep model, a non linear relationship between the activation energy and the current density J is assumed, leading to a non logarithmic time decay for J(t) as observed experimentally [22,23].

The great amount of data available on dissipative effects and the various theoretical treatments clearly show that the interplay between phase transitions in the flux line lattice and the picture of thermal depinning used for conventional superconductors, is not yet clarified.

In this paper we present results on pinning and dissipative effects in Bi-Pb-Sr-Ca-CuO samples (sintered and single crystals) investigated by measurements of :

a) hysteresis cicles ;

b) the time decay of the zero field cooled magnetization (M_{zfc});

c) the (H,T) irreversibility line .

The observed temperature dependence of the reduced pinning force, the crossover from a logarithmic to a non logarithmic time decay of M_{zfc} (for T>5OK) and the observed power law for the (H,T) line (0.30<n<0.40), are discussed in relation to the flux creep model, its extensions and the collective pinning theory .

2. Experimental

Sintered samples (powders and textured pellets) were prepared by a conventional ceramic method[24] (nominal composition: $Bi_2Pb_{0.4}Sr_2Cu_4O_x$). X-ray diffraction spectra and low field susceptibility measurements,performed with a commercial SQUID magnetometer and an AC susceptometer, confirmed that samples consist almost completely of the 2223 phase, with very small traces of the low T_c phase present in the powdered sample [24]. A single crystal of the 2212 phase ($0.98x0.95x0.16mm^3$; nominal composition: $Bi_{1.7}Pb_{0.3}Sr_2Ca_1Cu_2O_{8+x}$) grown by the flux technique [25] was also selected for the measurements.

Hysteresis loops [24] were measured up to 5 T at different temperatures (6K ÷ 50K) . The volume pinning force F_p has been evaluated using Bean's critical state model [26], in which the critical current density is proportional to the amplitude of magnetization cycles. Therefore :

$$\mathbf{F_p = J_c B \; \alpha \; \Delta M B} \qquad (1)$$

where ΔM denotes the difference in magnetization measured for increasing and decreasing field.

Plotting the reduced pinning force, $f = F_p/F_p^{max}$ (F_p^{max} is the maximum pinning force), as a function of the reduced field $h = H/H_{c2}^*$, curves of the type shown in fig.1 have been obtained in the temperature range investigated. For conventional type II superconductors the applied field H is scaled by the upper critical field H_{c2} , at which ΔM (and thus F_p) goes to zero. However in high T_c superconductors the magnetization becomes reversible at fields $H_{c2}^*(T)$ well below H_{c2} and therefore the choice of H_{c2}^* for the reduced variable h seems to be the more appropriate.

The time relaxation of the magnetization (M_{zfc}) was studied at different temperatures by cooling the sample down to a given temperature in zero field and then applying a field of 3 kOe, kept constant during the measurements. The time decay was recorded for 10^4 s. After each run was completed, the sample temperature was raised well above T_c in order to remove completely the trapped flux lines. The time decay was also studied as a function of field at 6K [24].

The irreversibility line was determined by DC and AC (mutual inductance bridge) susceptibility measurements at several applied magnetic fields (H_{max}= 10 kOe). In the former case the sample was cooled in zero field down to 5K and the susceptibility was measured with increasing the temperature (ZFC branch); then the sample was cooled down to 5K in the same applied field and the susceptibility was measured again (FC branch). The irreversibility temperature T_{irr} (H) is defined as the temperature at which the ZFC and FC branches split. In the latter case, T_{irr} is defined as the temperature where the AC loss (the imaginary part, χ'', of the AC susceptibility) shows a maximum. Both DC and AC measurements were performed with the DC field applied parallel to the c-axis of the samples (textured and single crystal).

3.1 Hysteresis cycles

It is well known that the behaviour of the f vs h curves may provide informations about the nature of the pinning processes operative in the material. Furthermore, the presence of a single pinning contribution operating in a certain temperature range, can be revealed by a scaling behaviour of f(h). Fig.1 shows the f vs h curves obtained from the hysteresis cycles on the powdered sample (the same measurements were performed also on the textured sample by applying the field parallel to the a-b plane.

We did not observe any scaling in the 18K - 35K range (fig.1a) suggesting that no common pinning mechanism operates in this temperature range. Moreover, data can not be fitted with any of the expressions derived in ref. 27 for f(h) assuming a single type of pinning. Nevertheless, the field dependence of f(h) at 45K and 50K (fig.1b) is well fitted with the law:

$$f\,(h) \propto h^{1/2}\,(1-h) \qquad (2)$$

This law can be derived in the framework of the Larkin - Ovchinnikov (LO) theory of collective pinning[18], in which randomly distributed weak pinning centers can give rise to pinning through elastic deformations of short- range ordered volumes of the flux line lattice. In Bi-based materials, pinning centers are known to be very effective at low temperatures (4.2K ÷ 20 K) as it results from the high J_c values widely reported in literature (see ref. 28, for instance, and references therein). As the temperature is raised and thermally activated flux motion becomes relevant, f(h) develops a maximum whose position shifts with temperature [28].

3.2 Relaxation

Relaxation data were preliminarly analyzed using the standard theory of flux creep [7,29], in which the pinning energy is assumed to vary linearly with the driving force:

$$U_e = U_o - BJV_c\,a \qquad (3)$$

where U_e represents the effective barrier to flux motion, U_o the pinning energy in absence of flux gradient, V_c the correlation volume and a an hopping distance. The linear approximation leads to the well known logarithmic decay of persistent currents (or equivalently the magnetization M), with a simple relationship between an apparent activation energy U_o^* and the measurable logarithmic creep rate:

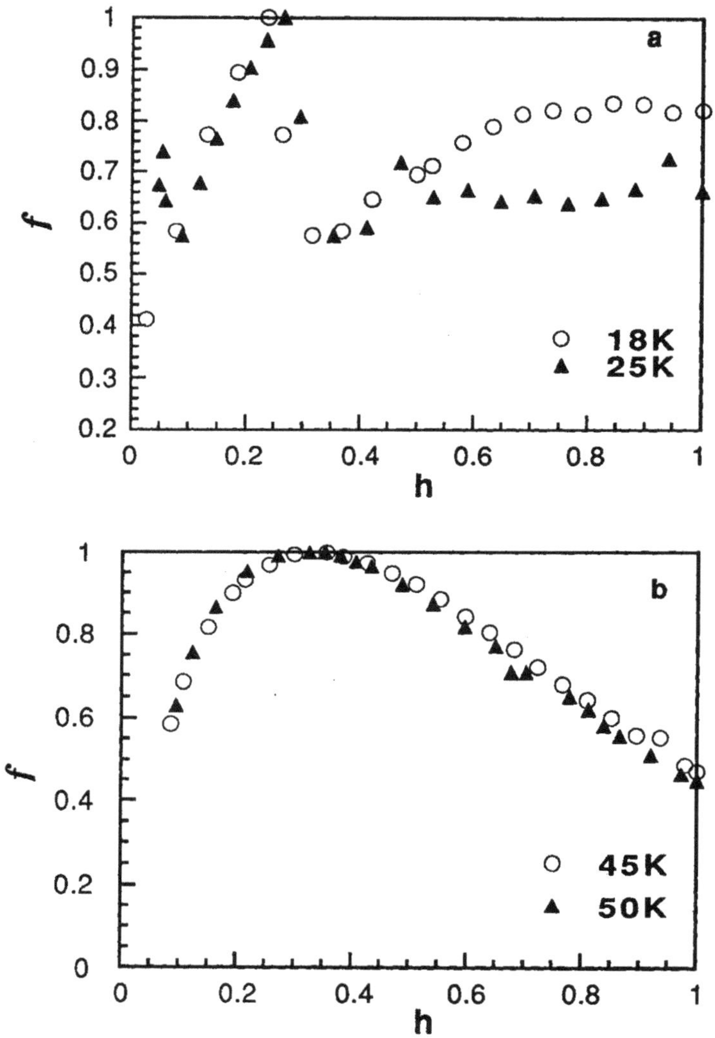

Fig.1 Reduced pinning force f as a function of reduced field at different temperatures for a powdered sample.

$$U_0^* = kT\,[(1/M_0)(dM/d\ln t)]^{-1} \qquad (4)$$

where M_0 is the first measured value of the magnetization.
In our case, logarithmic decays were observed between 6 and 50 K for fields below 10 kOe and U_0^* was extracted as a function of temperature (H_a= 3 kOe, fig.2) and field (at 6K). U_0^* showed a field dependence similar to that reported for the width of the hysteresis cycle at 6K [24].

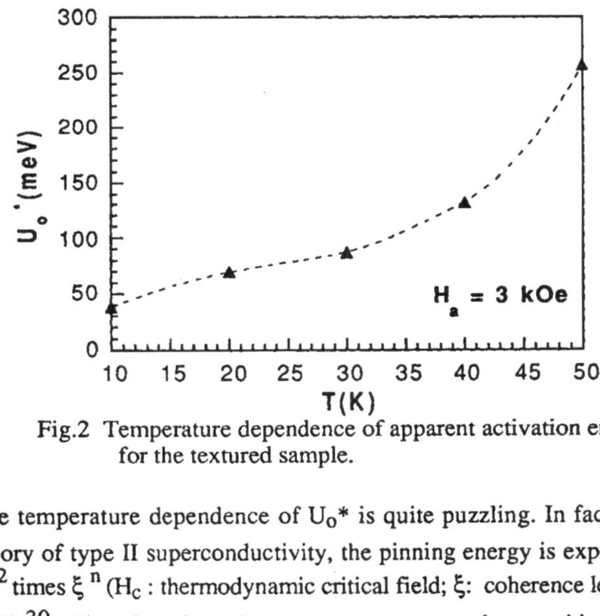

Fig.2 Temperature dependence of apparent activation energy U_0*
for the textured sample.

The temperature dependence of U_0* is quite puzzling. In fact, within the conventional theory of type II superconductivity, the pinning energy is expected to be proportional to H_c^2 times ξ^n (H_c : thermodynamic critical field; ξ: coherence length and n ranges between 1- 3) [30]. Therefore, in an homogeneous superconductor with one type of pinning center, the pinning energy is expected to rise from zero at T_c to some constant value at low temperature.On the other hand, U_0* is found to increase with temperature.
As originally proposed by Beasley et al. [31] and recently pointed out by Xu [32] and Weir[33], the monotonic increase of U_0^* with temperature may result from a non linear dependence of U_e on the critical current J. In fact, as it is shown in fig.1 of ref.31, the standard procedure for inferring U_0^* is equivalent to extract the intercept (U_0^*) of the tangent line to $U_e(J)$ (or equivalently $U_e(M)$) at the instantaneous value of J (M). As the temperature increases, smaller and smaller values of J (M) are sampled due to the activated flux motion,

resulting in the monotonic increase observed for U_o^*.

In order to elucidate the dependence of U_e on J (M), the rate equation for thermally activated flux motion[31] can be written in a suitable form, as pointed out by Maley et al. [34]:

$$U_e/k = -T \ln|dM/dt| + T \ln(B\nu\, a/\pi\, d) \quad (5)$$

where k is the Boltzmann constant, ν an attempt frequency and d an average sample thickness. Plotting the data in the form -T ln(dM/dt) vs - (M - M_{eq}) α Jd (M_{eq} is the equilibrium magnetization calculated at 3 kOe from hysteresis cycles), various U_e (J,M) curves were deduced (fig.3).

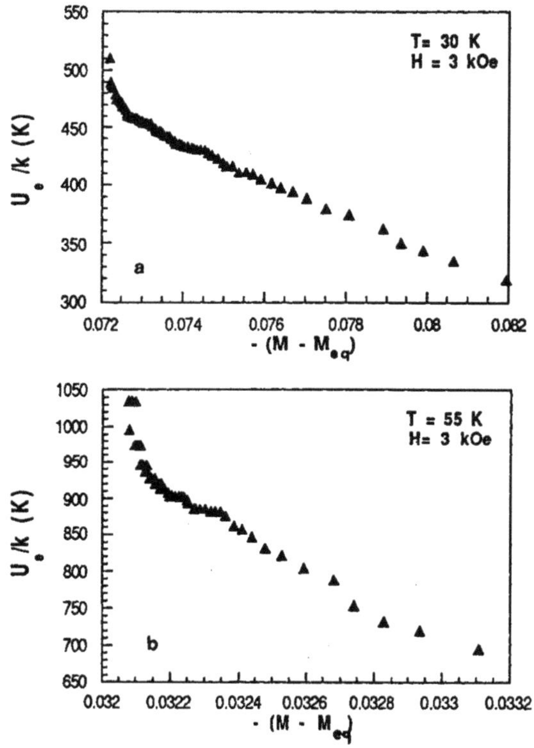

Fig.3 U_e/k (see text, eq.5) as a function of the instantaneous value of irreversible magnetization for the textured sample. (a): T= 30 K; (b): T= 55 K.

At 30K an almost linear regime can still be found in $U_e(M)$, except for a deviation at lower M (i.e. at longer times). Nevertheless, relaxation data at the same temperature (see fig.4a of ref.24) showed no significant deviations from the logarithmic behaviour. As the temperature raises, the deviation from linearity becomes more and more pronounced and above 50K the relaxation data (fig.4) exhibited a non logarithmic time dependence. The data recorded at 55K and 58K are well fitted with the form $M_{zfc}/M_o \propto [\ln(t)]^{-c}$ (c= 0.07 at 55K; c= 0.4 at 58K).

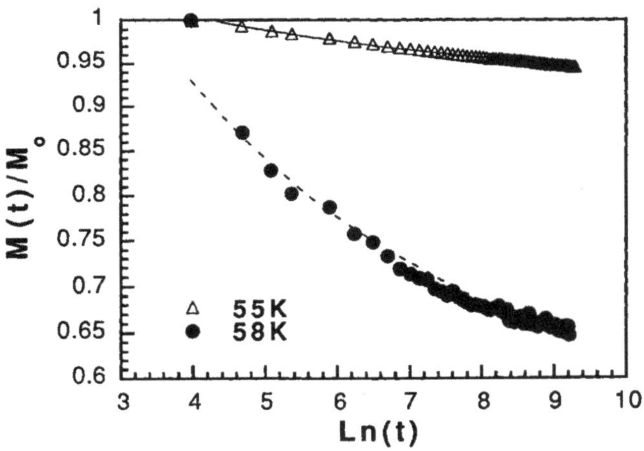

Fig.4 Time decay of M_{zfc} at 55K (Δ) and 58K (●). Solid and dashed lines are fittings with the law $[\ln(t)]^{-c}$

A non logarithmic time decay of the persistent currents (M) was predicted by Feigelman et al. [19,20] in a theory of flux creep based on the collective pinning. In their model, a power law was assumed for the $U_c(J)$ relationship:

$$U(J) = U_o\,(J_c/J)^{\beta} \qquad (6)$$

where U_o is a characteristic energy scale and the exponent α depends on the dimensionality of the flux line lattice and on the particular creep regime. The U(J) behaviour (6) gives rise to a non logarithmic form for the current (M) relaxation:

$$J(t) = J_c\,[(U_o/T)\ln(t/\tau)]^{-1/\beta} \qquad (7)$$

where τ is some attempt time.
The values of β deduced from relaxation data ($\beta_{exp} = 1/c$) are not in agreement with theoretical predictions, i.e.: in the three dimensional case $\beta_{th} = 1/7$ for creep of single flux lines ; $\beta_{th} = 3/2$ for collective creep of small flux bundles (transverse size smaller than the London penetration depth). It should be noted however, that a direct comparison between β_{th} and values inferred from relaxation data is quite difficult becuse β is expected to depend on the current density level (and thus also on temperature) considered, which is in turn affected by the actual pinning "environment" of the material.
The non logarithmic relaxation of persistent currents [22,32,35] is also characteristic of the vortex-glass theory developed by Fisher[12,13]; experimental evidences in favour of a superconducting glass state have been reported for YBCO by means of I vs V measurements [36-38]. Moreover, the anomalous temperature dependence of U_o^* could be also explained in a framework other than the collective pinning, assuming the existence of a distribution of energy barriers for indipendent relaxation processes, as proposed by Griessen et al. [39,40]. In this case the increase of U_o^* is due to the fact that increasing the temperature, stronger pinning sites are sampled within the experimental time window. It is always possible in principle, to find a suitable distribution of pinning energies to describe the behaviour of $U_o^*(T)$. The Gurevich's vortex percolation model [41,42], in which the effects of different pinning centers are coupled, can also account for the monotonic increase of U_o^* with temperature.

3.3 Irreversibility line

Figure 5a shows the field dependence of the reduced irreversibility temperature $t = T_{irr}(H)/T_c$ for all the samples investigated. A double logarithmic plot of variables 1-t and H yielded a straight line indicating that a power law, $1-t \propto H^n$ with $0.30<n<0.40$, is obeyed. The quasi de Almeida-Thouless law (n=0.66) has been found in YBCO single crystal[43] for fields up to 30 kOe, and BSCCO single crystals[44] only for fields below 500 Oe. Other values than 0.66 have been recently reported for n in the case of BSCCO single crystals[45] and (Bi-Pb)SCCO orientated tapes [46], for fields higher than 500 Oe. These results points to the existence of distinct pinning mechanisms in Bi-based systems and YBCO, although the DC definition of T_{irr} is affected by the experimental criterion (i.e. difference in magnetization) chosen to determine the splitting point of ZFC and FC branches, as also pointed out by Ota et al. [47]. Indeed, values between 0.36-0.46 are reported [47] for BSCCO crystal (H//c-axis), depending on the criterion adopted. Our values

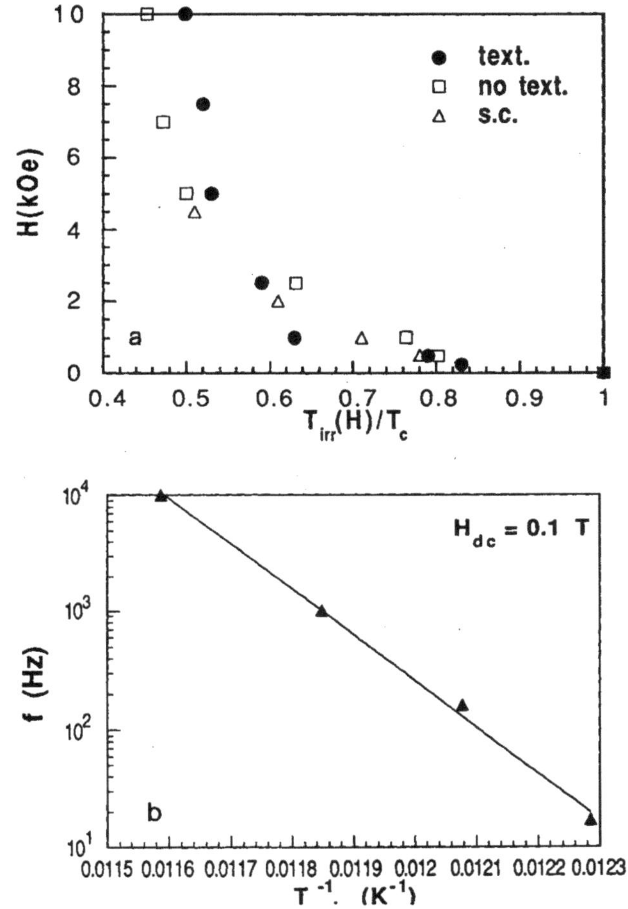

Fig.5 (a): field dependence of irreversibility temperature for sintered (textured and non textured) and single crystal; (b): frequency dependence for the textured sample.

of n are comparable with those reported in ref.47, corresponding to a criterion of $3 \cdot 10^{-9}$ Am^2. Fig. 5b shows a semilogarithmic plot of the frequency f versus T^{-1}_{irr} (T_{irr} is taken as the temperature at which a peak appears in χ''). As pointed out by Kes [48], such a plot is similar to a resistivity $\ln\rho$ vs T^{-1} plot, whereas the choice of a resistance threshold corresponds in our case to the choice of an AC measuring frequency. The linear dependence of ln(f) on T^{-1} reveals that the TAFF model [48] predicting at the irreversibility onset :

$$f = cD_0 \alpha \exp[-E_a/kT] \qquad (8)$$

where c is a geometric factor, D_0 the flux diffusion coefficient and E_a the activation energy for flux motion at a given field, can describe quite satisfactorily the AC susceptibility data.

4. Summary

We have investigated pinning and dissipative effects in Bi(Pb)-Sr-Ca-Cu-O samples by means of DC magnetization and magnetic relaxation measurements. As far as relaxation data are concerned, a description within the conventional flux creep picture (i.e. the assumption of a linear current dependence of the activation energy) works reasonably well only at low temperatures. Furthermore, the irreversibility (H,T) line investigated on the same compounds can not be described by the law $1\text{-}t \,\alpha\, H^{2/3}$ ($t = T_{irr}/T_c$), derived within the "giant" flux creep model [49]. An exponent n ranging between 0.30-0.40 was instead found .

At higher temperatures, being the non linearity of U_e vs J relationship more and more relevant, the Larkin - Ovchinnikov theory of collective pinning and its extension (Feigelman et al.) to high T_c superconductors could be the natural framework within both the observed behaviour of the reduced pinning force and the non logarithmic relaxation can be inserted. However, the interplay between the collective picture and other models such as Fisher's vortex glass, Griessen's distribution of pinning energies or Gurevich's percolation model, is not clear yet. Further experimental investigations are in progress.

Acknowledgements

The Progetto Finalizzato "Materiali Speciali per Tecnologie Avanzate" of Consiglio Nazionale delle Ricerche is acknowledged for financial supports.
One of us (A.M.T.) wishes to thank the Consorzio I.N.F.M. for the financial support.

1. T.K.Worthington, W.J.Gallagher, T.R.Dinger *Phys.Rev.Lett.* **59** (1987) 1160
2. A.C.Mota, A.Pollini, P.Visani, K.A.Muller, J.G.Bednorz *Phys.Rev.B* **36** (1987) 401
3. M.Tuominen, A.M.Goldman, M.L.Mecartney *Phys.Rev.B* **37** (1988) 548
4. Y.Yeshurun, A.P.Malozemoff, *Phys.Rev.Lett.* **60** (1988) 2202
5. P.L.Gammel, L.F.Schneemeyer, J.V.Waszczak, D.J.Bishop,*Phys.Rev.Lett.* **61**(1988) 1666

6. K.A.Muller, M.Takashige, J.G.Bednorz *Phys.Rev.Lett.* **58** (1987) 1143
7. P.W.Anderson, *Phys.Rev.Lett.* **9** (1962) 309
8. M.Tinkham, *Phys.Rev.Lett.* **61** (1988) 1658
9. T.T.M.Palstra, B.Batlogg, R.B.van Dover, L.F.Schneemeyer, J.V.Waszczak *Appl.Phys.Lett.* **54** (1989) 763
10. P.H.Kes, P.Berghuis, S.Q.Guo, B.Dam, G.M.Stallman, *J. of Less Comm.Met.* **151** (1989) 325
11. A.P.Malozemoff, T.K.Worthington, E.Zeldov, N.C.Yeh, M.W. Mc Elfresh, F.H. Holtzberg *"Strong Correlations and Superconductivity"*, Springer Series in Sol.St. Sci. vol. **89** (Ed. by H.Fukuyama, S.Maekawa, A.P.Malozemoff. Springer Verlag 1989) p. 349
12. M.P.A.Fisher, *Phys.Rev.Lett.* **62** (1989) 1215
13. M.P.A.Fisher, D.S.Fisher, D.A.Huse, *Phys.Rev.B* **43** (1991) 130
14. D.R.Nelson, *Phys.Rev.Lett.* **60** (1988) 1973
15. M.C.Marchetti, D.R.Nelson *Phys.Rev.B* **42** (1990) 9938
16. M.Tachiki, S.Takahashi *Sol.St.Comm.* **70** (1989) 291
17. B.Ivlev, N.B.Kopnin, *J. of Low Temp.Phys.* **77** (1989) 413; *Phys.Rev.Lett.* **64** (1990) 1828
18. A.I.Larkin, Yu N.Ovchinnikov, *J. of Low Temp.Phys.* **34** (1979) 409
19. M.V.Feigelman, V.B.Geshkenbein, A.J.Larkin, V.M.Vinokur, *Phys.Rev.Lett.* **63** (1989) 2303
20. M.V.Feigelman,V.B.Geshkenbein, V.M.Vinokur, *Phys.Rev.B* **43** (1991) 6263
21. V.M.Vinokur, M.V.Feigel'man, V.B.Geshkenbein *Phys.Rev.Lett.* **67** (1991) 915
22. P. Svedlindh, C.Rossel, K.Niskanen, P.Norling, P.Nordblad, L.Lundgren, G.V.Chandrashekar *Physica C* **176** (1991) 336
23. J.R.Thompson, Y.R.Sun, F.H.Holtzberg, *Phys.Rev. B* **44** (1991) 458
24. D.Fiorani, A.M.Testa, G.Calestani . *J. of Superconductivity* , in press.
25. G.Balestrino, U.Gambardella, Y.Liu, M.Marinelli, A.Paoletti, G.Paterno', *J. of Cryst. Growth* **92** (1988) 678
26. C.P.Bean, *Phys.Rev.Lett.* **8** (1962) 850
27. D. Dew Hughes *Phil.Mag.* **30** (1974) 293
28. J.Tenbrink, M.Wilhelm, K.Heine, H.Kranth, *IEEE Trans. on Magnetics* **27** (1991) 1239
29. A.M.Campbell, J.E.Evetts *Adv.Phys.* **21** (1972) 199
30. M.Tinkham, *"Introduction to Superconductivity"* (Mc Graw Hill, 1975)
31. M.R.Beasley, R.Labusch, W.W.Webb *Phys.Rev.* **181** (1969) 181
32. Y.Xu, M.Suenaga, A.R.Moodenbaugh D.O.Welch Phys.Rev.B **40** (1989) 10882
33. S.T.Weir, *Sol.State Comm.* **77**, 839 (1991)
34. M.P.Maley, J.O.Willis, H.Lessure, M.E.McHenry, *Phys.Rev.B* **42**, 2639 (1990)
35. J.R.Thompson, Y.R.Sun, F.H.Holtzberg *Phys.Rev.B* **44** (1991) 458
36. R.H.Koch, V.Foglietti, W.J.Gallagher, G.Koren, A.Gupta, M.P.A.Fisher *Phys.Rev.Lett.* **63** (1989) 1511
37. P.L.Gammel, L.F.Scneemeyer, D.J.Bishop *Phys.Rev.Lett.* **66** (1991) 953
38. T.K.Worthington, E.Olsson, C.S.Nichols, T.M.Show, D.R.Clarke *Phys.Rev.B* **43** (1991) 10538
39. C.W. Hagen, R.Griessen *Phys.Rev.Lett.* **62** (1989) 2857
40. R.Griessen *Phys.Rev.Lett.* **64** (1990) 1674
41. A.Gurevich *Phys.Rev.B* **42** (1990) 4857
42. A.Gurevich, H.Kupfer, C.Keller *Supercond.Sci.Technol.* **4** (1991) 591
43. Y.Yeshurun, A.P.Malozemoff, *Phys.Rev.Lett.* **60** (1988) 2202
44. B.D.Biggs, M.N.Kunchur, J.J.Lin, S.J.Poon, *Phys.Rev. B* **39** (1988) 7309

45. C.Quitmann, U.Ebels, P.C.Splittgerber-Hunnekes, G.Guntherodt *Physica B* **165-166** (1990) 1143
46. Y.Yamada, S.Murase, K.Yamamoto, Y.Kamisada, *Cryogenics* **30** (1990) 581
47. S.B.Ota, G.P.Rapson, P.A.J. de Groot, *Physica B* **165-166** (1990) 1145
48. C.J. van der Beek, P.Kes *Phys.Rev.B* **43** (1991) 13032
49. Y.Yeshurun, A.P.Malozemoff, T.K.Worthington, R.M.Yandrofski, L.Krusin-Elbaum,T.R.Dinger, G.V.Chandrasekar, *Cryogenics* **29** (1989) 258

INTERGRANULAR FLUXONS IN HTSC

R. Vaccarone and M. Gaspari †
I.N.F.N., Sezione di Genova
Via Dodecaneso 33, 16146 GENOA, ITALY

ABSTRACT

The low transport current in granular High T_c superconductors is undoubtely attributed to the presence of weak links between grains. Magnetization and critical current measurements at very low field indicates that this low transport current behaves as expected from a critical state model.

In this paper we examine the flux structures which can be present in such granular systems, at low magnetic fields, by using a two-dimensional model of the material, which is assumed to consist of S/C grains connected by weak links showing Josephson properties. We show that the discreteness of the the system is the cause of flux pinning and that by using very simple dinamical equations it is possible to study the critical state and the irreversible hysteresis which arouse in the sample under a varying magnetic field.

By using a very sensitive magnetometer based on a SQUID sensor we have detected the presence of the individual fluxons sitting in the samples and we have shown that even their movement in the sample can be detected.

1. Introduction.

Most of the High T_c superconductors are easily prepared in the form of granular aggregates by a sintering process. In such systems the grains have good superconducting properties and below some low critical field H_{c1} present a complete Meissner effect. The surface of separation between the grains is the weak point for the transport current, showing a critical current orders of magnitude lower than the grains. Anyway the material as a whole still shows superconducting properties. The junctions between the grains are superconducting weak links and show Josephson character, as nicely demonstrated in [1]. Save in the materials with very high density, a number of voids is left between the grains and these holes play an important role in determining the low field behaviour of these materials. The system can be indeed modeled as an array of SQUIDs, or as a network of Josephson junctions[2].

The system is truly tridimensional and with a very high disorder. We expect that in such a complex network nor the current can follow a regular straight path nor the field can be described by a two-dimensional map.

However in this paper we use simple two-dimensional regular models which will contain the fundamental physical features.

† In partial fullfillment of his thesis work

A linear array of Josephson junctions has been studied by Yamashita et al. [5] by means of a mechanical analog. Pinning has been detected but the critical state has not been searched. Networks of junctions in two dimensions have been studied by many authors [6,7], but all of them have neglected the effect of the self field and by this choice they where not able to obtain information about the existence of a critical state in the system.

2. Two-dimensional SQUID array.

A granular superconductor can be described by a two-dimensional array of superconducting grains connected by Josephson Junctions. We analize here a system with point-like junctions and with holes of a finite size (which of course can be well described as a SQUID array).

The grains are arranged at the corners of a square mesh and the junctions that connect adjacent grains are oriented along the coordinate axis.

We will label the grains by two indices i, j running respectively along the x and y axis; to any grain we can associate two junctions, starting from it and connecting it to the adjacent grain on the right, on x direction, or to the upper one, in the y direction; the unitary cell will contain also the hole on the right and above the i, j grain.

The equation wich describe the system can be written in terms of the phase difference across the junctions in x directions φ_x or in y direction φ_y. In the same way the current has two components I_x and I_y while the magnetic field b is directed along the z axis.

A static equation for the system can be obtained in the following way:

We consider a closed square path encircling an hole, with the four corners in the center of the grains i, j , $i+1, j$, $i+1, j+1$ and $i, j+1$, and crossing the junctions which connectes the four grains together. By integrating along this path the equation which relates the current density in the superconductor to the vector potential and the space variation of the phase of the wave function

$$j(r) = eh\ (\ \nabla\varphi - e/hcA(r))\,\psi_0{}^2 \tag{1}$$

we obtain a relationship between the phase difference across the junctions and the flux enclosed in the hole.

$$\varphi_{x,i,j} + \varphi_{y,i+1,j} - \varphi_{x,i+1,j+1} - \varphi_{y,i,j} = 2\pi\frac{\Phi_{i,j}}{\Phi_0} \tag{2}$$

A second equation can be obtained by considering the field change on the two sides of the slab formed by subsequent junctions in the same direction (which in this model are infinite in the z direction)

$$\Delta_x b = \mu_0 \cdot I_x \tag{3a}$$

$$\Delta_y b = -\mu_0 \cdot I_y \tag{3b}$$

where I is the current in the junctions for unit lenght along z.

The junction current must be related to the phase difference according to the Josephson equation

$$I_x = I_{xm} \cdot \sin(\varphi_x) \tag{4a}$$

$$I_y = I_{ym} \cdot \sin(\varphi_y) \tag{4b}$$

From equations 2, 3, 4 we get the following system of difference equations:

$$\Delta_x(\Delta_x \varphi_y - \Delta_y \varphi_x) = \beta \cdot \sin(\varphi_y) \tag{5a}$$

$$\Delta_y(\Delta_x \varphi_y - \Delta_y \varphi_x) = -\beta \cdot \sin(\varphi_x) \tag{5b}$$

The field in the holes will be given by

$$b_{i,j} = \frac{\Phi_{i,j}}{S_{hole}} = \frac{\Phi_0}{S_{hole}} \cdot \frac{\varphi_{x,i,j} + \varphi_{y,i+1,j} - \varphi_{x,i+1,j+1} - \varphi_{y,i,j}}{2\pi} \tag{6}$$

and the grain size will enter only in determining the average field in a cell

$$\langle b \rangle_{i,j} = b_{i,j} \frac{S_{hole}}{(S_{hole} + S_{grain})} \tag{7}$$

The equations given above in particular cases can be put in the form of differential equations for a continuous medium, but in doing this we will loose the possibility to find pinning. Another important point must be given about the vector fields. While I_s has null divergence in the continuous limit, the same is not true for φ_s.

The difference equation 5 can be solved with suitable boundary conditions and a solution representing a single fluxon can be found. If a square region of the lattice of sufficient size is considered, the solution will hardly depent on the boundary conditions, and the fluxon can be taken as insulated. We will describe now the caracteristics of such an object. The field profile in the holes of the lattice is shown in fig. 1.

We can define the size of the fluxon as the radius of the circle which contains half a flux quantum. We call this lenght Λ_Φ which is found to be almost exactly

$$\Lambda_\Phi = \frac{\pi}{\sqrt{\beta}} = \pi \cdot \lambda_j \tag{8}$$

The current density decays at large radii (where an average over many junctions in azimutal direction is meaningful) as

$$\langle j \rangle \propto K_1(r\sqrt{\beta}) \tag{9}$$

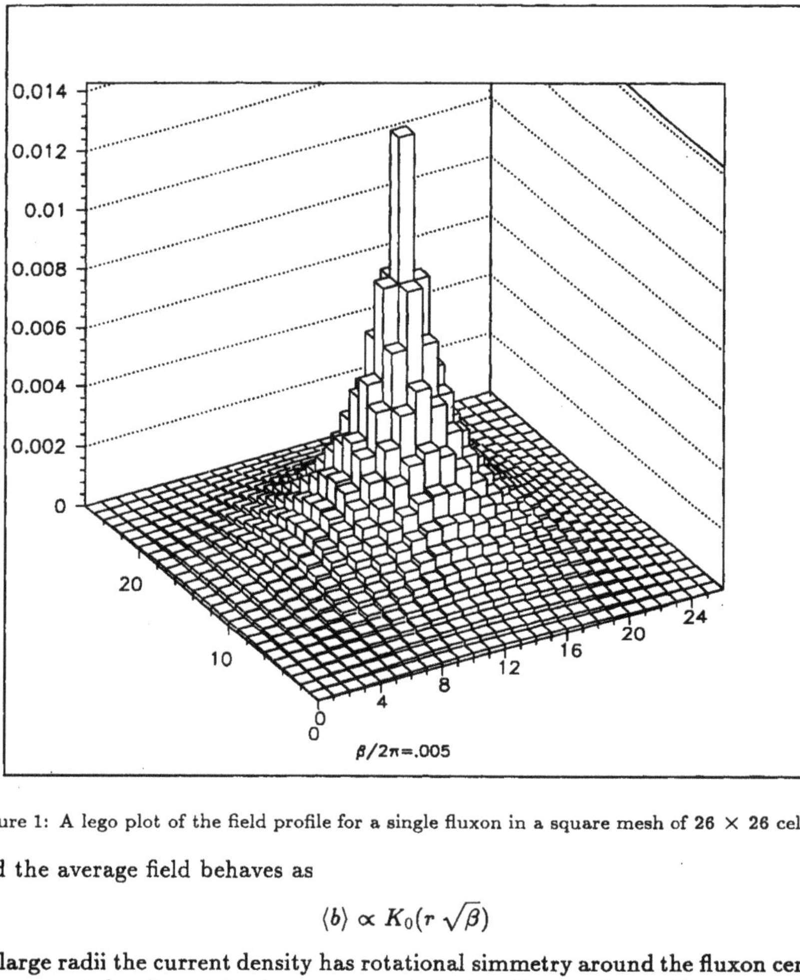

Figure 1: A lego plot of the field profile for a single fluxon in a square mesh of 26 × 26 cells.

and the average field behaves as

$$\langle b \rangle \propto K_0(r\sqrt{\beta}) \tag{10}$$

At large radii the current density has rotational simmetry around the fluxon center. The current density attains is maximum at the junctions enclosing the central hole, irrespective of the magnitude of the parameter β.

3. Fluxon pinning and interaction.

The normalized energy density in the array can be calculated easily by adding together the magnetic term ε_m due to the field in the holes and the term ε_J due to the current flowing in the junctions

$$\varepsilon_{i,j} = \varepsilon_{m\,i,j} + \varepsilon_{J\,i,j} = {b_{i,j}}^2/2 + (1 - cos(\varphi_{x\,i,j})) + (1 - cos(\varphi_{y\,i,j})) \tag{11}$$

The energy of an insulated fluxon is obtained by summing over the cells of the whole mesh the energy density.

In our reduced units the self energy varies with λ_j as $1/\lambda_j^2$.

As in the onedimensional case[4,8] we can consider the state with the fluxon centered in the hole or over a junction (i.e. shared between two adjacent holes). The fluxon energy will be different in the two cases and this difference gives the energy barrier to overcame in moving across the junctions network. The barrier energy as a function of β is given in fig. 2. Because periodic boundary conditions are used, the network size gives also the fluxon spacing.

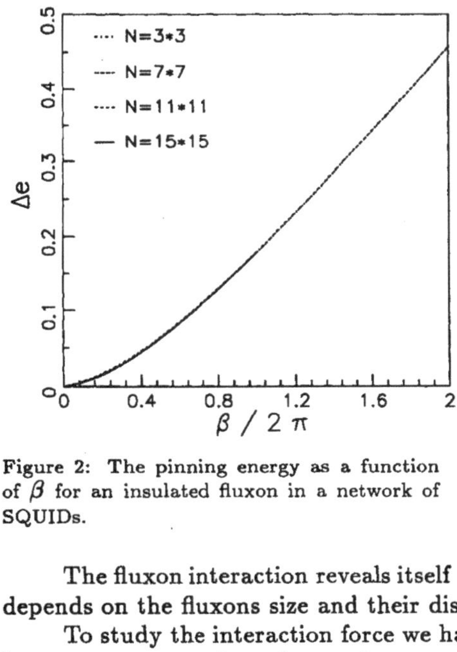

Figure 2: The pinning energy as a function of β for an insulated fluxon in a network of SQUIDs.

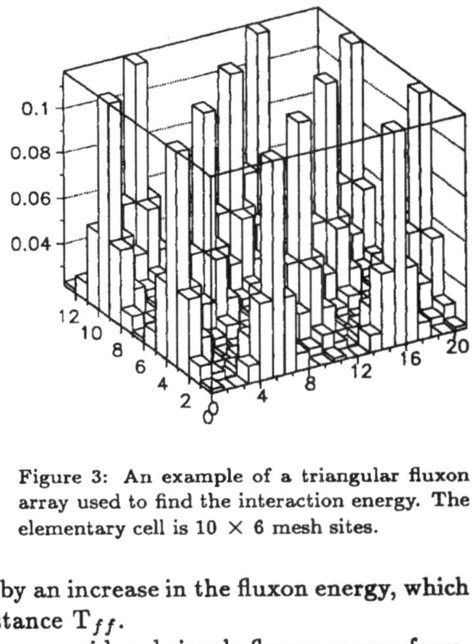

Figure 3: An example of a triangular fluxon array used to find the interaction energy. The elementary cell is 10 $\times$ 6 mesh sites.

The fluxon interaction reveals itself by an increase in the fluxon energy, which depends on the fluxons size and their distance T_{ff}.

To study the interaction force we have considered simple fluxon arrays forming a square or a triangular mesh.

The interaction energy depends on λ_j / T_{ff} with a quadratic law being valid when $\lambda_j \gg 1$ and the spacing is greater than $2\,\lambda_j$.

In conclusion we can see the fluxon as a particle with self energy (rest mass), sitting in a periodic potential, and interacting with other fluxons.

4. The dinamical equation and the magnetization cycles.

From the system of difference equation wich describe our SQUID array it is possible to obtain field profiles with a gradient which is linear on the average, but shows even the individual fluxons sitting in the sample.

The use of a dinamical equation allows us to verify that the system behaves according to the critical state model, and to obtain magnetization cycles or $VvsI$ plots.

In two dimensions the discrete Sine-Gordon equations, with a dissipative term representing the normal conductivity of the weak links are:

$$\Delta_x(\Delta_x \varphi_y - \Delta_y \varphi_x) = \gamma \frac{\partial^2 \varphi_y}{\partial t^2} + \beta \cdot \sin(\varphi_y) + \eta \frac{\partial \varphi_y}{\partial t} \qquad (12a)$$

$$\Delta_y(\Delta_x \varphi_y - \Delta_y \varphi_x) = \gamma \frac{\partial^2 \varphi_x}{\partial t^2} - \beta \cdot \sin(\varphi_x) + \eta \frac{\partial \varphi_x}{\partial t} \qquad (12b)$$

The coefficients γ , β and η are given by:

$$\gamma = \mu_0 C S_{hole} \qquad (13a)$$

$$\beta = \frac{2\pi \mu_0 S_{hole} I_m}{\Phi_0} \qquad (13b)$$

$$\eta = \frac{\mu_0 S_{hole}}{R} \qquad (13c)$$

where C and R are the junction capacitance and shunt resistance and I_m is the maximum Josephson current; S_{hole} is the hole area and μ_0 S_{hole} is equivalent to the self inductance of the loop encircling the hole. Both I_m, C, R and L are calculated per unit length along the field direction.

We obtain a resistance $R = 10^{-5}\Omega/m$, starting from a typical resistivity of $1\,m\Omega\,cm$. The capacitance C is about 1. 10^{-8} F for an oxide barrier 50 Å thick, but can be orders of magnitude lower for S-N-S junctions. If we consider a square mesh of grains and we impose as boundary conditions an external field changing in time on the peripheral cells, the field in the sample evolve according to a critical state model, and the total flux linked with the square domain shows a typical hysteresis cycle. In the present simulation we have neglected the finite size of the junctions and the magnetization is expected to follow the Bean model[3]. A magnetization cycle obtained for a square mesh of 14 $\times$ 14 cells at $\beta/2 \cdot \pi = 1.3$, with high damping and zero junction capacitance ($\eta = 1$, $\gamma = 0$) is shown in Fig. 4. The external field H_{ext} is normalized to the field which gives a flux quantum in an hole (of unit area). The magnetic field profile has a slope near one and the sample is almost fully penetrated. The cycle has an overall shape which agrees with the Bean model with large steps superimposed on it. At any step a large number of fluxons enter in the sample or leave it. Indeed at high βs on the surface of the sample we find a onedimensional flux profile, from which single fluxons enucleates, moving inside where a truly two-dimensional arrangment is present. Another example is given in Fig. 5. The $\beta/2\pi$ value is lower than unit and the fluxons will have a larger size and larger separation.

This graph shows first of all that even at low β the pinning exists and the critical state is in effect. The shape of the curve cannot be compared with a Bean

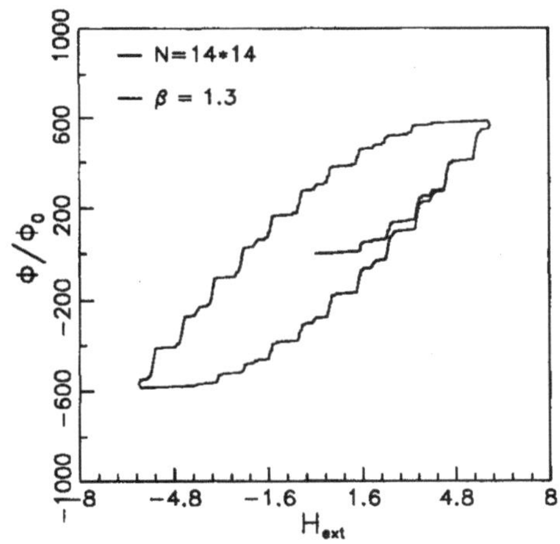

Figure 4: The hysteresis cycle obtained in a SQUID network with high pinning (β = 1.3).

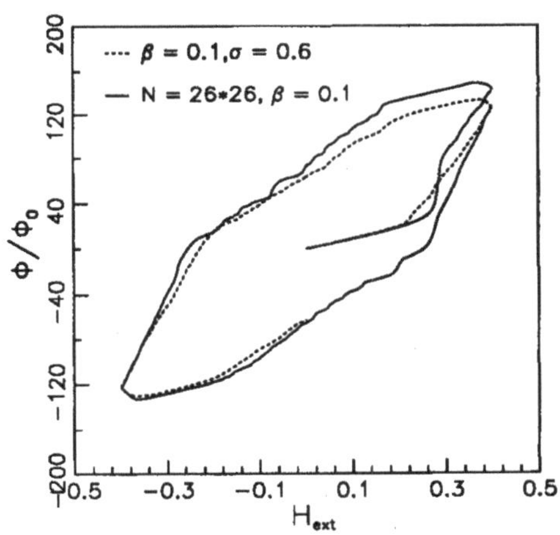

Figure 5: A system with low pinning and uniform (continuous line) or statistically distributed (dashed line) junction currents.

cycle, because single fluxon effects are important. In this case the steps due to surface shielding are reduced and the movement of the flux profile in the bulk can be seen better.

The results for a SQUID network with Junction currents distributed statistically with a variance of 0.6 are shown on the same graph. The main point we want to stress is that the cycle amplitude, i.e. the critical current in the system, is almost unchanged. Indeed a granular system has pinning due to the presence of voids even in absence of any statistical distribution of its properties.

5. The magnetometer

A simple SQUID magnetometer has been built to measure the magnetization of small samples of granular High T_c Superconductor in the low field range where single fluxon effects are expected to show up.

The magnetometer is designed to allow fast recording of magnetization cycles in magnetic fields ranging from 0.1 μT to a few mT on samples with a minimum section area of 10^{-9} m^2. At the actual stage the measurements can be performed at a fixed temperature of 4.2 K.

The magnetometer is housed in a glass dewar, and both the SQUID probe, the coil assembly and the sample are immersed in liquid He. Liquid Nitrogen surrounds the Helium criostat and a Mumetal shield at room temperature is used to reduce the external fields to about 3 μT.

The flux measuring devices is a commercial SQUID detector (BTi 330X

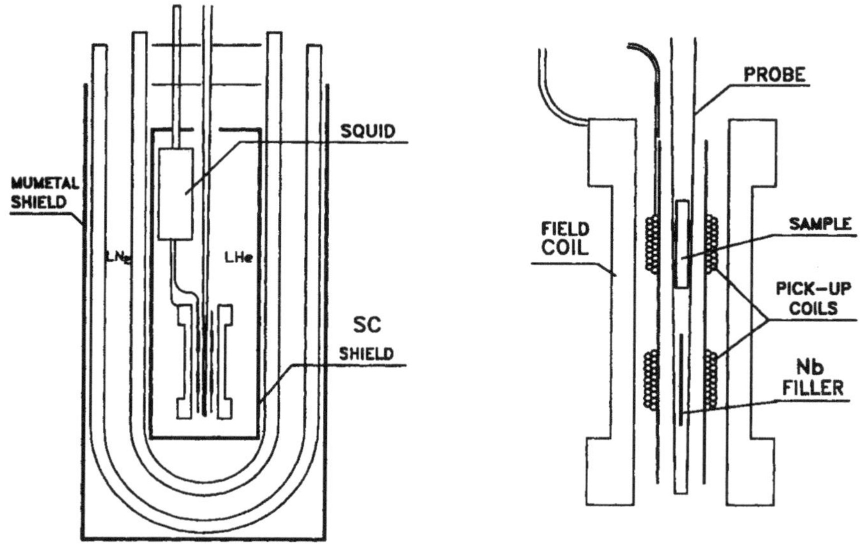

Figure 6: Schematic view of the magnetomether: general view (left), expanded view of the magnetic field coil, gradiometer and sample area (right).

Series SQUID measurement system with HYBRID SQUID sensor and MFP probe) coupled to the sample by a superconducting flux transformer. Two pick-up coils wound in opposite directions are used, to form a first order gradiometer. The coils are made with NbTi wire, in two layers of 13 turns each, on a fiberglass coil former of 3.9 mm diameter. The center to center distance of the coils is 16 mm. The transformer as a flux transfer ratio of 1 / 14.

The field coils is 60 mm long with an inner diameter of 7 mm. The winding is made with copper wire, 32660 turn/m, with correction coils 4.5 mm long at the ends.

The experimental area is surrounded by a Superconducting Lead shield.

A general view of the cryostat is given in Fig. 6 (left). On the right we show an expanded view of the detection assembly. It is possible to see the two opposite pick-up coils containing the sample and a few short wires of Nb which are used for a first order compensation of the gradiometer umbalance. The use of such a filling allows us to get compensation of the flux transformer at $\sim 1/3000$ starting from the as-done astatic coils with an umbalanced flux of $\sim 1/200$.

6. Magnetometer operation

With this magnetometer it is possible to obtain magnetization cycles in a fast and straightforward way.

The samples can be changed at any time by inserting them in a top-loading probe. During cool down the sample reaches its transition temperature in a low field zone, inside the Mumetal shield.

The sample is then exposed to a triangular periodic magnetic field (usually 6 to 18 cycles of the same amplitude) generated by a digital source of high accuracy and low noise. A cycle is completed in two and a half minutes.

The magnetic flux is recorded (1000 points per cycle) and an averaging is done over 4 points. When appropriate a further averaging is performed later if many cycles of the same amplitude have been acquired.

The hysteresis cycles are usually measured under a sequence of increasing magnetic fields, to simplify the magnetic history of the sample.

7. Experimental data.

We measured a few samples of YBCO and BISCCO. We present here in detail only the results obtained in an YBCO sample we call VAC IV.

The sample is cut from a wire with Ag matrix, which is left in place for the measurements.

The sample has an outer diameter of 0.5 mm including the Silver sheet and the YBCO core has a diameter of 0.34 mm

The sample lenght is about 4 mm

The starting material has an average grain size $= 10\mu m$

The critical current densities for the newly prepared sample where: $J_c =$ 242A/cm^2 @ 77. K and $J_c = 2350.$ A/cm^2 @ 4.2 K.

The present measurements comes after one year of ageing of the sample and the critical current density at 4.2 K is reduced to $\sim$ 400 A/cm^2

The reduced values of the transport properties are anyway useful to us and have been chosen on purpouse to simulate at 4.2 K the behaviour usually expected at liquid Nitrogen temperature, keeping low the effects coming from thermal fluctuations in the material.

The magnetic history of the sample has been analized starting as usual from low amplitude cycles, but we present first the series of hysteresis loops obtained at higher fields. In fig. 7 you can see the close similarity of the Φ vs H we have measured with the expectation of the Bean model which is shown in fig. 8. A good agreement is found if we assume a critical current density of 180 A/cm^2 and a filling factor of 95 %.

In Fig. 9 we show four hysteresis cycles taken at lower fields. In this graph you can note that the amplitude of the cycle, and in particular the difference between the residual fluxes at $H_{ext} = 0$ on increasing and decreasing fields is not an integer multiple of the flux quantum. This can seem strange at first glance but this behaviour can be easely explained by recalling that even in very simple structures, which in a sense are the ingredients from which the granular samples

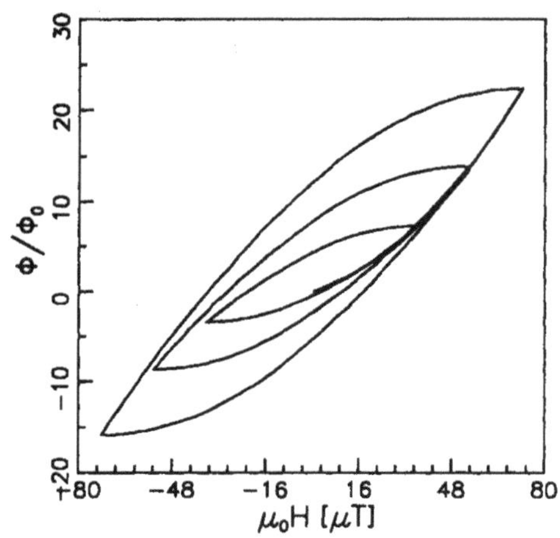

Figure 7: Three magnetization cycles at field amplitude ranging from 40 to 75 μT.

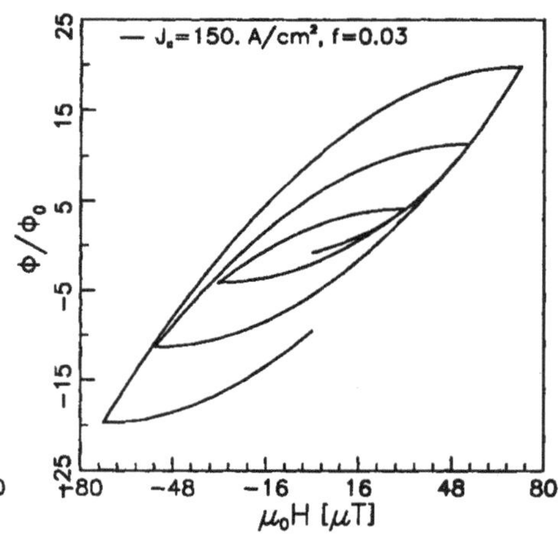

Figure 8: Magnetization cycles obtained by the Bean model.

are made, like a single superconducting loop closed by a Josephson Junction or an uniform Josephson Junction, the internal flux can be nonzero and non integer at null external field. In fig. 10 we show how the internal flux at zero external field varies in a SQUID as a function of the loop area (symbols) and in an uniform Junction as a function of its lenght.

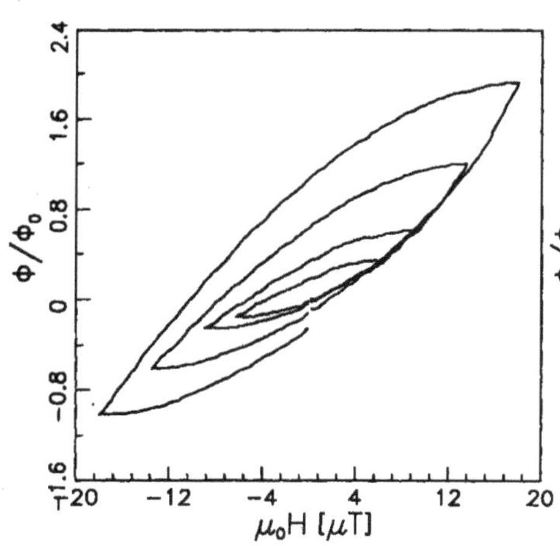

Figure 9: Hysteresis cycles measured in sample VAC IV at intermediate amplitude.

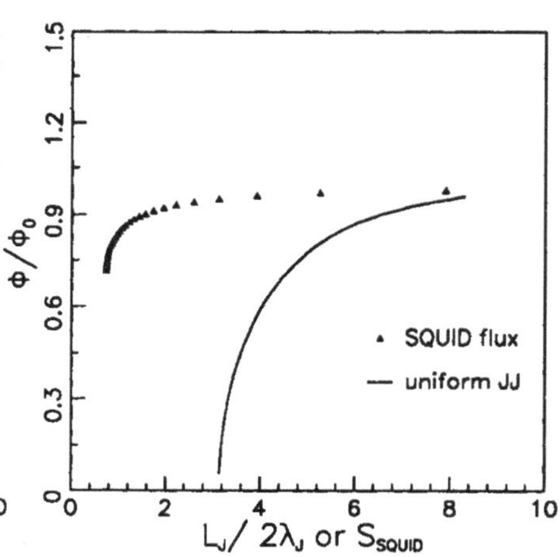

Figure 10: The internal flux in a SQUID as a function of its area (a.u.), and in an uniform Junction as a function of its relative lenght.

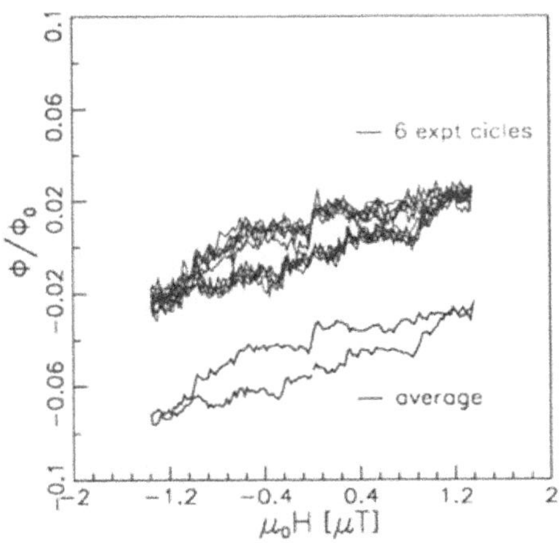

Figure 11: Six experimental cycles measured at very low field amplitude and their average.

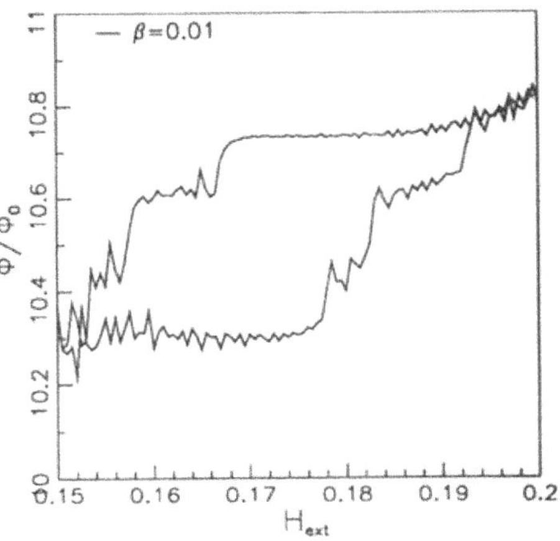

Figure 12: Non-integral flux steps in a two-dimensional simulation of a sample with $\beta/2\pi$=0.01

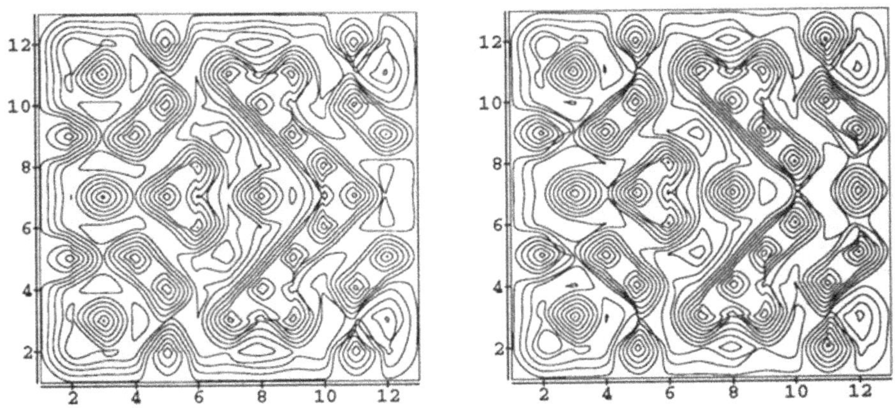

Figure 13: The magnetic field distribution in the sample at two different field levels. The fluxon on the center of the right side is moved by one mesh site.

The first cycles which show hysteresis were obtained at magnetic fields of the order of 1 μT. A sequence of six cycles measured at such a field level is shown in fig. 11. The irreversibility of the behaviour is clearly seen and is confirmed by the averaged cycle shown in the picture. The cycle presents a few steps, of which the central one is the more apparent, with a flux change of a few percent of the flux quantum.

Of course we cannot attribute such a step to the entrance of a fluxon in the sample.

We decided at this time to have a closer look at the simulation analizying a system with a $\beta/2\pi = 0.01$, performing small cycles around a nonzero external field after having trapped a few fluxons inside. The results of such a simulation are shown in fig. 12 in which flux steps of $\sim 0.1\Phi_0$ can be seen. The larger amplitude of the step compared with the experimental ones can be attributed to the fact that we use a 2-D simulation, and in this case we did not even introduced a statistical distribution.

Corresponding to this small steps the field map in the sample shows clearly the movement of the fluxons sitting near the boundary of one cell, as shown in fig. 13, which corresponds to the central flux jump in decreasing field in fig.12.

The origin of the jumps in the experimental data, in spite of their smaller size, which can be due to a partial movement in the third direction (along the field), can be explained by the same effect.

8. Conclusions

The use of a new magnetometer of very high sensitivity, tailored to the specific needs of the measurement of magnetization in superconductors at low fields, has allowed us to detect in high T_c superconducting samples the presence of fluxons and their movement from one pinning center to another.

We have shown that the pinning in granular superconductors come out directly from the discretness of the system, formed by superconducting grains connected by Josephson Junctions, while the statistical distribution of the properties of the material changes only slightly its macroscopic behaviour. The magnetic history of the sample is determined by the critical state model.

Intergranular fluxons can be spread over several cells of the network of junctions, if the junctions have low critical current, and still be pinned and follow a critical state model.

The hysteresis cycles can show at low fields, for small samples made by low current materials, quantum structures (due to individual fluxons) mixed with the usual macroscopic behaviour.

While a quantitative agreement between the experiment and our numerical model is hardly possible, because of the simplified treatment, based on a 2-D model with pointlike junctions, qualitavely we were able to explain all the features seen experimentally.

References

1. P. Chaudhari, J. Mannhart, D. Dimos, C. C. Tsuei, J. Chi, M. M. Oprysko and M. Scheurmann, *Phys. Rev. Lett.* **60** (1988) 1653
2. V. Calzona, M.R. Cimberle, C. Ferdeghini, F. Pupella, M. Putti, C. Rizzuto,

A. Siri and R. Vaccarone, *Cryogenics* **30** (1990) 569
3. C. P. Bean, *Rev. Mod. Phys.* **36** (1964) 31
4. R. Vaccarone, *Proc. 3rd National Meeting on High Temperature Superconductivity*, Feb. 12-14, 1990, Genoa, Italy, p. 346
5. T. Yamashita and L. Rinderer, *J. Low Temp. Phys.* **21** (1975) 153
6. A. Giannelli and C. Giovannella, *Proc. 3rd National Meeting on High Temperature Superconductivity*, Feb. 12-14, 1990, Genoa, Italy, p. 358
7. C. J. Lobb, D. W. Abraham and M. Tinkham, *Phys. Rev. B* **27** (1983) 150
8. F. Parodi and R. Vaccarone, *Phisica C* **173** (1991) 56

RESISTIVE MEASUREMENTS IN BiSrCaCuO FILMS: ANISOTROPY AND DIMENSIONALITY

M.Giura and R.Marcon

Dipartimento di Fisica - Università di Roma La Sapienza - P.le A.Moro, 2 - I 00185 Roma, Italy

We present the angular and temperature dependences of a critical magnetic field $H^*(\vartheta)$ defined as the onset of the reduced d.c. resistivity ρ/ρ_n vs H at a given temperature for highly c-oriented 2212 BiSrCaCuO films. ϑ is the angle between the direction of the external magnetic field and the a-b plane.The experimental results put into evidence a crossover between a two-dimensional (2-D) and an anisotropic three-dimensional (3-D) behavior by varying the temperature. This crossover has the same experimental characteristics as those observed in the artificially prepared layered superconductors for the H_{c2} upper critical field. The early studies carried out on these multilayered systems can be applied to the high-T_c copper oxides. Anisotropy is also observed at microwave frequencies.

In this report we will show a set of experimental results concerning the anisotropy properties of the Bi:2212 films. The anisotropy is studied by means of measurements of dissipative losses both in d.c. and at microwave frequencies when an external magnetic field is present. The relevant physical parameter is the angle θ between the a-b plane (the film plane) and the direction of the magnetic field H.

The preminence of this argument lies in two respects:

a) it should be possible to clarify the dimensionality of the superconductors carriers;

b) it should allow to better understand the mechanisms involved in the dissipation in high-T_c superconductors, in particular the role of the fluxons (flux motion, flux-creep), of the Josephson junctions (intrinsic or extrinsic), of the slippage, etc..

With reference to point a), because of the difference between the values of the coherent length ξ in the a-b plane and along the c-axis [$\xi_{\parallel} = \xi_{ab} \simeq 30$ Å, $\xi_{\perp} = \xi_c \simeq 1$ Å], and because of the small value of $\xi_{\perp}$ with respect to the distance between the Cu-O_2 planes, it is natural to think that these new superconductors show the same phenomena as the artificially prepared layered traditional superconductors (e.g. Nb/Cu) [1] when the coherent length is comparable to the layer thickness and the interlayer spacing. The physical quantity suitable to characterize these phenomena is the upper critical field H_{c2}. In fact the experimental results on artificially layered systems are well described in the framework of the extended Ginzburg-Landau (GL) theory for the H_{c2} calculation.

Depending on the value of d, the layer thickness, and s, the interlayer spacing, two regimes have been found for the angular dependence of the critical field H_{c2}:

i) 2-D behavior when $d < \xi_\perp < s$. In this case the Tinkham model obtained in the limit $d \to 0$ is valid and $H_{c2}(\vartheta)$ is given by solving the equation [2]

$$\left|\frac{H_{c2}(\vartheta)\sin\vartheta}{H_{c2\perp}}\right| + \left[\frac{H_{c2}(\vartheta)\cos\vartheta}{H_{c2\parallel}}\right]^2 = 1 \quad , \tag{1}$$

with ϑ the angle between the external magnetic field and the layer plane. $H_{c2\perp}$ and $H_{c2\parallel}$ are the H_{c2} value orthogonal and parallel to the a-b plane respectively . From Eq.(1) one obtains

$$\left|\frac{dH_{c2}}{d\vartheta}\right|_{\vartheta=0} = \frac{H^2_{c2\parallel}}{2H_{c2\perp}} \neq 0 \quad , \tag{2}$$

and the curve $H_{c2}(\vartheta)$ shows a cusp at $\vartheta = 0$, as observed experimentally in layered Nb/Cu. [1]

ii) 3-D anisotropic behavior when $\xi_\perp > s$. In this case the superconducting layers are coupled via Josephson or proximity effects. For a Josephson coupling, Lawrence and Doniach have shown that $H_{c2}(\vartheta)$ is given by [3]

$$H_{c2}(\vartheta) = \frac{\phi_o}{2\pi\xi_\parallel^2(T)\left[\sin^2\vartheta + \left(\frac{m}{M}\right)\cos^2\vartheta\right]^{1/2}} \quad , \tag{3}$$

where ϕ_o is the flux quantum, m and M are the equivalent effective masses respectively perpendicular and parallel to c-axis. From Eq.(3) one finds

$$\left|\frac{dH_{c2}}{d\vartheta}\right|_{\vartheta=0} = 0 \qquad (4)$$

and the curve $H_{c2}(\vartheta)$ shows a round maximum at $\vartheta = 0$.

According to the temperature dependence of the GL coherence length, a crossover between i) and ii) regimes should be found varying the superconductor temperature. Precisely, lowering the temperature one should go from a Lawrence-Doniach to a Tinkham regime, as previously found in Nb/Cu (Fig.1). However, in high-T_c superconductors it is possible to obtain the experimental behaviors of $H_{c2\parallel}$ and $H_{c2\perp}$ as a function of the temperature only in a small range below T_c, owing to the high value of dH_{c2}/dT.

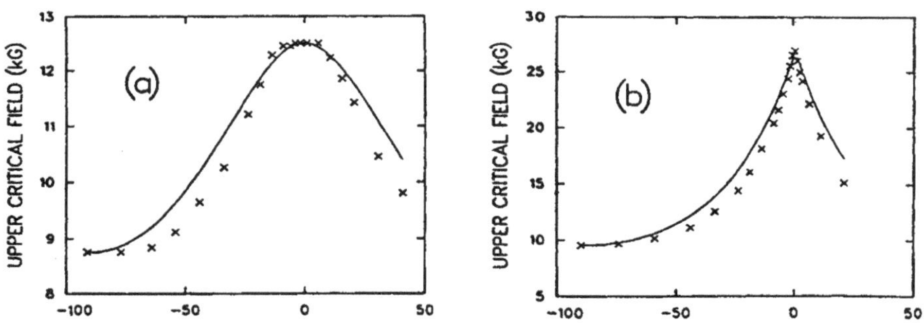

Fig.1 Angular dependence of critical fields for Nb/Cu superlattice showing (a) anisotropic 3D behavior ($d=D_{Nb}$ = 23Å, $s=D_{Cu}$ = 23Å and T=1.17K), and (b) 2D behavior ($d=D_{Nb}$=172Å, $s=D_{Cu}$ =333Å, and T=1.17K). [From Ref.(1)].

Here we present the angular dependence of the critical magnetic field $H^*(\vartheta)$, defined as the onset of the reduced resistivity ρ/ρ_n *vs* H at a given temperature, for a samples of BiSrCaCuO. The experimental definition of the field H* follows from the Fig.2, where the typical behavior of the normalized resistivity ρ/ρ_n as a function of the magnetic field H is shown. The field H^* for the onset of dissipation is clearly detectable in the limit of the accuracy of our measurements (5 nV with I=10μA).

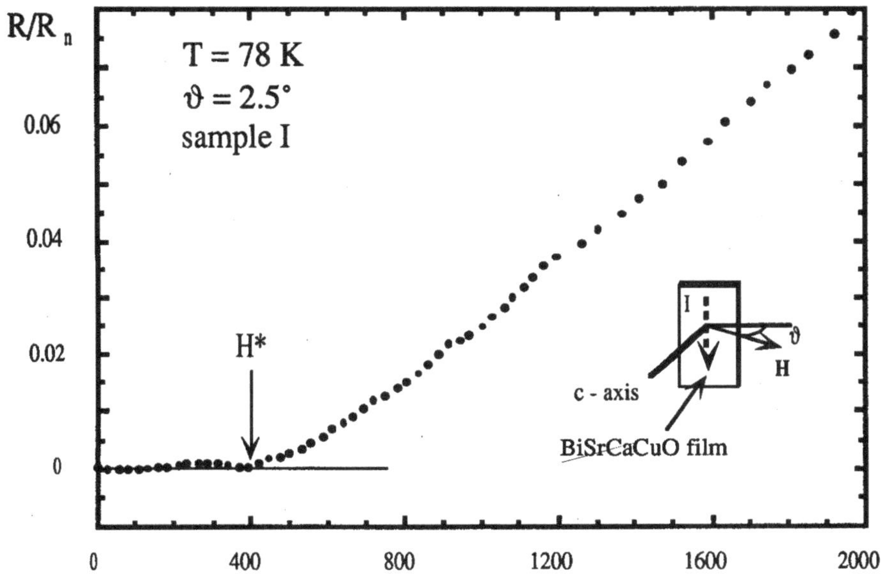

Fig.2 Typical behavior of the normalized resistance R/R_n and determination of the field H* for the onset of the resistance. R_n is the normal-state resistance.The sample arrangement with respect to the magnetic field **H** and the probing current **I** is also sketched. [From Ref.(9)].

The measurements of resistivity are made with the four-probe d.c. resistance method; they are completely reversible and independent on the direction of the current. Samples are highly c-axis oriented epitaxial film of 2212 BiSrCaCuO grown on $LaGaO_3$ (sample I) and $NdGaO_3$ (sample II) substrates with a mosaic spread less than 0.1 degrees [4]. The film thickness is about 1 μm, and the zero-resistance temperature at $H = 0$ is 79.1 K. The normal-state d.c. resistivity at the onset of the transition is $\rho_n = 90$ μΩ cm. The sample holder was adjusted and fixed so that the a-b plane lies in a vertical plane. The magnetic field, supplied by a traditional electromagnet, can rotate in a horizontal plane. The rotation angle ϑ was measured to be accurate to 0.1 degree.

As it can be seen from Fig.3, $H^*_\perp$ (T) lies in the irreversibility line founded from measurements with different technique [5]. In this sense H* can be interpreted as a critical field. In order to investigate the analogy with layered traditional superconductors, the critical field H^* *vs* ϑ has been measured. A cusp for $\vartheta = 0$ in the $H^*(\vartheta)$ curve seems to be present when the temperature is well below the transition, as shown in Fig.4 for T = 76.6 K. This cusp-like behavior is present also in measurements of many physical properties: critical currents [6], resistivity [7-9], magnetization [10] and microwave absorption [11].

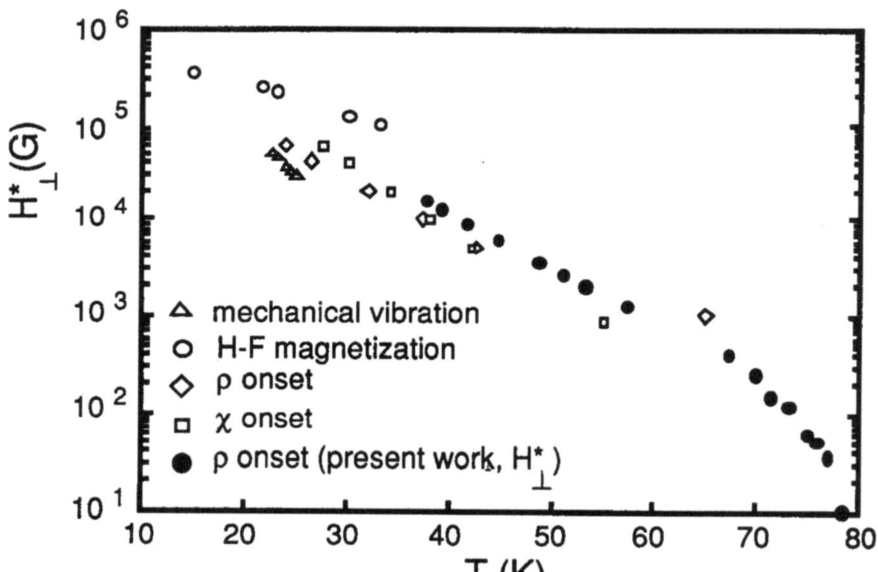

Fig.3 Onset fields when the external magnetic field is perpendicular [$H^*_\perp(\vartheta)$, full dots] to the a-b plane as a function of the temperature. Data of the irreversibility line obtained by means of other techniques are reported for comparison.

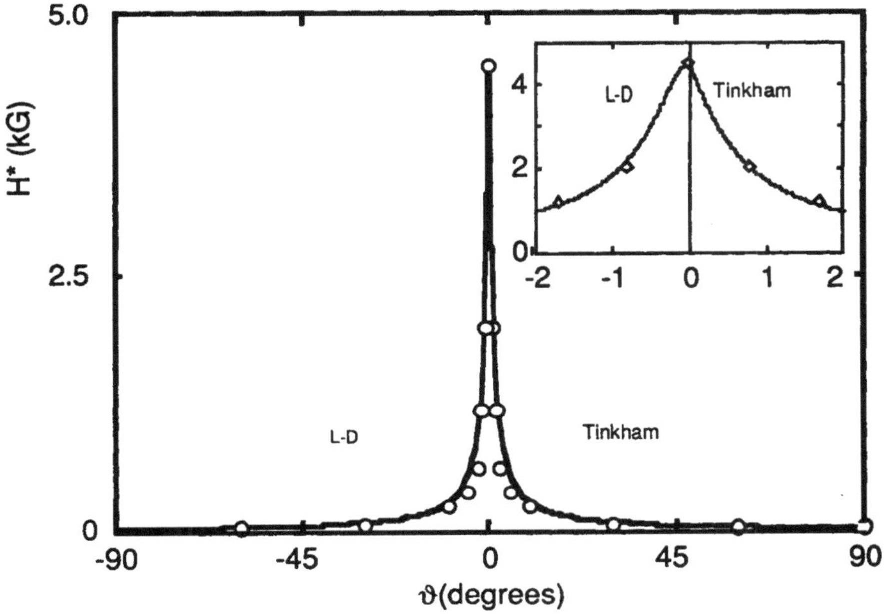

Fig.4 Open circles refer to the angular dependence of $H^*(\vartheta)$ at T=76.6K. The fits by means of Eq.1 for $\vartheta > 0$ and by means of Eq.(5) for $\vartheta < 0$ (continuous lines) are reported. ϑ is the angle between the a-b plane and the direction of the external magnetic field. The insert shows that it is necessary to perform many measurements in the range $-1.0^\circ \leq \vartheta \leq 1.0^\circ$, in order to discriminate the 2-D Tinkham from the anisotropic 3-D Lawrence-Doniach (L-D) model. [From Ref.(8)].

In Fig.4 the fits of the measurements are reported: for $\vartheta > 0$, we have used the Tinkham theory expressed by Eq.(1), and for $\vartheta < 0$, the Lawrence-Doniach equation rewritten in the form [3]

$$\left[\frac{H_{c2}(\vartheta)\sin\vartheta}{H_{c2\perp}}\right]^2 + \left[\frac{H_{c2}(\vartheta)\cos\vartheta}{H_{c2\parallel}}\right]^2 = 1 \quad . \tag{5}$$

For the fit of Fig.4 we have placed $H^*_{\parallel}$ and $H^*_{\perp}$ instead of $H_{c2\parallel}$ and $H_{c2\perp}$ in Eqs.(1) and (5), as previously explained. As one can see, because of the strong anisotropy, the fit of the angular-dependent data for angles not strictly close to $\vartheta=0$, using the current models given by theory, does not allow to discriminate between the 2-D Tinkham formula and the anisotropic 3-D Lawrence-Doniach model. Hence, in order to distinguish the two regimes accurate measurements must be carried out in a narrow range around $\vartheta = 0$. Furthermore, since in all superconductors the coherence length $\xi(T)$ diverges as the temperature approaches the transition temperature T_c, in high-T_c systems some crossover from two-dimensional (2-D) to three-dimensional (3-D) regime should occur below T_c and it must experimentally come into sight.

In Fig.5 the data of $H^*(\vartheta)$ for $T = 76$ K and $T = 78$ K in the angular range $-1.0^\circ \leq \vartheta \leq 1.0^\circ$ are shown: at $T = 78$ K the cusp is substituted by a round maximum. In the same figure, with dashed and continuous lines the theoretical curves are reported. It is clear that at $T = 76$ K the cusp shows that the behavior is Tinkham-like, while at $T = 78$ K the round maximum proves that the behavior is Lawrence-Doniach-like. Therefore between 76 and 78K there is a crossover between a 2-D and an anisotropic 3-D behavior. Further measurements have given for the crossover temperature T* the value T*=77.8K. Analogous experimental results have been obtained in a 2212 BiSrCaCuO epitaxial film grown on $NdGaO_3$ substrate (sample II). The crossover temperature for this sample is 80K, with T=80.4K for the zero-resistance temperature at H=0.

The results show that the angular dependence of H* is completely analogous to that of H_{c2} in the artificial layered superconductors. Then the knowledge of H* can take the place of H_{c2} in order to test the anisotropic properties of high-T_c systems. The results for the crossover reduced temperature $t^* = T^*/T_c$ are $t^* = 0.984$ for sample I and $t^* = 0.996$ for sample II .

The crossover reduced temperature may be extimated considering that the crossing point demarking the anisotropic 3-D from the 2-D (or quasi 2-D) behavior is given by the condition $\xi_\perp(t^*) = s/\sqrt{2}$ [12] with $\xi_\perp(t^*) = \xi_\perp(0)\ (1-t^*)^{-1/2}$. In the case of BiSrCaCuO, in our temperature range the coupling between the neighbouring Cu-O_2 planes (interplane distance s_1 = 3.35 Å) is strong enough to consider them as one superconducting sheet, therefore the distance between the superconducting layers is the half lattice constant $s \cong 15$ Å [13]. Because the extrapolated zero-temperature length along the c-axis is $\xi_{\perp}(0) \cong 1$ Å , one obtains $t^* \cong 0.99$ which is in good agreement with the previous data.

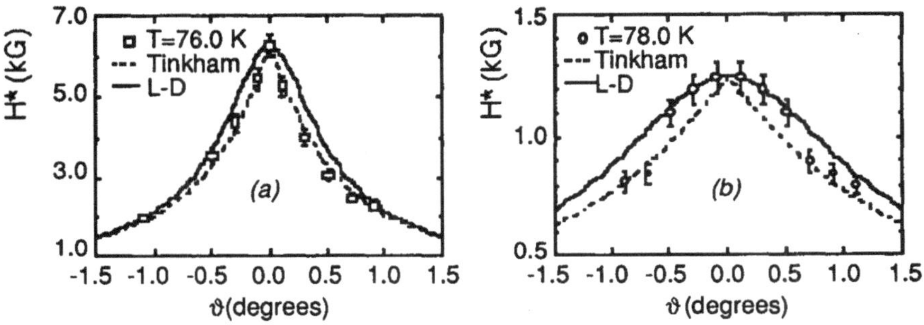

Fig.5 The behavior of $H^*(\vartheta)$ at two different temperatures in a narrow angular range around $\vartheta = 0$ ($-1.0° \leq \vartheta \leq 1.0°$) is shown. A crossover between a 2-D and an anisotropic 3-D behavior going from (a) to (b) is present: a cusp is clearly detectable at ϑ=0 for T=76.0 K, while $dH^*(\vartheta)/d\vartheta=0$ at ϑ=0 in the curve at T=78.0 K . For each temperature the fit of the experimental results by means of the two theoretical models [Eqs.(1) and (5)] is reported for comparison. [From Ref.(8)].

The presence of a crossover temperature in the ϑ-dependence of H* should be coupled to a crossover in the T-dependence of H* around ϑ=0 . Therefore, we have carried out a set of measurements of $H^*_{\parallel}$ as a function of the temperature T. The data are reported in Fig.6 both for sample I and for sample II. As it can be seen, in connection to the values T*, found by means of the ϑ-dependence method, a variation of the slope of the function H*(T) is found.

In summary of the a) point, we have shown that the ϑ-dependence of H* is analogous to that of the upper critical field H_{c2} found in the artificially prepared layered superconductors. This is a puzzling result, because H* is the onset field of the motion of fluxons and not the critical field of the disappearance of superconductivity. Therefore, theoretical work is needed in order to justify both the ϑ-dependence and the

T-dependence of the onset field H*. However, the results are experimentally clear and the measurements of H* seem to offer an alternative method to H_{c2} for the investigation of the anisotropy and the dimensionality of high-T_c copper oxides.

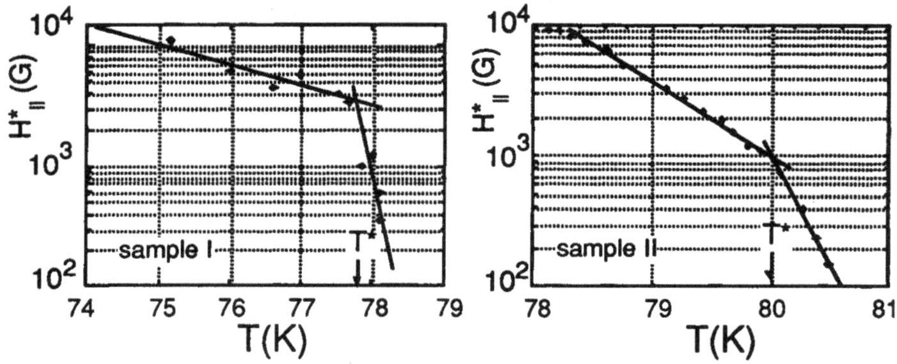

Fig.6 T-dependence of H*(ϑ=0) for sample I and II. The arrows indicate the crossover temperature T*. Continuous lines are guides for the eye. [From Ref.(9)].

Regarding to the second point b), we present now the magneto-absorption measurements. At microwave frequencies (21-23 GHz) it is not possible to detect the onset field H* when measurements of surface resistance are performed because the resistance is immediately developed upon application of the field. In Fig.7 $R_S(H)-R_S(0)$ the variation of the surface resistance as a function of the magnetic field is reported at a fixed value of temperature and at various angles ϑ [14]. Again an anisotropic behavior is present and the absorption shows a Hsinϑ scaling for ϑ >4°, as it can be seen in Fig.8. This scaling is in agreement with the Kes *et al.* [15] model: when the magnetic field is applied parallel to the Cu-O_2 planes (ϑ=0), it penetrates completely the material which becomes "magnetically transparent"; whereas for an arbitrary field orientation (ϑ finite) only the perpendicular component to a-b plane (Hsinϑ) gives rice to a vortex lattice. This model has been successfully applied to the d.c. resistance [16], critical current [17] and torque measurements [18]

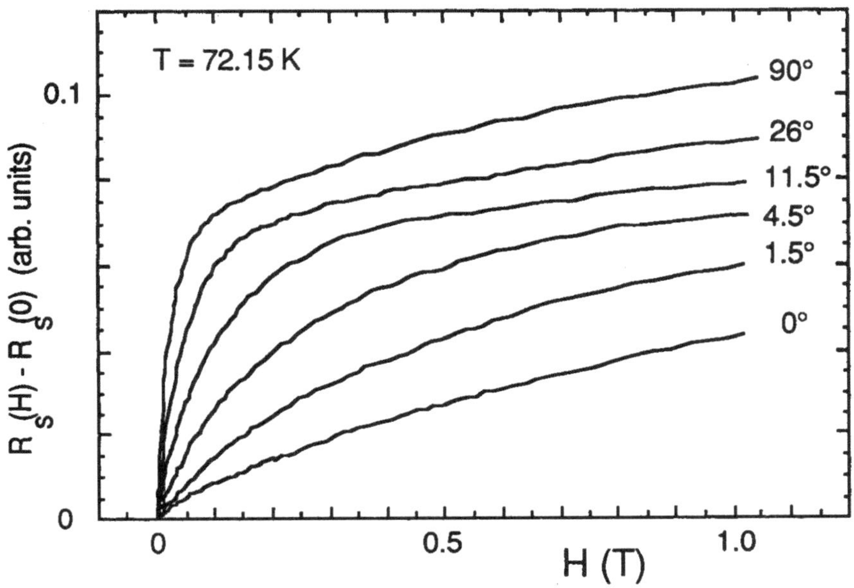

Fig.7 Typical behavior of the microwave surface resistance for a given temperature at various angles ϑ [From Ref.(14)].

More recently, d.c. resistance measurements in 2212 BiSrCaCuO thin films carried out by Raffy *et al.* [19] show that the data can be perfectly scaled with H$\sin\vartheta$ between $\vartheta_{crit} < \vartheta < \pi/2$, where ϑ_{crit} is of order of some degrees ($\vartheta_{crit} \simeq 2°$). Analogous measurements, with a high angular resolution ($\simeq$ 0.02 degrees) have been performed by Iye *et al* [20]. They find that the H$\sin\vartheta$ relation is well obeyed for $\vartheta >$ 5°. In the region $\vartheta \simeq 0°$ at high current density (J=2000 A/cm^2) and at low temperatures (T< 50K) a novel feature, i.e. an enhancement of the resistance below 0.5° , is observed. This effect is also observed in our measurements at microwave frequencies for $-0.4° < \vartheta < 0.4°$:

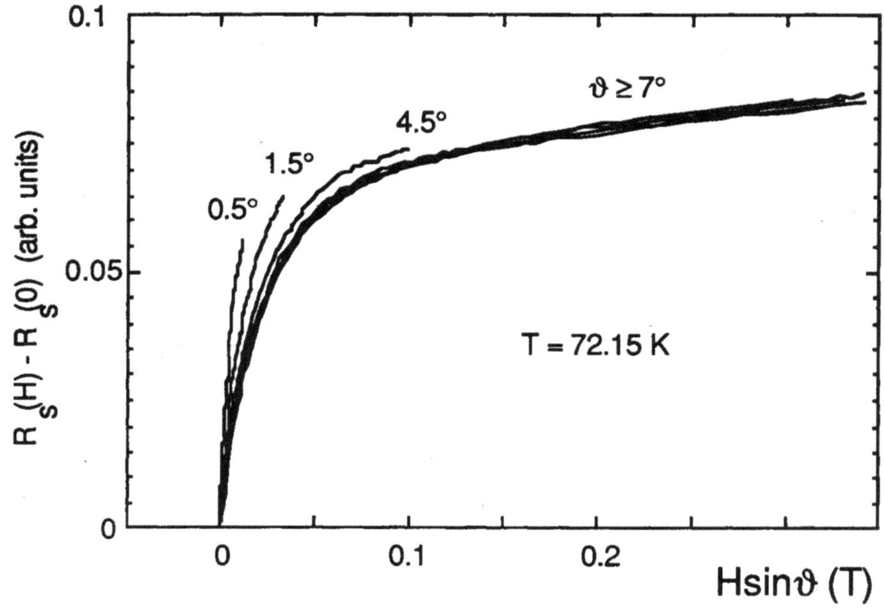

Fig.8 Scaling of the curves of Fig.7 as a function of Hsinϑ. In figure only the part below Hsinϑ = 0.3 T is shown. [From Ref.(14)].

For field near parallel to the $Cu\text{-}O_2$ planes D.H.Kim *et al.* [21)] rule out the suggestion that the non validity of the Hsinϑ scaling for $\vartheta < 5°$ is due to as small misalignment of the field with respect to the $Cu\text{-}O_2$ planes [15)]. A Josephson coupling model is considered to understand the results at small angle ϑ. Iye et al [(20)] tentatively ascribed the enhancement of the d.c. dissipation at $\vartheta \sim 0°$ and with high currents ($J = 2000\ A/cm^2$) to the thermal fluctuations which can cause the creation of flux kinks due to local migration of flux lines. Therefore the theoretical picture of the whole problem is far to be clear and experimental investigations need to understand this very important aspect of HTC superconductors.

In summary, we have reported d.c. measurements of the angular dependence of the field $H^*(\vartheta)$ for the onset of dissipation in 2212 BiSrCaCuO films. The experimental results show that the $H^*(\vartheta)$ behavior is completely analogous to that of the upper critical field $H_{c2}(\vartheta)$ in artificial layered conventional superconductors. The main result is the measurement of the T* crossover temperature from a 2-D to an anisotropic 3-D behavior.

REFERENCES

1. Chun, C.S.L., Zheng, G.-G., Vincent, J.L. and Shuller, I.K., Phys.Rev.B 29, 4915 (1984)
2. Tinkham, M., Phys.Rev. 129, 2413 (1963); in Superconductivity edited by P.R.Wallace (Gordon and Breach, Science Publishers (New York, 1969), pp.371-424; P.E.Harper and M.Tinkham, Phys.Rev. 172, 441 (1968).
3. Lawrence, W.E. and Doniach, S., in *Proceedings of the 12th International Conference on Low Temperature Physics* (Kyoto, Japan, 1970) edited by E.Kanda (Kiegaku, Tokyo, 1971) pp.361-362.
4. Balestrino, G., Foglietti, V., Marinelli, M., Milani, E., Paoletti, A. and Paroli, P., Appl. Phys.Lett. 57, 2359 (1990).
5. Kadowaki, K., Li, N.J., de Boer, F.R., Frings P.H. and M.Franse, J.J., *LT 19 Satellite Conference on High Temperature Superconductivity* - 13-15 August 1990 Cambridge U.K..
6. Roas, B., Schulz, L. and Saemann-Ischenko, G., Phys.Rev.Lett. 64, 479 (1990); Schmitt, P., Kummeth, P., Schultz L. and Saemann-Ischenko, G., Phys.Rev.Lett. 67, 267 (1991).
7. Palstra, T.T.M., Batlogg, B., Schneemeyer, L.F., van Dover, R.B. and Waszczak, J.V. , Phys.Rev.B 38, 5102 (1988) ; Raffy, H.,Labdi, S., Laborde, O. and Monceau, P., Physica B 165-166 , 1423 (1990) ; Juang , J.Y., Cutro, J.A., Rudman, D.A., van Dover, R.B., Schneemeyer, L.F. and Waszczak, J.V., Phys.Rev.B 38, 7045 (1988) ; Naughton, M.J., Yu, R.C., Davies, P.K., Fischer, J.E., Chamberlin, R.V., Wang, Z.Z., Jing, T.W., Ong, N.P. and Chaikin, P.M., Phys.Rev.B 38, 9280 (1988).
8. Fastampa , R., Giura, M., Marcon, R. and Silva, E., Phys.Rev.Lett. 67, 1795 (1991).

9. Marcon, R., Fastampa, R., Giura, M. and Silva, E., Europhys.Lett. 16, 757 (1991).
10. Tuominen, M., Goldman, A.M., Chang, Y.Z. and Jiang, P.Z., Phys.Rev.B 42, 412 (1990).
11. Fastampa, R., Giura, M., Marcon, R. and Silva, E., unpublished.
12. Bulaevskii , L.N., Ginzburg, V.L. and Sobyanin, A.A., Zh. Esp. Teor. Fiz. 94, 355 (1988) [Sov. Phys. JETP 68, 1499 (1988)]; Bulaevskii, L.N., ibid. 64, 2241 (1973) [ibid. 37, 1133 (1973)].
13. Tarascon, J.M., Le Page, Y., Barboux, P., Bayley, B.G. , Green, L.H., McKinnon, W.R., Hull, G.W., Giroud, M. and Hwang, D., Phys.Rev.B 37, 9382 (1988) .
14. Silva, E., Giura, M., Marcon, R., Fastampa, R., Balestrino, G., Marinelli, M., Milani, E., (unpublished).
15. Kes, P.H., Aarts, J., Vinokur, V.M. and van der Beek, C.J., Phys.Rev.Lett. 64 , 1063 (1990) .
16. Iye, Y., Nakamura, S. and Tamegai, T., Physica C 159, 433 (1989) .
17. Farrell, D.E., Bonham, S., Foster, J., Chang, Y.C., Jiang, P.Z., Vandervoort, K.G., Lam, D.J. and Kogan, V.G., Phys.Rev.Lett. 63 , 782 (1989) .
18. van den Berg, J., van der Beek, C.J., Kes, P.H., Mydosh, J.A., Menken, M.J.V. and Menovsky, A.A., Supercond. Sci. Technol. 1, 249 (1989) .
19. Raffy, H., Labdi, S., Laborde, O. and Monceau, P., Physica B 165-166, 1423 (1990).
20. Iye, Y., Tamegai, T. and Nakamura, S., Physica C 174, 227 (1991).
21. Kim, D.H., Gray, K.E., Kampwirth, R.T. and McKay, D.M., Phys.Rev.B 42, 6249 (1990).

FLUX PINNING ENHANCEMENT BY FAST NEUTRON IRRADIATION IN $YBa_2Cu_3O_{7-\delta}$ SINGLE CRYSTALS

H.W. WEBER

Atominstitut der Österreichischen Universitäten, A-1020 Wien, Austria

ABSTRACT

Neutron irradiation experiments on YBCO single crystals will be reviewed. Since all the measurements on these crystals are made by magnetometry, a thorough reexamination of the measuring process in a SQUID magnetometer is needed to establish the appropriate evaluation conditions for the critical current densities and the irreversibility lines. This will be used to discuss the results of a sequential neutron irradiation program on several YBCO single crystals. The data show considerable enhancements of $J_c^{a,b;H//c}$ and $J_c^{c;H//a,b}$, shifts of the anisotropic irreversibility lines to higher fields and temperatures as well as a reduction of magnetization relaxation rates. All of these observations indicate the presence of "strong" pinning centers, similar to precipitate pinning in the case of melt textured bulk YBCO or pinning by amorphous tracks in heavy ion irradiated YBCO single crystals.

1. Introduction

The critical current densities within the grains of high-T_c ceramics or within high-quality (non-granular) single crystals are severely limited by inadequate defect structures for flux pinning and, as a consequence, by thermally assisted flux flow[1], magnetization relaxation[2] and complete depinning[3] of the flux lines ("irreversibility line"). Furthermore, the highly anisotropic nature of the intrinsic superconductive parameters, such as the upper critical field H_{c2}[4], as well as obviously anisotropic flux pinning conditions in the "as-grown" state represent serious obstacles for an application of these materials, especially at the desired operating temperature of 77 K. On the contrary, rather high values of J_c have been reported for epitaxial thin films of $YBa_2Cu_3O_7$ (~ 5 x 10^{10} Am^{-2}, 77 K, zero field), if the field is applied parallel to the basal plane. Such differences in J_c, which may reach one or two orders of magnitude, can be ascribed only to the existence of totally different defect structures for flux pinning, a conjecture which has been substantiated recently by the observation of large numbers of screw dislocations[5] in some epitaxial films.

Modification of the original defect structure in bulk high-T_c materials is therefore of paramount importance. The most remarkable progress, also in view of practical applications, has been achieved by a special melt texturing process for YBCO (melt-powder-melt-growth[6]), in which a substantial amount of normal conducting precipitates of the $Y_2Ba_2CuO_5$-phase is introduced into the 123-matrix. The resulting bulk samples are not only significantly textured, but also show high J_c-values, reduced relaxation rates and considerably enhanced irreversibility lines. The only other technique, which has led to similar beneficial effects, consists of an in-situ modification of the defect structure by particle irradiation[7]. It has turned out that the results of such an irradiation treatment depend sensitively on the type and energy of the irradiating particles, i.e. the distribution of recoil energies produced by a single event, and that all of the above mentioned beneficial effects, which are taken as a signature of strong flux pinning centers, occur only after fast neutron[7] and GeV heavy ion irradiation[8,9].

In the present contribution, I will concentrate on the main features of fast neutron induced property changes in single crystals of $YBa_2Cu_3O_{7-\delta}$, i.e. the changes of T_c, J_c and the irreversibility line, as revealed by sequential irradiation experiments on a few well characterized single crystals. Additional information obtained by electron microscopy on the nature, size and volume concentration of the radiation-induced defects[10,11] as well as the results of a 5 K-irradiation experiment will provide us with a solid basis to explain the data in a conclusive way.

2. Experimental

Because of the small size of high-T_c single crystals, essentially all of the information on their irreversible (i.e. $J_c \neq 0$) properties is based on the analysis of magnetic moments measured as a function of applied field in SQUID or vibrating sample magnetometers. The usual evaluation procedure consists of several steps. Following the conversion of magnetic moments into magnetizations, the magnetization data for increasing and decreasing fields at *fixed external field* are used as input parameters in the Bean model[12], in order to calculate J_c. Modifications of the original model have been developed, which account for the finite geometry of the sample[13] as well as for the possibility of anisotropic critical current distributions[14,15]. In view of strong demagnetization effects in very thin single crystals and because of the failure of conventional demagnetization corrections[14] in the case of very large demagnetization factors (D > 0.7), many published J_c-values are affected by considerable evaluation errors. As pointed out, e.g., in Ref. 16, more detailed considerations on the field and current distributions within these extreme sample geometries are mandatory. The second major problem with this kind of experiments, which cannot be removed because of its physical nature, is related to distinct magnetization relaxation effects, i.e. a continuously "changing" critical state. Hence, the results of a magnetization experiment are clearly dependent on the timing of the instrument, which may be particularly slow in SQUID magnetometers equipped with superconducting high field magnets.

Most of the above mentioned evaluation problems can be solved by reconsidering the actual measuring process in the SQUID magnetometer[17]. The measured magnetic moment can be considered as being the dipole moment of a network of current loops within the superconductor. Consequently, it can be used to calculate the magnitude of the currents flowing in the plane perpendicular to the field direction in a Bean-model-type way (J_c = const.). This is currently being done for situations, where the fully penetrated state has been reached. Furthermore, based on this value of J_c, the average *induction B within the sample*, i.e. the quantity needed for all evaluations in terms of flux lattice spacings and bulk pinning forces, can be calculated. Contributions of the reversible magnetization can be detected easily by comparing the J_c's calculated for the two branches of the magnetization loop. Demagnetization corrections are incorporated by replacing the external field H in the usual relation $B = \mu_0 (H + m/V)$ by the effective field $H_{eff} = H + DM$, i.e. $B = \mu_0 [H + (1 - D) m/V]$, and using either the measured demagnetization factor (from the Meissner slope) or the calculated one for the appropriate sample geometry[18]. These corrections for the induction B are generally small, except for low fields where the moments are large. This leads to an unambiguous assessment of the J_c-B dependence, when the sample is magnetized parallel to the c-direction ($J_c^{a,b;H//c}$) under the assumption of a homogeneous current distribution in the basal plane. The most significant alteration of this approach compared to the original Bean model for infinite geometries lies in the fact, that field (i.e. induction) independent J_c's do *not* lead to a *constant slope* of the flux density gradient in these finite sample geometries.

For H//a,b, a similar approach incorporating the anisotropic Bean model with two differently steep flux density gradients can be taken. Since the critical current densities involved in this case are $J_c^{c;H//a,b}$ and $J_c^{a,b;H//a,b}$, which are both unknown a priori, information on these quantities obtainable from the angular dependence of J_c in thin films[19] and some computer modelling have been used. In the thin film work, the critical current densities flowing along the a- or b-directions can be assessed both for H//a,b and for H//c. The results show that the ratio $J_c^{a,b;H//a,b}/J_c^{a,b;H//c}$ is always larger than 2, sometimes even very much larger. Hence, we assume that *one* of the two flux density gradients in the magnetization experiment with H//a,b is very steep corresponding, e.g., to $J_c^{a,b;H//c}$, 2 or 5 x $J_c^{a,b;H//c}$, or to infinity. It turns out that the results determined under these assumptions are practically identical, i.e. that the measured magnetization hysteresis in entirely determined by the *less steep* flux density gradient, i.e. by $J_c^{c;H//a,b}$. Hence, $J_c^{c;H//a,b}$ can be assessed without further assumptions. The only critical current density which is not accessible to a magnetization experiment on presently available YBCO single crystals is therefore $J_c^{a,b;H//a,b}$, i.e. the critical current density along the a- or b-directions for H//a,b.

With regard to the measurement of the irreversibility line in a SQUID magnetometer, the following procedures can be employed. Firstly, the irreversibility field H_{irr} can be determined from the disappearance of the magnetization hysteresis at constant temperature. Secondly, the irreversibility temperature T_{irr} can be determined from the appearance of hysteresis in fc/zfc measurements at fixed applied field. Thirdly and preferably, in both of these operating conditions of the magnetometer the original response curves of the SQUID can be analyzed. Since the crystal is moved slowly between two pick-up coils in a slightly inhomogeneous magnetic field, which is equivalent to the superposition of a small "ac ripple field" onto the dc field, the response curve of the SQUID is slightly distorted (Fig. 1) by the action of the "ac field" on pinned flux lines, if hysteretic magnetization conditions prevail. The absence of these distortions represents clear evidence for reversible magnetization conditions on a relatively fast time scale ($\delta t \sim 35$ s) and with good experimental resolution ($J_c < 10^4 - 10^5$ Am^{-2}). It can be seen from Fig. 1, that an application of this procedure immediately removes the usual ambiguities associated with the determination of the irreversibility point from the other two procedures mentioned above.

3. Critical Current Densities

Neutron irradiation experiments on conventional superconductors[20], even on compounds which are sensitive to disorder, have shown that the transition temperature T_c is only slightly reduced at low fluence levels ($\leq 10^{22}$m^{-2}, E > 0.1 MeV), whereas significant T_c-depressions occur later on. This behavior is basically confirmed for the high-T_c compounds investigated so far in a systematic way. As an example, the T_c-dependence on fluence is shown for an YBCO single crystal in Fig. 2. Hence, in the low fluence regime ($\leq 4 \times 10^{21}$m^{-2}) radiation-induced changes in the magnetization curves can be related primarily to *changes in the defect structure* responsible for flux pinning rather than to changes in the intrinsic superconductive parameters.

Most of our recent systematic work on fast neutron damage in YBCO has been done on a different single crystal (HW01, a ~ b ~ 1.2 mm, c = 47 μm) with a very sharp T_c at 91.1 K. At high temperatures (e.g. 77 K), clear indications of an anomalous pinning feature, which is often referred to as "fishtail" magnetization[21], have been observed.

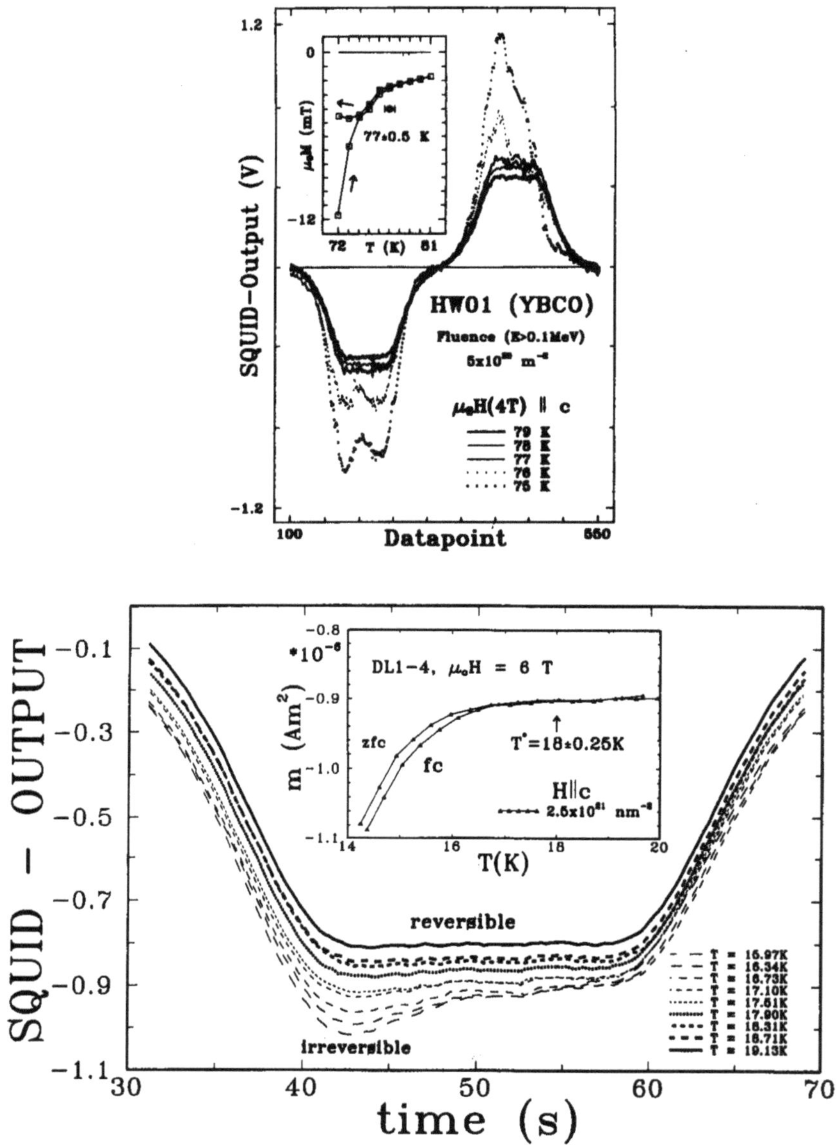

Fig. 1: SQUID output voltage as a function of time. a) Full scans at fixed magnetic fields, but different temperatures near the irreversibility temperature. b) Close-up, taken from similar measurements on an irradiated BiSCCO single crystal.

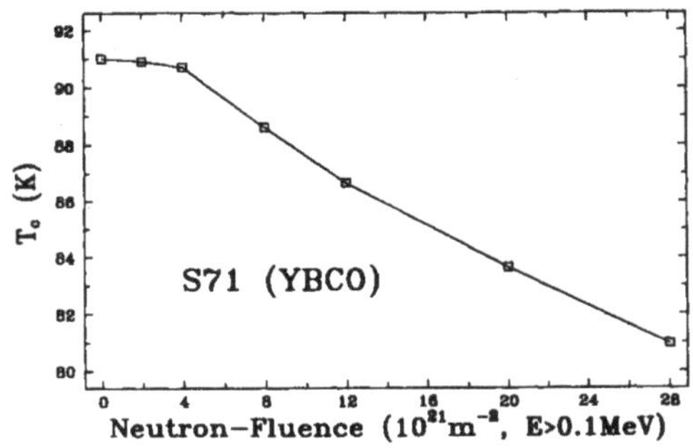

Fig. 2: Transition temperature T_c of single crystal S 71 as a function of fast neutron fluence.

The results of magnetization measurements carried out prior to irradiation and at four different fluence levels up to 4 x 10^{21} m^{-2} (E > 0.1 MeV) are summarized in Fig. 3, where the magnetic moments at 5 and 77 K are plotted versus field for both orientations of the crystal (H//c, H//a,b). The observed radiation effects are strikingly different at these two temperatures as well as for the two field orientations. At 5 K and for H//c, the width of the hysteretic magnetization loops increases continuously with neutron fluence. For H//a,b, we observe only a small increase at the lowest fluence and larger enhancements later on. At 77 K, J_c increases by a factor of about 100 at low fields (H//c), but disappears for fields $\geq$ 3 T due to thermal activation of flux lines.

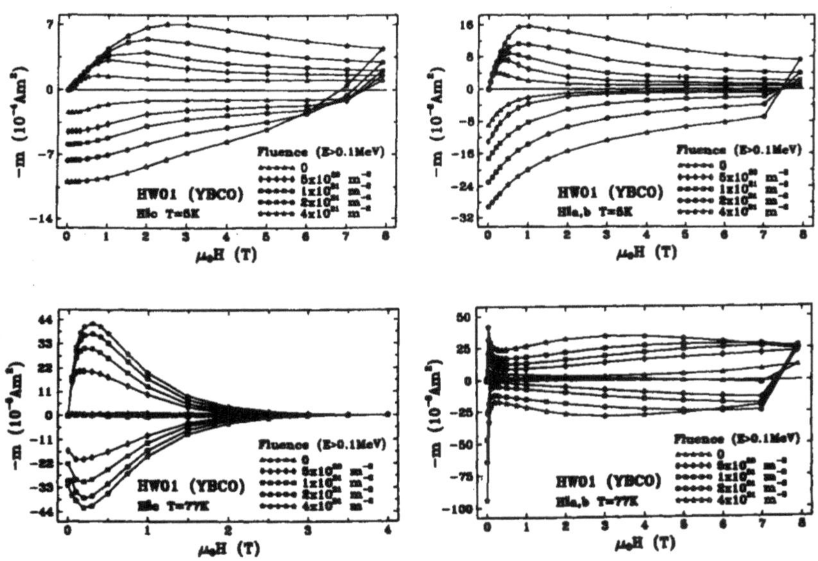

Fig. 3: Magnetic moments versus external field for single crystal HW01 at various neutron fluences (T = 5 K, 77 K). Left: H//c, right: H//a,b

On the contrary, large enhancements are observed for H//a,b at high fields. Since the defects produced by neutron irradiation and their distribution are completely isotropic, these observations can only be explained by proposing (a) the existence of two distinctly different pinning mechanisms for H//c and H//a,b, respectively, (b) the occurrence of radiation-induced modifications of the pre-irradiation defect structure, and (c) the appearance of strong additional pinning centers.

The results for the critical current densities obtained according to the procedures outlined in section 2, are shown in Fig. 4. It should be noted that the magnitude of $J_c^{a,b}$, i.e. the critical current density in the basal plane, is very large at 5 K in the irradiated state ($4 \times 10^{21} m^{-2}$) and approaching results reported for thin films. On the other hand, J_c^c is about two orders of magnitude smaller, but enhanced more significantly by the irradiation. At 77 K, the most significant enhancements of $J_c^{a,b}$ occur at inductions below 1 T, followed by a drop towards the irreversibility line. However, because of the fluence dependence of this line (cf. section 4), significant J_c-values are observed at higher fields in the irradiated state, e.g. ~ $10^8 Am^{-2}$ at 3 T. The "fishtail" character of the data seems to be completely eliminated as a consequence of the irradiation. The field dependence of J_c^c is rather complicated at 77 K and seems to be influenced to a larger extent by the pre-irradiation pinning features. It should be noted, however, that the field dependence of J_c^c is much less pronounced in the irradiated state (except for the anomalous enhancement near B = 0, which is not understood).

High fluence irradiations are not yet available for this single crystal. However, results obtained on other single crystals[22,23] (cf. also Fig. 1) demonstrate that the critical current densities even at 5 K start to decrease at fluence levels of $\geq 1.2 \times 10^{22} m^{-2}$ for H//c and of $\geq 2 \times 10^{22} m^{-2}$ for H//a,b. More drastic effects occur at these fluence levels at high temperatures because of the significant depression of T_c. Hence, both this change of intrinsic superconductive parameters *and* the formation of an unfavorable defect structure observed at this fluence level by electron microscopy[11] are responsible for limitations of further J_c-enhancements by fast neutron irradiation in YBCO.

4. Irreversibility Lines and Magnetization Relaxation

The disappearance of flux pinning at certain fields and temperatures by thermal activation of the flux lines, as well as the strong relaxation effects at lower fields and temperatures which are obviously caused by insufficiently deep pinning potentials, are of course subjects of major concern in view of potential applications of high temperature superconductors. Whereas depinning occurs at Tesla-fields and rather close to T_c in YBCO, the depinning field $\mu_0 H_{irr}$ already falls below ~ 1 T at 25 K in BiSCCO single crystals. In both materials the irreversibility lines are clearly anisotropic[23,24] and always occur at higher fields and temperatures for H//a,b.

As has been shown previously[25,26], the position of the irreversibility line can be changed significantly by neutron irradiation. However, many interesting features of these shifts, which may shed some light on the physics involved, can only be detected by studying the fluence dependence in a systematic way. Results of this kind are summarized for YBCO in Fig. 5. For H//c, the irreversibility line shifts at first to *lower* fields and temperatures, then back again and significantly beyond the initial position at a fluence of $4 \times 10^{21} m^{-2}$. On the other hand, a large shift to *higher* fields and temperatures is observed for H//a,b at the lowest fluence and barely changes up to a fluence of $2 \times 10^{21} m^{-2}$. At the next fluence level a reversal of this trend sets in. This completely different response for H//c and for H//a,b, respectively, points again to the existence of different flux pinning mechanisms as mentioned in section 3. In fact, details of this fluence dependence can be understood qualitatively in terms of a recent model[23,27] involving at least three different flux pinning mechanisms in irradiated YBCO

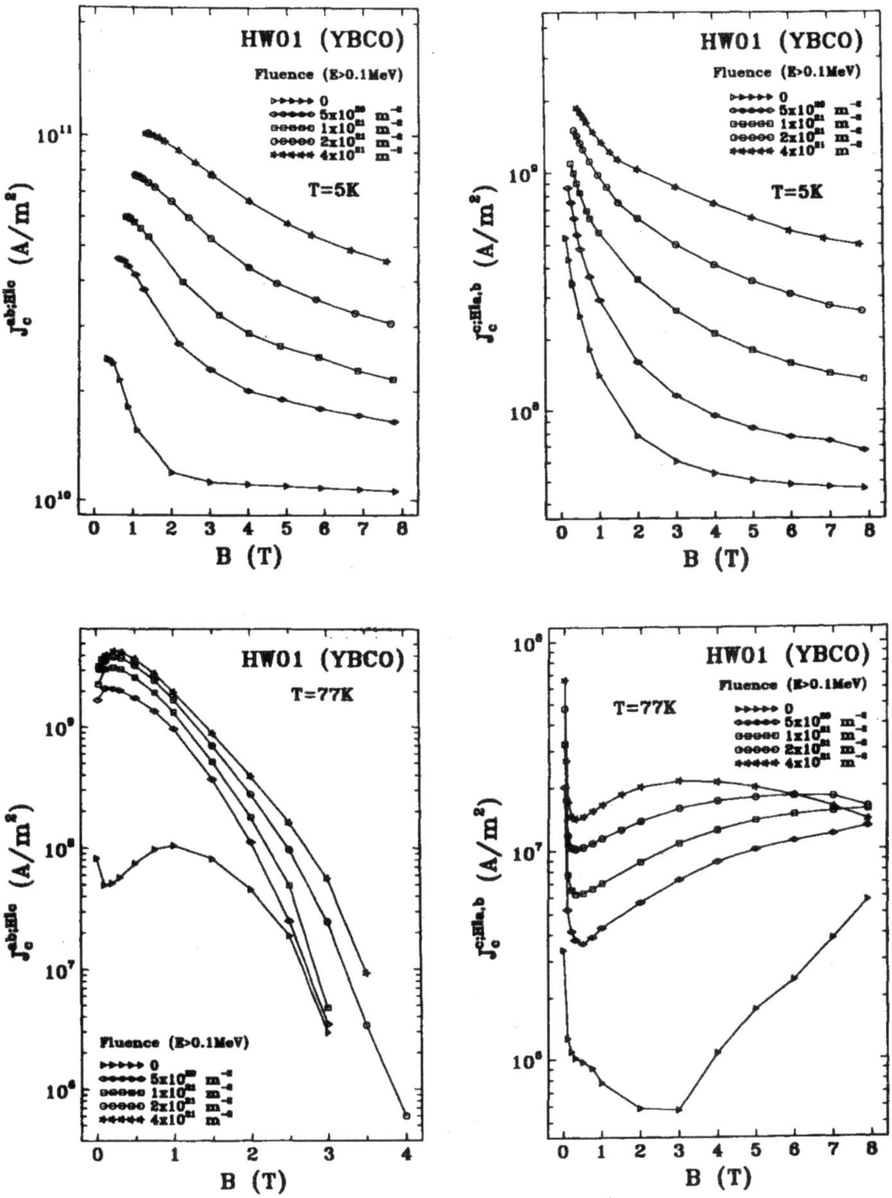

Fig. 4: Critical current densities at 5 K and 77 K calculated from the data of Fig. 3.
Left: H//c ($J_c^{a,b}$), right: H//a,b (J_c^{c}).

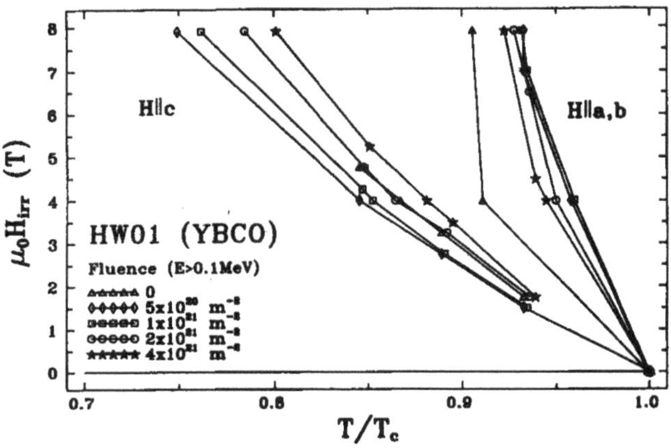

Fig. 5: Irreversibility lines for an YBCO single crystal (HW01) at various neutron fluences and for H//c and H//a,b, respectively.

single crystals (cf. also below). According to these considerations the initial reduction of the irreversibility line for H//c is attributed to a large decrease in the number of pre-irradiation defects, while the concentration of strong radiation-induced pins is still insufficient. This trend is reversed at higher fluence levels because of the increase in cascade concentration. For H//a,b on the other hand, already a few large defects can change the depinning behavior significantly because of the quasi one-dimensional current flow in the presence of intrinsic pinning[28]. With regard to the temperature dependence of H_{irr} we find reasonable agreement with predictions based on a simple activation model[3], i.e. $H_{irr} \propto (1 - t)^{3/2}$ with $t = T/T_c$, but only for H//c, thus confirming again the differences in the pinning/depinning processes for these two field orientations.

As a final remark, first results on changes in the magnetization relaxation rates observed in the irradiated state by torque magnetometry[29] should be mentioned. The presently available data refer to two experiments made on YBCO single crystals which were irradiated to fluences of 0.5 and 2 x 10^{21} m^{-2}, respectively. The measured normalized relaxation rates following magnetization to a field of 1.5 T are in both cases drastically reduced and nearly temperature independent up to ~ 60 K. At higher temperatures, an increase of the relaxation rate is observed in complete opposition to the behavior in the unirradiated state. This difference in radiation response (low versus high temperature) has already been emphasized in section 3. With regard to the relaxation behavior, it must be borne in mind that the relaxation is of course sensitive to *all* the defects in the crystal and that the removal of a significant amount of weak pinning centers as a consequence of the irradiation may explain the observed effects in a regime where the flux line lattice is soft and where flux lines pinned by defects with deep pinning potentials do not contribute to the measured relaxation at all.

5. Discussion and Conclusions

As outlined in the previous sections, the main features of radiation-induced property changes in single crystals consist of considerable J_c-enhancements, shifts of the irreversibility lines to higher fields and temperatures and reductions of relaxation rates. All of them indicate the presence of *strong* flux pinning centers. The nature of these defects is currently being investigated by electron microscopy[10,11]. Individual defect cascades with diameters of 5 - 10 nm have been identified in YBCO single crystals. At a fluence of $2 \times 10^{21} m^{-2}$ the density of cascades was found to be ~ $10^{22} m^{-3}$ (which corresponds to an average defect spacing of 46 nm), in another crystal irradiated to $8 \times 10^{21} m^{-2}$ it was ~ $4 \times 10^{22} m^{-3}$, nicely consistent with the previous result. At higher fluences ($2 \times 10^{22} m^{-2}$), significant cascade overlapping and large disordered areas have been observed. Hence, the size and the density of these defects are very well suited for strong flux pinning at low fluences ($\leq 10^{22} m^{-2}$) but no longer at higher fluences. Damage calculations have shown that the fraction of high-energy recoil atoms leading to cascade formation is very large in the case of fast neutron irradiation. Low energy recoils are of course produced by a certain neutron spectrum as well, which will lead to smaller defects (point defects), point defect clusters and some local release of oxygen. Hence, *weak* pinning centers (e.g. oxygen deficient lattice cells), which are characteristic of the pre-irradiation defect structure, may be *removed* through oxygen recovery or direct replacement by a large cascade or a point defect cluster.

This picture, i.e. the interplay between a gradual build-up of strong pinning centers and a simultaneous (but more rapid) removal or weakening of the pre-irradiation defect structure, is qualitatively adequate to explain the J_c-enhancements, the observed reduction and subsequent increase of the irreversibility line as well as the temperature dependence of the relaxation rates for H//c. For H//a,b, however, flux pinning by the intrinsic pinning mechanism must be taken into account in addition to the above mentioned mechanisms, leading to quasi one-dimensional flux line motion *along* the Cu-O layers. A significant weakening of this mechanism is expected only at high fluence levels where the long range order of Cu-O planes is severely disturbed, such as in the single crystal irradiated to $2 \times 10^{22} m^{-2}$ and investigated in the electron microscope (Fig. 1 of Ref. 11). In fact, the results presented for YBCO in figures 3 and 5 (H//a,b) confirm that the first signs of a trend reversal just appear at the highest fluence level investigated so far, i.e. $4 \times 10^{21} m^{-2}$.

In conclusion, it should be pointed out that the overwhelming contribution of cascades and larger point defect clusters to the observed J_c-enhancements has been demonstrated recently[30] by a low temperature irradiation experiment. In this experiment, a single crystal was irradiated at ~ 5 K to a fluence of $8 \times 10^{21} m^{-2}$ (E > 0.1 MeV) and J_c measured prior to and following warm-up to room temperature. Whereas a significant recovery of T_c is observed as a consequence of the room temperature anneal, which can be attributed to the annealing of point defects, especially oxygen, the critical current densities remain nearly unaffected by this thermal cycle. According to Fig. 6, J^c (which is roughly proportional to the loop widths) increases by ~ 2500 % (T = 5 K, $\mu_0 H$ = 4 T) following low temperature irradiation and measured prior to warm-up. After room temperature annealing, where most of the point defects have disappeared, J_c is reduced by only ~ 20 % of its largely enhanced value. This confirms the predominant role of cascades and larger point defect clusters in the flux pinning process of irradiated YBCO crystals and supports the above conjecture, that pinning centers of a *different nature* ("strong pinning") are responsible for the overall property changes summarized in the previous sections.

Acknowledgements

I wish to express my gratitude for the dedicated efforts of my group, in particular Dr. F.M. Sauerzopf, M.C. Frischherz, Dr. W. Kritscha and H.P. Wiesinger. The fruitful cooperation with Drs. G.W. Crabtree and M.A. Kirk (Argonne) as well as with Prof. R. Griessen and J.G. Lensink (Amsterdam) is acknowledged. This work has been supported in part by Fonds zur Förderung der Wissenschaftlichen Forschung, Wien, under project no. 7970.

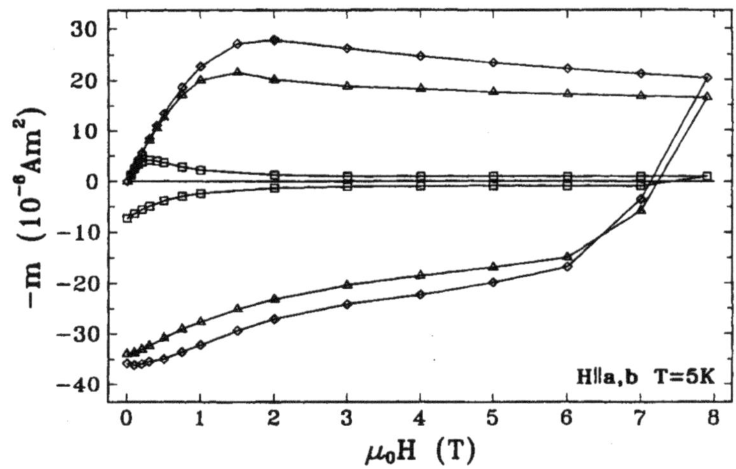

Fig. 6: Magnetization of an YBCO single crystal irradiated at low temperature to a fluence of 8 x 10^{21} m^{-2}. □ □ □ ... prior to irradiation, <> <> <>...following irradiation without warm-up, ΔΔΔ...following irradiation and warm-up to room temperature (H//a,b; T= 5 K).

References

1. P.H. Kes, J. Aarts, J. van den Berg, C.J. van der Beek and J.A. Mydosh, *Supercond. Sci. Technol.* **1** (1989) 242
2. C. W. Hagen and R.P. Griessen, in *Studies of High Temperature Superconductors*, ed. A.V. Narlikar (Nova Science Publishers, New York, 1989), Vol. 3, p. 159
3. A.P. Malozemoff, T.K. Worthington, Y. Yeshurun, F. Holtzberg and P.H. Kes, *Phys. Rev. B* **38** (1988) 7203
4. U. Welp, W.K. Kwok, G.W. Crabtree, K.G. Vandervoort and J.Z. Liu, *Phys. Rev. Lett.* **62** (1989) 1908
5. J. Mannhart, D. Anselmetti, J.G. Bednorz, C. Gerber and D.G. Schlom, *Supercond. Sci. Technol.*, in press
6. M. Murakami, M. Morita, D. Doi and K. Miyamoto, *Jap. J. Appl. Phys.* **28** (1989) 1189
7. H.W. Weber and G.W. Crabtree, in *Studies of High Temperature Superconductors*, ed. A.V. Narlikar (Nova Science Publishers, New York, 1991), Vol. 9
8. L. Civale, A.D. Marwick, T.K. Worthington, M.A. Kirk, J.R. Thompson, L. Krusin-Elbaum, Y. Sun, J.R. Clem and F. Holtzberg, *Phys. Rev. Lett.* **67** (1991) 648
9. W. Gerhäuser, H.W. Neumüller, W. Schmidt, G. Ries, O. Eibl, G. Saemann-Ischenko and S. Klaumünzer, *Physica C* **185-189** (1991) 2339
10. M.C. Frischherz, *Ph.D. Thesis* (TU Wien), 1992
11. M.A. Kirk, M.C. Frischherz, J.Z. Liu, L.R. Greenwood and H.W. Weber, *Phil. Mag. Lett.* **62** (1990) 41
12. C.P. Bean, *Rev. Mod. Phys.* **36** (1964) 31

13. A.M. Campbell and J.E. Evetts, *Critical Currents in Superconductors* (Taylor & Francis, London, 1972)
14. F.M. Sauerzopf, H.P. Wiesinger and H.W. Weber, *Cryogenics* **30** (1990) 650
15. E.M. Gyorgy, R.B. van Dover, K.A. Jackson, L.F. Schneemeyer and J.F. Waszczak, *Appl. Phys. Lett.* **55** (1989) 283
16. M. Daeumling and D.C. Larbalestier, *Phys. Rev. B* **40** (1989) 9350
17. H.P. Wiesinger, *Ph.D. Thesis* (TU Wien), 1992
18. J.A. Osborn, *Phys. Rev.* **67** (1945) 351
19. R.M. Schalk, H.W. Weber, Z.H. Barber, C.E. Davies, J.E. Evetts, R.E. Somekh and D.H. Kim, *Supercond. Sci. Technol.* **4** (1992) S 224
20. H.W. Weber and B.S. Brown, in *Concise Encyclopedia of Magnetic and Superconducting Materials*, ed. J.E. Evetts (Pergamon Press, Oxford, 1992), in press
21. M. Daeumling, J.M. Seuntjens and D.C. Larbalestier, *Nature* **346** (1990) 332
22. F.M. Sauerzopf, H.P. Wiesinger, W. Kritscha, H.W. Weber, G.W. Crabtree and J.Z. Liu, *Phys. Rev. B* **43** (1991) 3091
23. F.M. Sauerzopf, H.P. Wiesinger, H.W. Weber and G.W. Crabtree, *Adv. Cryog. Eng.* **38**, in press
24. W. Kritscha, F.M. Sauerzopf, H.W. Weber, G.W. Crabtree, Y.C. Chang and P.Z. Jiang, *Physica C* **179** (1991) 59
25. W. Kritscha, F.M. Sauerzopf, H.W. Weber, G.W. Crabtree, Y.C. Chang and P.Z. Jiang, *Europhys. Lett.* **12** (1990) 179
26. H. Küpfer, C. Keller, R. Meier-Hirmer, K. Salama, V. Selvamanickam and G. Tartaglia, *IEEE Trans. Magn.* **MAG-27** (1991) 1369
27. F.M. Sauerzopf, H.P. Wiesinger, H.W. Weber, G.W. Crabtree, M.C. Frischherz and M.A. Kirk, *Supercond. Sci. Technol.* **4** (1992) S 105
28. M. Tachiki and S. Takahashi, *Solid State Comm.* **70** (1989) 291
29. J.G. Lensink, R. Griessen, H.P. Wiesinger, F.M. Sauerzopf, H.W. Weber and G.W. Crabtree, *Physica C* **185-189** (1991) 2287
30. H.P. Wiesinger, F.M. Sauerzopf, H.W. Weber, H. Gerstenberg and G.W. Crabtree, to be published

Proton irradiation and pinning in $Ba_2YCu_3O_{7-\delta}$

R. B. van Dover
E. M. Gyorgy
A. E. White
L. F. Schneemeyer
R. J. Felder
J. V. Waszczak

AT&T Bell Laboratories
Murray Hill, NJ USA

ABSTRACT

Proton irradiation of $Ba_2YCu_3O_{7-\delta}$ single crystals increases the critical current markedly, but has only a small effect on the pinning energy. We explore some of the materials issues which lie behind this result. While oxygen point defects may account for some of the pinning (both before and after irradiation), we conclude that other defects are more important, and that these are probably cation point defects.

INTRODUCTION

Flux pinning has been an important scientific and technological issue for high-temperature superconductors ever since they were discovered. Since flux pinning necessarily involves spatial inhomogeneity, it is easier to isolate and study pinning effects in clean single crystals, where the initial level of pinning can be quite weak. By irradiating such crystals we can deliberately introduce defects of various sorts and study the interaction of vortices with these defects. This interaction can proceed through two primary mechanisms:[1] the attractive magnetic interaction of the vortex with the image field associated with the defect, or core energy reduction, in which the energy of the vortex in the superconductor is lowered when the normal core of the vortex is coincident with a region of the superconductor with a reduced pair potential, *i.e.*, a defect. The latter mechanism is expected to dominate for the small defects created by light-ion irradiation, and most discussions of pinning have taken this for granted. It may be necessary to consider magnetic interactions, however, for the case of neutron or heavy-ion irradiation, where large collision cascades can be induced at low fluences. Other complications are the possibility that vortices with weak core pinning might nevertheless be pinned by interactions with other vortices (collective pinning) and the possibility that the anisotropy of the coherence length and penetration depth might qualitatively affect the pinning interactions (pancake vortices).

In this work we have focussed on the effects of proton irradiation on pinning in single crystal $Ba_2YCu_3O_{7-\delta}$. The use of protons has some experimental advantages, as will be described, and also ensures that the vast majority of induced defects (at the fluences investigated here, which range up to $\phi t=3\times10^{17}$ cm^{-2}) will be point defects, which should lead to a fairly weak core pinning interaction. Note that pinning by point defects is negligible in conventional superconductors, partly because the coherence length in such materials is in the range 2-5 nm (rather than the .1-1 nm range in $Ba_2YCu_3O_{7-\delta}$) but also because the pairing interaction is rarely as sensitive to disorder as in the cuprates. The defects induced by proton irradiation produce reproducible effects both from sample to sample within our lab, and also in comparisons between results reported by various research groups.

The interaction between an isolated vortex and a pinning center can be characterized as a potential well in the free energy of the system. Two parameters are particularly significant in this picture—the depth of the well, or pinning energy U, and the maximum restoring force ($\propto n\nabla U$, where n is the density of pinning centers) which is related to the critical current density. It is important to keep a clear distinction between these two parameters. As we will see, the critical current can be increased by a large factor by proton irradiation, while the effect on the pinning energy is much more subtle.

EXPERIMENTAL

In this work we report on measurements of $Ba_2YCu_3O_{7-\delta}$ single crystals which were grown by the CuO-flux method[2] in yttria-stabilized zirconia crucibles. Typical crystal dimensions were 1 × 1 × 0.02 mm. Thin crystals were chosen to ensure that the proton-induced damage would be uniform through the depth of the crystal. Energy dispersive x-ray florescence and Auger spectroscopy allow us to put an upper limit of 0.1 atomic per cent on Zr contamination from the crucible, a level which would have a negligible impact on any of the properties discussed in this report. Crystals were chosen from five separate batches differing slightly in BaO:Y_2O_3:CuO proportions, with no systematic effect on the results. The crystals were annealed in static CO_2-gettered oxygen for three weeks to six months, again with no systematic effect detected on the results. These crystals had a moderate density of twins (that is, they were not deliberately detwinned) but examination of twin boundary pinning[3] in proton irradiated crystals suggests that there is no interaction between the two sources of pinning (e.g., that induced defects do not migrate to twin boundaries and increase or decrease their pinning).

The resulting crystals may be described as "high quality", although that term is ill-defined. They do typically have a low normal-state resistivity, sharp resistive transition, sharp flux-expulsion curve, and low critical current before irradiation. We have found a systematic correlation that crystals which initially have the highest critical currents achieve a lower optimized value after irradiation, compared to those which start out with the lowest critical currents.[4] The T_c of the crystals varies from 89-92 K; we have ruled out inadequate oxygen annealing as a variable (some of the 89 K crystals were annealed 6 months) and tentatively ascribe the variation as due to a homogeneous nonequilibrium oxygen defect distribution.

The samples are irradiated with 3.5 MeV protons in a Van de Graff facility at a flux of 0.01-1 $\mu A/cm^{-2}$. The crystals are attached to a liquid-nitrogen-cooled Cu stage during irradiation, and warmed to room temperature before any measurements. A comparison of the properties of crystals irradiated at the highest and lowest fluxes shows no effect; we conclude that even at the highest flux there is no significant sample heating. At this incident energy the calculated depth of implantation for protons in $Ba_2YCu_3O_{7-\delta}$ is 55 µm. The damage profile (calculated using TRIM-91 code) is fairly flat to a depth of about 40 µm. Figure 1 shows the results of a Monte Carlo[5] calculation of the energy loss due to recoils in a typical crystal, 20 µm thick. Apart from statistical noise, the damage is quite uniform. The Monte Carlo calculation also indicates that roughly 80% of the recoils result in only 1 or 2 atomic vacancies, confirming the expectation that point defects are the dominant form of induced damage. While it is possible that these point defects could subsequently migrate to form larger clusters, the observation that there is no change in twin boundary pinning means either that the defects are neither attracted nor repelled to twin boundaries, or that they are not free to move. The latter possibility seems more likely. Experiments on thin films irradiated at low temperature and measured before and after warming show that some defects do anneal out during the initial warmup[6], although this may be enhanced by thermal expansion mismatch. We have found that the critical current of our irradiated crystals is identical for measurements made within a few hours of irradiation and those made after sitting six months at room temperature, again suggesting that defect migration is not a major consideration.

Critical currents were inferred from magnetization measurements; in this report we will exclusively discuss the values obtained with the current flowing in the **a-b** plane, *i.e.*, where the magnetic field is applied along the **c** axis. The critical current is inferred from the magnetization using the critical state model with an appropriate geometrical factor:[7] we assume that the

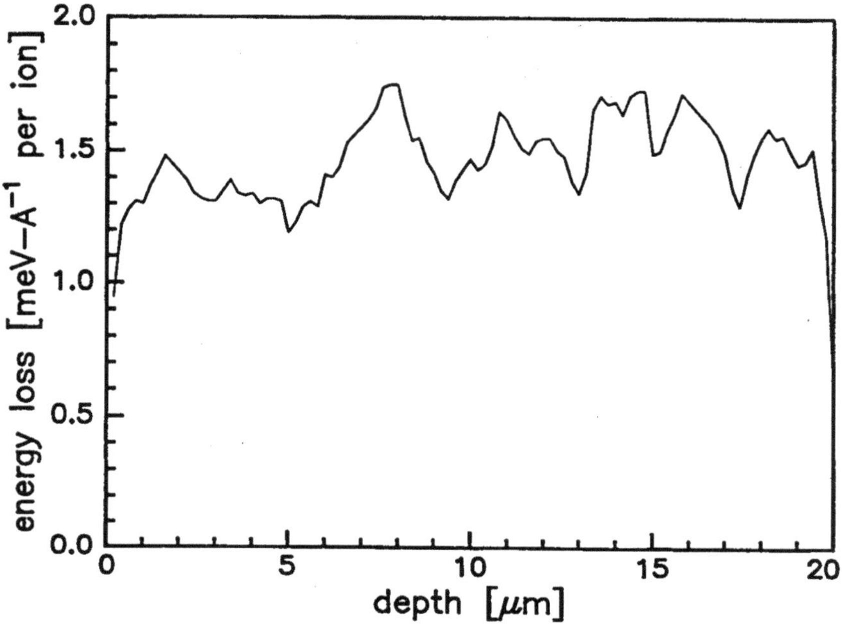

Figure 1. Energy lost to recoils through the depth of a 20 μm crystal of $Ba_2YCu_3O_{7-\delta}$, as calculated using TRIM-91 code. The damage profile is extremely uniform except for thin layers at the top and bottom surfaces.

induced current circulates uniformly throughout the crystal. This is reasonable because we see no signs of granularity in our crystals.

Transport measurements were made on crystals which had evaporated Ag contacts annealed in flowing oxygen at 400 °C. These crystals were thermally anchored to sapphire substrates using Stycast 2850 (Al_2O_3-filled epoxy) in order to avoid heating at high currents. The resistive transitions were measured in magnetic fields from 0-9 T (H || c) using a typical current density of 25 A-cm^{-2}. The results obtained at 2.5-250 A-cm^{-2} were indistinguishable from each other, implying an ohmic behavior in the regime investigated.

RESULTS and DISCUSSION

Proton irradiation causes a large change in the critical current of these crystals. Because the effects of various pinning centers and mechanisms are basically additive, the difference between the critical current before and after irradiation will be largest for the crystals which initially have the lowest critical current, *i.e.*, lowest density of pinning centers. In order to reliably examine the effect of the radiation-induced pinning centers, therefore, it is necessary to begin with high-quality crystals. The magnitude of the change in critical current in our crystals is illustrated in Figure 2,

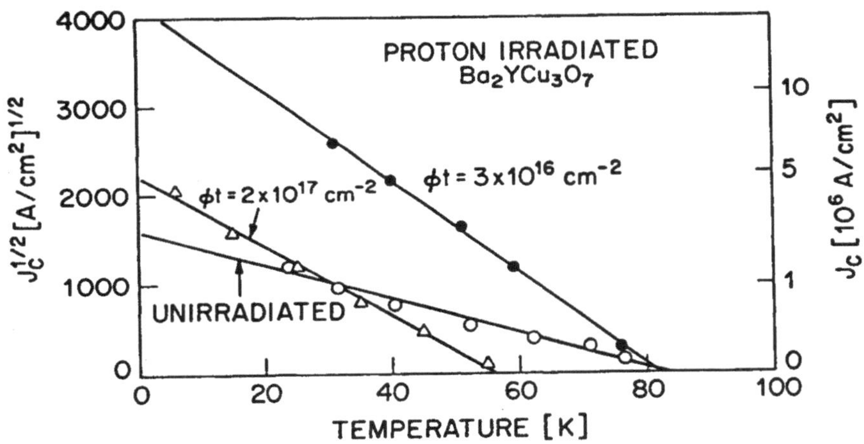

Figure 2. Temperature dependence of the critical current inferred from magnetization measurements at B=1 T. Note that the vertical scale is $J_c^{1/2}$, a scale which yields a linear behavior. The y- and x-intercepts are referred to as $J_c(0)$ and T^*, respectively. The proton fluence which gives the highest critical current is $\phi t = 3\times10^{16}$ cm^{-2}.

in which we plot the critical current measured at 1 T versus temperature. Note that the vertical axis is actually $J_c^{1/2}$, a transformation which consistently gives a linear behavior versus temperature in the regime we have investigated.

Extrapolating the critical current to T=0 gives $J_c(0)$, while the temperature at which $J_c(T)$ extrapolates to 0 is T^*. Note that T^*, which is indirectly related to the so-called irreversibility temperature, is virtually unaffected by proton damage up to a fluence $\phi t=3\times10^{16}$ cm^{-2}, even though $J_c(0)$ increases 20-fold. The value of T^* is 82.5 K at 1 T, which is consistent with other estimates of the irreversibility line. At higher fluences, both T_c and T^* decrease due to an overall reduction in the order parameter when the density of defects exceeds one per unit cell. For proton-irradiated $Ba_2YCu_3O_{7-\delta}$ this occurs at about $\phi t=10^{17}$ cm^{-2}. The value of $J_c(0)$ decreases proportionally due to the reduction in the maximum modulation of the order parameter (which is responsible for pinning).

The fact that T^* does not change for fluences below 3×10^{16} cm^{-2} implies that the pinning energy is essentially unchanged. We interpret this to mean that defects produced by irradiation are similar to the preexisting defects, that they do not interact, and that there is little change in the overall pinning mechanism.

The magnetization of a typical sample, which was given the optimal[8] dose $\phi t=3\times10^{16}$ cm^{-2} is shown in Figure 3 for nine temperatures, where the vertical axis is translated into the inferred critical current density. The data at low fields (<.25-1 T, depending on temperature) are not corrected for self-field effects, and should not be taken seriously. Perhaps the most striking feature of these curves is the almost complete absence of field-induced pinning,[1] for which the critical current *increases* with field. That behavior is commonly seen in unirradiated samples of $Ba_2YCu_3O_{7-\delta}$—both single crystal and polycrystalline—and is attributed to pinning by oxygen-deficient regions[4,9] (which might be oxygen point defects). In contradiction to the previous paragraph, this implies that the nature of either the induced defects or the pinning mechanism has changed.

The critical current at low temperatures and relatively low fields is extremely high in these irradiated crystals. Certainly it is comparable to the value found in the best (unirradiated) thin films under the same conditions. Following the logic of Hylton and Beasley,[10] and assuming that the fields are low enough to

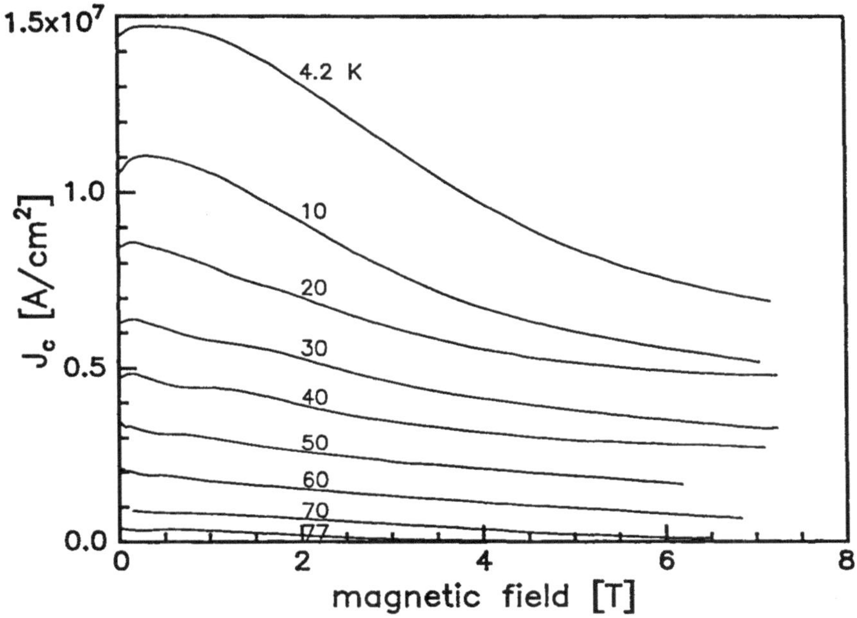

Figure 3. Critical current (inferred from magnetization) of a typical $Ba_2YCu_3O_{7-\delta}$ crystal, irradiated with protons fluence of 3×10^{16} cm^{-3}. Note that the critical current is essentially a monotonically decreasing function of the magnetic field. Except at the highest temperatures, the fields are not high enough to determine the location of the peak in pinning force ($J_c \cdot B$).

give isolated vortex pinning, we conclude that a large fraction of each vortex must be intercepted by defects, *i.e.*, that the occasional large defect clusters which are produced by proton irradiation are not sufficient to account for the magnitude of the increase of the critical current.

In order to look for possible changes in the pinning energy, and learn about the effect of oxygen defects on pinning in $Ba_2YCu_3O_{7-\delta}$, we have found it useful to look at the results of transport measurements. As with any measurement of pinning energy, there is ambiguity[11] in interpreting the results, but we feel that by concentrating on the flux-creep regime we can make meaningful qualitative inferences. The approach described below is completely empirical and makes no assumptions about specific pinning models.

We have found that the resistive transition of almost all of our crystals, at various levels of irradiation, and in magnetic fields from 1 to 9 T, can be described by a single functional curve, which is scaled vertically to the normal state resistivity and on the horizontal axis to a single scaling temperature (and T_c). Figure 4a

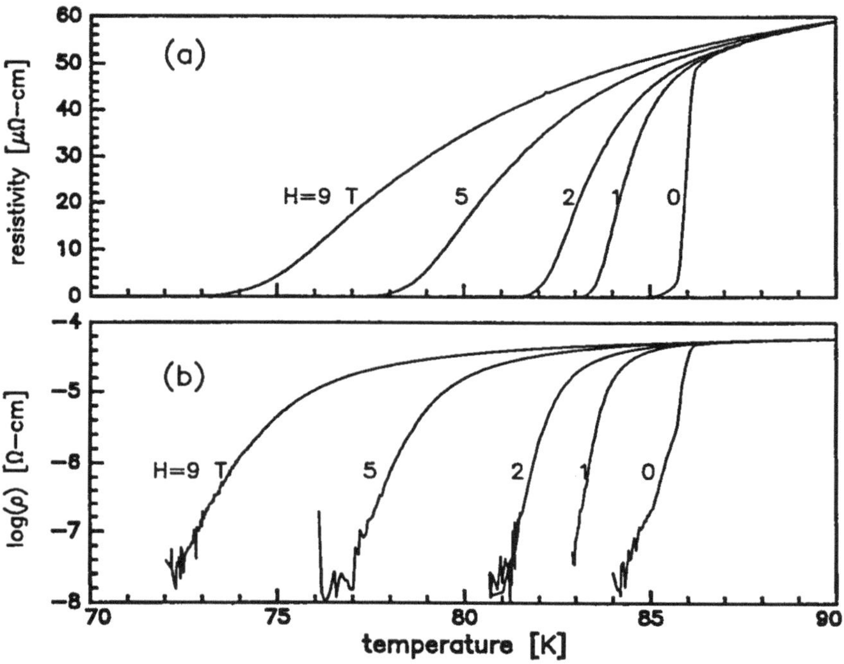

Figure 4. (a)Resistive transition of a typical crystal ($\phi t = 3\times10^{16}$ cm^{-2}) in various magnetic fields. (b) Same data on a logarithmic vertical axis. Note the similarity in the shapes of the curves.

shows a typical series of resistive transitions, taken in static magnetic fields (H $\parallel$ c) from 0 to 9 T. The data are comparable to those reported in the literature.[12] In Figure 4b we show the same data on a logarithmic scale, so that the low resistance behavior can be seen clearly. Although the resistivity can be seen to decrease roughly exponentially with temperature, it does not accurately follow an activated form (because of the inherent temperature dependence of U(T) near T_c). The shapes of the curves for $1 \leq B \leq 9$ T are extremely similar. This can be seen in Figure 5a, in which the data for all of the fields have been collapsed onto one curve by scaling the horizontal axis of each curve: $X = (T_c^{mf} - T)/(T_c^{mf} - T_s(B))$, where T_c^{mf} is the mean-field T_c (obtained by extrapolating the fluctuation conductivity and taken as independent of the magnetic flux density B), and T_s is the scaling temperature, chosen to be in the "activated" regime, arbitrarily at the temperature at which $R = R_n/100$. The scaling temperature is obviously field dependent, and is found to vary roughly as $T_s \propto B^{2/3}$. Finally, Figure 5b shows the result for seven different crystals, at the various fields. In this figure the resistances are normalized to the value above T_c to allow comparison. The consistency of the shape of the R(T) curve from sample to sample is remarkable.

We are thus able to describe the entire resistive transition by a single independent parameter, T_s. If there were a concensus theory which described the shape of the resistive transition, we could presumably relate the scaling parameter directly to a pinning energy. But the detailed shape of the curve is not well-understood, so we limit ourselves to using the parameter T_s to compare the behavior of various samples. We argue that the magnitude of T_s is inversely related to the pinning energy: a smaller pinning energy implies a broader transition, and so a larger T_s.

In Figure 6 we show how this parameter varies with fluence for various crystals. Data for which the same crystal was irradiated sequentially are connected by solid lines. Notice first the overall consistency in the behavior, *i.e.*, that the data are linear with a small statistical variance. A significant trend can be seen as well: the effective pinning energy increases slightly (T_s

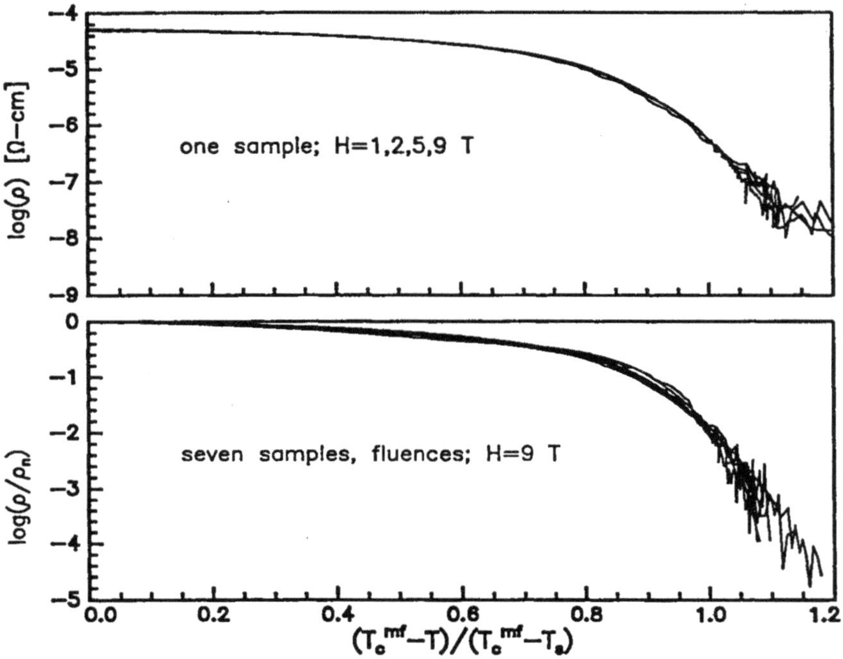

Figure 5. (a) The top graph illustrates the same data as in Figure 4, but with the data for each field ($1 \leq H \leq 9$ T) scaled linearly in temperature (T_c -T), so as to collapse the curves. This demonstrates that the curves are of identical shape for this sample. (b) The bottom graph is a composite of data (specifically, the 9 T data) for four different crystals at three fluences (*i. e.*, multiple fluences for some samples). This demonstrates that the curve shape is a consistent property of the material.

decreases) with fluence. The effect is small, consistent with our inability to resolve a change in the irreversibility temperature T^* at 1 T, as described above. The effect is also consistent with information about the pinning energy which can be inferred from magnetization decay measurements, as reported by Civale, et al.[13]

A variation in pinning energy of this magnitude is consistent with two

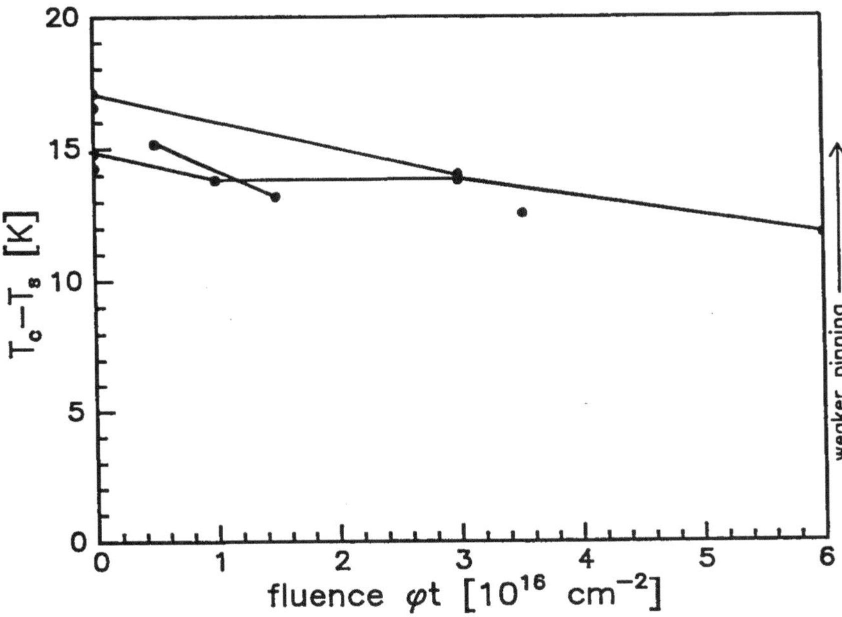

Figure 6. Scaling temperature T_s versus fluence for various crystals. A slight overall trend is discovered. The scaling temperature is a measure of how far below T_c it is necessary to cool the sample in order to reduce the resistivity by a certain factor. Thus it is an inverse measure of the pinning energy (larger $T_c - T_s \rightarrow$ lower pinning energy).

possibilities: that the induced defects are not the same as the original defects in the superconductor (which are most likely oxygen point defects), or that they are the same and that the change in inferred pinning energy is a result of the increased density of pinning centers (for example, an increased density of pinning centers along the length of a vortex might increase its tilt or shear modulus).

In order to shed light on this issue we have examined the behavior of a crystal where the density of oxygen point defects was deliberately increased by removing a small amount of oxygen. This was done by annealing the crystal at 450-500 °C for 50-100 hours, long enough to establish equilibrium in 1 atm. flowing O_2. It was not possible to directly analyze such small samples for their oxygen deficiency (δ), and rather than attempt to infer the value from data on bulk samples we present the results as a plot of scaling temperature versus the mean-field transition temperature, T_c^{mf}. In this way we can directly compare the high-temperature-quenched oxygen-deficient samples[14] with those which have been irradiated. We note, nevertheless, that a decrease in T_c of 3 K corresponds roughly to an increase of δ by .02 under these conditions,[15] which is equivalent to an increase in the oxygen defect density of 1.3×10^{20} cm^{-3}. This is close to the calculated vacancy density (assuming the standard 20 eV energy to create a displacement) for samples irradiated at a fluence of 3×10^{16} cm^{-2}, which is 1.4×10^{20} cm^{-3}.

The data of Figure 7 include one datum for the highly-oxygen-deficient phase $Ba_2YCu_3O_{6.6}$, which has been shown to have a markedly lower pinning energy than $Ba_2YCu_3O_{6.95}$. This datum is from Reference 16, scaled to 9 T for comparison. It is immediately apparent from Figure 7 that even small levels of oxygen deficiency dramatically lower the pinning energy rather than increasing it. There is no indication that these oxygen defects, *per se*, act as pinning sites.

This is not to say that oxygen point defects can never act as pinning centers. It is not possible to rule out oxygen defect clustering, and indeed evidence of this has been seen in other crystals.[9] But it seems likely that oxygen point defects do not dominate the pinning behavior in proton-irradiated $Ba_2YCu_3O_{7-\delta}$. TEM studies of irradiated crystals do not indicate the presence of extended defects at the density needed to explain the extremely high critical currents we obtain.[17] Therefore we suggest that it is actually cation point defects that dominate the pinning behavior.

A similar suggestion has been offered to account for pinning in $Ba_2YCu_3O_{7-\delta}$ thin films grown by the so-called *in situ* approach.[18] In the thin films this interpretation is supported by the observation that the c-axis lattice parameter, c_0, is dilated in the films from its nominal bulk value of

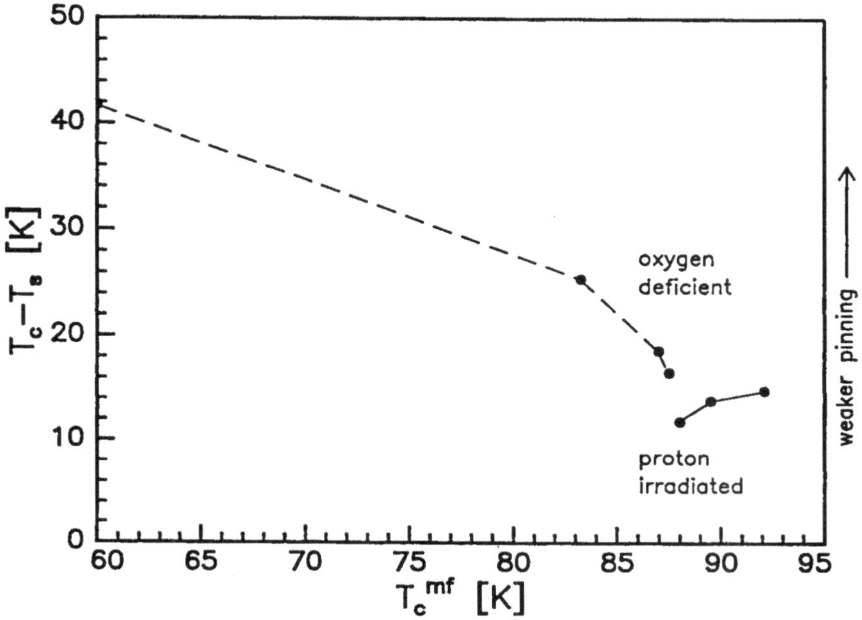

Figure 7. Scaling temperature versus mean-field T_c for a crystal which was proton irradiated (solid line) and a crystal from which a small amount of oxygen was removed (dashed line). Also shown (leftmost point on the dashed line) is the value for a crystal with the stoichiometry $Ba_2YCu_3O_{6.7}$, the so-called "60 K" oxygen phase, inferred from Reference 16.

11.67 by 0.02-0.1 A. However, in our crystals the value of c_0 is unchanged (±0.01 A) after irradiation. This may be due to the fact that thin films incorporate cation defects in a slightly different manner (*e.g.*, Ba-for-Y antisite disorder[18]) than the presumably random interstitials generated by irradiation. Our experiments to verify the presence of cation defects by examining the diffuse scattering between Bragg peaks were inconclusive.

CONCLUSIONS

The question of what defects are responsible for the increased pinning in proton-irradiated $Ba_2YCu_3O_{7-\delta}$ is difficult to resolve unambiguously, because efficient pinning can be obtained on a scale impossible to inspect directly, e.g., by HRTEM. The evidence suggests that only point defects are present at a sufficient density to account for the extremely high critical currents observed. As in unirradiated material, oxygen point defects may account for some of the pinning, but proton irradiation is likely to produce just as many cation point defects, which is consistent with the increase in effective pinning energy in the radiation-damaged samples.

This conclusion has profound implications for the preparation of strongly-pinned bulk material: a moderate concentration of some impurity which could substitute on a cation site might strongly increase the critical current. Experiments to date have achieved only moderate success,[19] for reasons that are not understood. However, one requirement for optimized pinning is that the substitution be uniform (with no clustering at the atomic scale) and at a level of approximately 2 mole per cent overall. Perhaps there are stringent limitations of the chemistry which must be overcome to achieve critical currents in bulk $Ba_2YCu_3O_{7-\delta}$ which are comparable to optimized proton-irradiated samples.

REFERENCES

1. For an extended discussion, see A. M. Campbell and J. E. Evetts, Advances in Physics **21**, 199 (1971).

2. L. F. Schneemeyer, J. V. Waszczak, T. Siegrist, R. B.van Dover, L. W. Rupp, B. Batlogg, R. J. Cava, and D. W. Murphy, Nature **328**, 601 (1987).

3. Observed using the technique described by E. M. Gyorgy, R. B van Dover, L. F. Schneemeyer, A. E. White, H. M. O'Bryan, R. J. Felder, and J. V. Waszczak, and W. W. Rhodes, Appl. Phys. Lett. **56**, 2465 (1990).

4. R. B. van Dover, E. M. Gyorgy, L. F. Schneemeyer, A. E. White, S. Glarum, R. J. Felder, and J. V. Waszczak, Proc. Mat. Res. Soc. **169**, 911 (1990).

5. F. J. Ziegler and J. P. Biersack, *The Stopping and Range of Ions in Solids* (Pergamon, New York 1985).

6. G. C. Xiong, H. C. Li, G. Linker, and O. Meyer, Phys. Rev. **B38**, 240(1988).

7. E. M. Gyorgy, R. B. van Dover, K. A. Jackson, L. F. Schneemeyer, and J. V. Waszczak, Appl. Phys. Lett. **55**, 263 (1989).

8. R. B. van Dover, E. M. Gyorgy, A. E. White, L. F. Schneemeyer, R. J. Felder, and J. V. Waszczak, Appl. Phys. Lett. **56**, 2681 (1990).

9. M. Daeumling, J. M. Seuntjens, and D. C. Larbalestier, Nature **346**, 332 (1990).

10. T. Hylton and M. R. Beasley, Phys. Rev. **B41**, 11669 (1990).

11. For a discussion of some of the ambiguities involved in flux creep measurements, see D. O. Welch, IEEE Trans. Mag. **27**, 1133 (1991).

12. T. T. M. Palstra, L. F. Schneemeyer, R. B. van Dover, J. V. Waszczak, and B. Batlogg, Appl. Phys. Lett. **54**, 763(1989).i

13. L. Civale, A. D. Marwick, M. W. McElfresh, T. K. Worthington, A. P. Malozemoff, F. H. Holtzberg, J. R. Thompson, and M. A. Kirk, Phys Rev.

Lett., **65**, 1164 (1990).

14. R. B. van Dover, L. F. Schneemeyer, R. J. Felder, and J. V. Waszczak, Physica **C** (in press).

15. J.-M. Tarascon, P. Barboux, B. G. Bagley, L. H. Greene, W. R. McKinnon and G. W. Hull in *Chemistry of High-Temperature Superconductors*, D. L. Nelson, M. S. Whittingham, and T. F. George, eds (American Chemistry Society, Washington DC, 1987), p. 198; J.-M. Tarascon, L. H. Greene, B. G. Bagley, W. R. McKinnon, P. Barboux, and G. W. Hull, in **Novel Superconductivity**, S. A. Wolf and V. Z. Kresin, eds. (Plenum, 1987) p. 705.

16. T. T. M. Palstra, B. Batlogg, R. B. van Dover, L. F. Schneemeyer, and J. V. Waszczak, Phys. Rev. **B41**, 6621 (1990).

17. C. H. Chen (private communication).

18. V. Matilasevic, P. Rosenthal, K. Shinohara, A. F. Marshall, R. H. Hammond, and M. R. Beasley, J. Mater. Res. **6**, 682 (1991).

19. E.g., R. Wordenweber, K. Heinemann, G. V. S. Shastry, and H. C. Freyhardt, Physics **C162-164**, 1601 (1989); K. A. Delin, T. P. Orlando, E. J. McNiff, Jr., S. Foner, R. B. van Dover, L. F. Schneemeyer, and J. V. Waszczak, Proc. Mat. Res. Soc. (to be published).

PROTON IRRADIATION ENHANCEMENT OF IMPLANTED LAYERS ON CERAMIC YBCO BULK

G. Castagno(#), E. Mezzetti(*), B. Minetti(*), F. Periale
Dipartimento di Fisica, Politecnico di Torino, Italy
(#) I.N.F.M.
(*) I.N.F.N.

D. Andreone, R. Gerbaldo
Istituto Elettrotecnico Nazionale "Galileo Ferraris", Torino, Italy

F. Abbattista, M. Vallino
Dipartimento di Scienza dei Materiali, Politecnico di Torino, Italy

R. Cherubini, G. Moschini
Laboratori Nazionali di Legnaro, I.N.F.N., Legnaro (Padova), Italy

ABSTRACT

In this work the possibility to radiatively modulate the performance of ceramic high T_c materials is investigated and experimentally proven.
Surface layers, about 200 µm deep, in different YBCO samples, have been irradiated with 6.5 MeV protons at fluence of $10^{17} p/cm^2$.
A complete set of transport measurements has been performed before and after proton implantation: the goal was to find the main physical properties and the chief parameters which affect and limit the critical current density enhancement obtained in the surface thick layer, by irradiation.
Several irradiated samples coming from two different irradiation - runs show, at 77K and zero field, a critical current density enhancement up to values greater than an order of magnitude. This increase is due to the layer of implanted protons which short-circuits the unaffected underlaying bulk. The irradiation affects also the critical current density vs. temperature curves which show a steeper decrease with temperature.
All results are discussed, both in terms of Thermally Assisted Flux Flow (TAFF), and of Kosterlitz-Thouless vortex-antivortex theory.

1. - INTRODUCTION

The studies concerning the effects of radiation induced defects on the performances of high T_c superconductors are very valuable to understand the most important aspects of the basic structures for controlling the superconducting behaviors. In particular they can give an insight into the relationship between defects and some driving superconducting properties, as critical current density J_c.

The role of defects in controlling the superconducting properties of high T_c materials has, up to now, two different aspects; firstly the oxygen vacancy ordering or, perhaps, superordering [1,2,3] dominated by oxygen points defects in the Cu-O chains or the local structures of interstitial oxygen defects in $La_2 C_u O_{4+\delta}$ or $Ba_{1-x} K_x Bi O_3$ are responsible for the high T_c superconductivity [4,5,6,7,8]. On the other side many crystal defects can interact with their flux lines and pin the motion so that the values of J_c are determinated by the flux pinning strength [9].

Moreover in syntered samples, the grain boundaries act as a special kind of defects [10]: some boundaries may be "naturally" oxygen deficient in a fully oxygenated specimen [11]; such a boundary is likely to be weakly superconducting. These characteristics strongly affect the macroscopic properties of polycrystalline superconductors. The destruction of homogeneity of the crystal leads to a suppression of the order parameter along the grain boundaries. Depending on the strength of the suppression, grain boundary effects range from being beneficial for the macroscopic properties of the superconductors [12,13,14,15] to being destructive for the macroscopic coherence of the material. Generally, the pinning strength reaches the maximum as the average diameter of the pinning centers is approximately equal to the coherence length of the superconductor [16]. In high T_c materials, however, the coherence length is estimated to be on the order of 10 A [17,18]. Thus, the size of appropriate pinning centers in the high T_c materials is reduced to the atomic lattice spacing.

What can be expected when radiation induced defects interact with pre-irradiation structures?

The expected effects are:

1) radiation damage due to destruction of intragranular (microscopic) and/or intergranular (macroscopic) coherence [19,20];
2) radiation enhancement caused by an enhancement in pinning phenomena [21,22,23]. It is to be underlined that the impact of these effects must be evaluated taking into account the optimum pinning size;
3) radiation enhancement caused by intragranular or intergranular optimization of the coherence between "orderer" regions [24].

In particular, as mentioned before, the high critical current density in YBCO is usually associated with strong flux pinning by oxygen vacancies and generally with suitable modifications of the original defect structure. Between a few other techniques (as for example melt-powder-melt-growth [25]) the most remarkable progress has been effectively obtained after neutron or ion irradiation of both YBCO and BISCO single crystals [26]. Extensive transmission electron microscopy (TEM) data reveal a highly defected microstructure in the irradiated samples [27]. For what concerns ceramic high T_c materials, the problem is of paramount importance but the results are up to now scarce and rather ambiguous, due to the possibility of unpredictable interactions between induced and pre-irradiation defect structure. It should be emphasized that the ion induced defects are of defect-cascade type resulting from an ion impact [23] and their size is approximately 50 A, that is larger than the predicted optimal one.

We began investigation bulk ceramic samples of YBCO and BISCO, irradiated by monoenergetic 6.5 MeV protons. The fluence of the primary beam was about 10^{17} p/cm^2 . A complete set of transport measurement has been performed before and after proton implantation. In fact, for what concerns the evaluation of the critical current density J_c, transport measurement allow to obtain unambiguous results. The average critical current has been strongly enhanced but in one case also weakened, probably because of the unfavorable interaction with the pre-irradiation defects structure. A systematic detailed investigation of the dependence of the critical current density enhancement on the fabrication parameters and on irradiation conditions is, up to now, at very early stage. However the possibility to enhance the superconductive properties of the irradiated surface layers (about 200 μm thick) over the unaffected bulk has been experimentally proved.

2. - EXPERIMENTAL PROCEDURE

The samples analyzed are syntered ceramic YBCO pellets, (preparation details are given in [28]), and nominally stoichiometric lead doped BSCCO 2223, prepared by Pirelli S.p.A. via a modified citrate routing. The specimen of BISCO was prepared by cold pressing powders in the shape of a bar (whose cross section is about 3 mm^2) and by syntering on air to 840 K for 120 hours. The resistive T_c was 105 K, whereas the critical current density J_c at 77 K, zero field, was poor (about 50 A/cm^2).

The pellets of YBCO, were properly double-T shaped for critical current-density measurements and successively syntered at 950 ^oC and then annealed at 400 ^oC.

The current was forced through a constriction less than 0.5 mm thick. Values of J_c at 77 K range from 50 to 80 A/cm^2. I-V characteristics were measured while passing step-current pulses through the sample. The duration of pulse (typically 0.5 sec) was selected in order to avoid significant heating of the sample. A 50 msec acquisition rate was employed for automatic data recording.

Irradiation experiments, at room temperature, were performed in vacuum at the APT scattering chamber, [29], of the 7 MeV Van de Graaf CN accelerator of INFN Laboratori Nazionali di Legnaro. For proton irradiation a 6.5 MeV beam at fluence of about 10^{17} p/cm^2 was used.

Here we present results concerning two different irradiation runs. For the first run all the sample were simultaneously irradiated and the primary proton beam was scattered through a 120 $\mu g/cm^2$ thick gold foil, so that the fluence on the specimens was about 10^{13} p/cm^2. In the second run, the samples were successively exposed to the primary beam. Proton beam fluence was determined by considering charge integration of the proton spectrum recorded at 45 degree by a Silicon Surface Barrier

Detector (1000 μm depth). The proton beam affects a surface layer on the bulk about 200 μm thick.

3. RESULTS

3.1 - First run:

As before mentioned, in the first run the samples were simultaneously irradiated at a fluence of 10^{13} p/cm^2. For two out three proton irradiated specimens, an enhancement of the critical current density was found at 77 K. The first one, (sample #9, YBCO) showed a transition from a semiconducting behavior to a superconducting one. This fact is reported here only for completeness and will be discussed only after further experimental work. Sample #6, YBCO, as well as BISCO 2, showed an increase of the critical current density. For one sample (#11, YBCO) the expected enhancement was not found. On the contrary, the deeply damaged surface introduced an average degradation of superconductive properties of the whole specimen, totally similar to the one obtained by low fluence neutron irradiation [19]. this behavior could be attributed to unfavorable interaction with pre-irradiation defects. Also the variation of the I-V characteristic show the presence of a damaged surface layer and reflects different response to the irradiation. For sample #6 and #11 the widths of R vs. T transition curves were slightly affected, while the measured heights remained nearly unchanged, due to presence of large regions of undamaged material.

To stress the low resistance behavior of the two quoted specimens, the data are plotted on a logarithmic scale as a function of 1/T (Fig.1). This Arrhenius plot shows that, below 10^{-4} ohm, at 1% of the normal state resistance regime, dominated by thermally assisted flux flow,[9,30,31] is differently affected by the radiation. We assume an average thermal activation energy U, (temperature independent) in the framework of Anderson's general expression for the electric field in a limiting case at $J \ll J_c$.

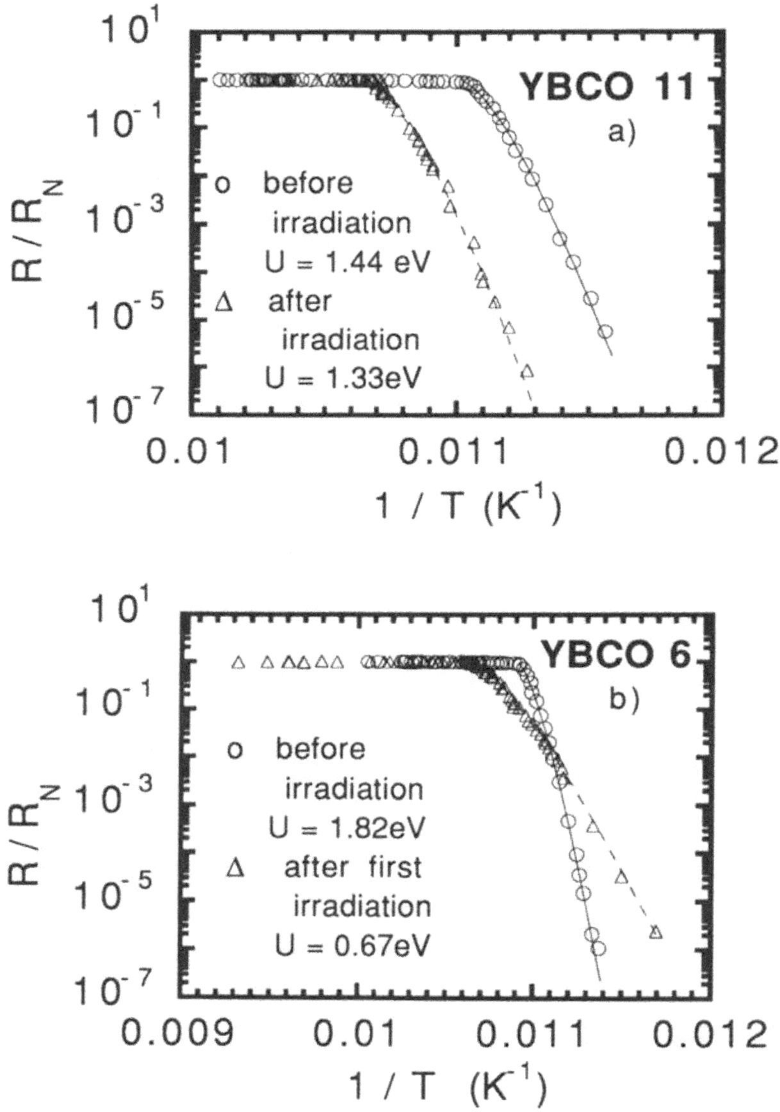

Fig.1: Arrhenius plot of the electrical resistivity before and after irradiation. a) sample #11; b) sample #6

We fit the normalized resistance, following the model proposed by Brandt [9] as

$$R/R_n = C(U/kT)\exp(-U/kT)$$

where C and U may be interpreted as phenomenoligical parameters. Several microscopic interpretations of these parameters have been proposed [32]. They depend on the considered pinning models.

In our particular experimental conditions, these parameters are useful as terms of comparison between an average jump rate of flux lines (or correlated volumes or "bundles" of flux lines) over pre-irradiation and after-irradiation pin structure. The constant values found in this work could be interpreted as temperature averaged values of a more general activation energy U(T).

In Figures [1a] and [1b] the values obtained in zero field before and after irradiation are reported. The damaged sample (YBCO #11) presents a nearly constant value of the energy barrier U, while the specimen (YBCO #6) shows a non expected strong decrease.

We argue that a larger activation energy stems from the presence of disordered extended defects. In contrast, the dominant "good" pins in improved sample are speculated to be implanted, well distributed, point-defects, due to extensive cross substitution resulting in much smaller activation energies, i.e. we think that in this case a difference in the dominant defect structure is the origin of the difference in pinning energies.

The I-V characteristics at temperatures near the superconducting transition exhibit power dependence $V=AI^{a(T)}$ in the low current region, which is usually interpreted on the basis of the Kosterlitz-Thouless model (KT) [33]. The I-V characteristics of the most improved sample #6, before and, respectively, after irradiation are shown in Fig. 2a and 2b. The critical current density enhancement, is localized in the upperlaying paths (about 200 μm thick) that short-circuits the underlaying not irradiated bulk. In the upperlaying layer the critical current density enhancement is certainly stronger than the improvement of the current in the sample. In fact, taking into account that

$$J_c/J_{co} = (\sigma_o I_c)/(\sigma I_{co})$$

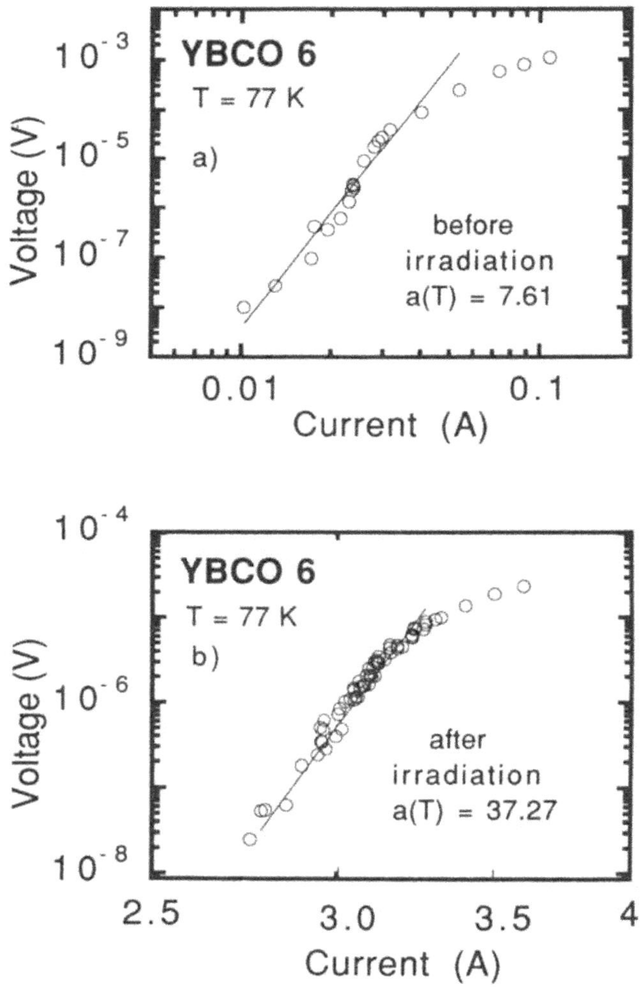

Fig.2: I-V characteristic fitted with a power-law function $I^{a(T)}$ in the low current region.for
a) sample #6, before irradiation
b) sample #6, after irradiation

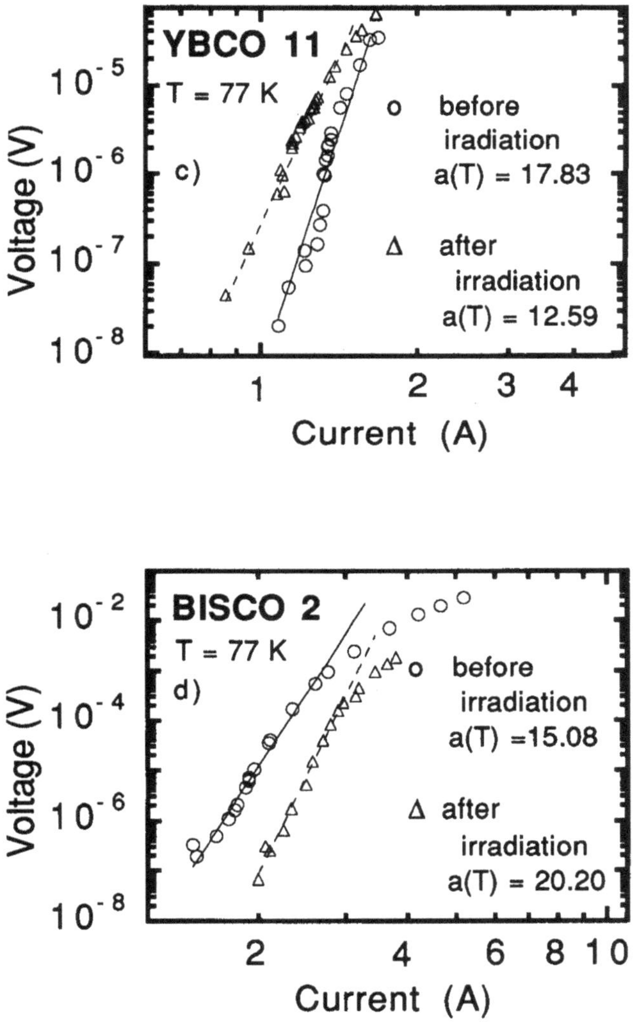

Fig.2: I-V characteristic fitted with a power-law function $I^{a(T)}$ in the low current region for
c) sample #11 before and after irradiation
d) sample BISCO 2 before and after irradiation.

where σ_0 and σ are the unaffected and irradiated cross sections, respectively, the critical current enhancement has a magnitude order grater than 10^2, being $\sigma_0/\sigma > 2$.

In Fig.2c the I-V characteristics at 77K of the damaged sample #11 are shown.

On the basis of the power law fit used in the low current regime, we observe a noticeable increase of the exponent for sample #6, while for sample #11 a decrease is found. In the framework of the KT theory an increase of a(T) is due to the increased strength of the interaction between vortex and antivortex. In the opposite situation, the dissociation of the vortex-antivortex pairs, whose unbinding gives rise to resistive dissipation, is enhanced.The positive results are confirmed also by the BISCO sample. In Fig. 2d the I-V characteristics at 77K before and after irradiation are shown. The logarithmic fit, shows that the exponent a(T) ranges from 15 to 20 and the critical current enhancement is about 3/2. Taking into account that the thickness of the cross section of the bulk is about 2 mm, and the implanted layer is about 200 μm deep, the enhancement of the critical current density has at least a magnitude order of 15.

The critical currents were also measured for several temperatures in the range 77-90K. The temperature dependence seems to be characterized by an exponential factor [34]. The exponential fits $J_c(T)=J_c(77)\ \exp(-b/T)$, before and after irradiation, respectively, are given in Figures 3a and 3b.

The $J_c(T)$ values ($T > 77K$) for sample #11 surprisingly seems to be slightly enhanced by the irradiation, while for sample #6 a depression of the performance is observed.

Qualitatively we can observe that in our samples the affected and unaffected regions are not equally weighted, so that, as the temperature increase, the superconducting upperlaying percolation pathways do not dominate any more, and the behaviors are more affected by self-field effects and generally by a more disordered situation.

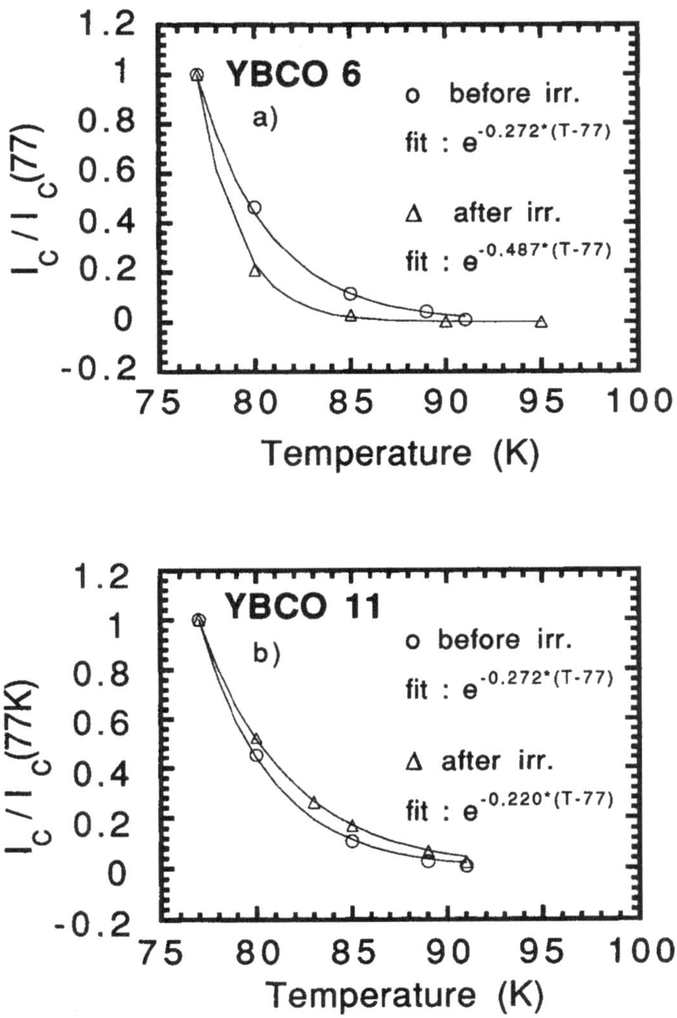

Fig.3 Normalized critical currents vs temperature curves with exponential fit
a) sample #11 before and after irradiation
b) sample #6 before and after irradiation

3.2 - Second run

In the second run the samples were successively exposed to the primary 6.5 MeV beam at fluence of 10^{17} p/cm^2. The results concern a successive irradiation of sample #6, and the irradiation of a new sample #22, whose superconductive properties have been compared with those of a twin sample R1 taken as reference.

Sample #6, again exposed to the flux, has reacted with a "damage" with respect to the previous irradiation, while all the parameters: critical current density, vortex-antivortex strength-coupling, activation energy assumed values intermediate between the unirradiated conditions and previous-irradiation values.

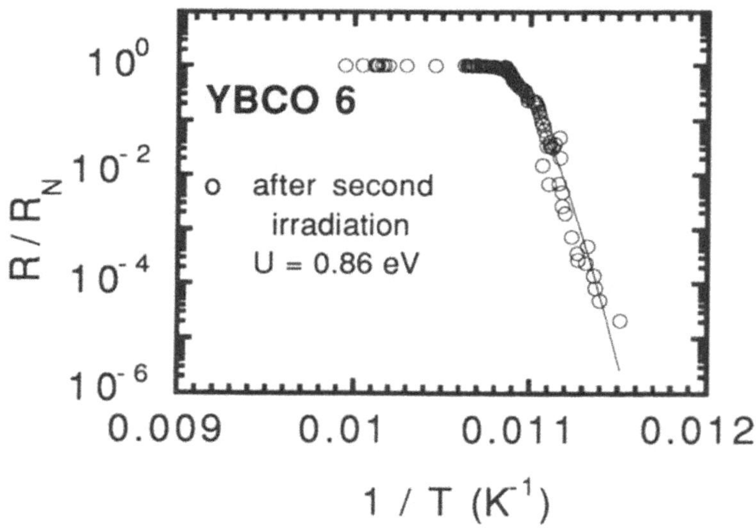

Fig.4 Arrhenius plot of the sample #6 after the second irradiation

The Arrhenius plot of the sample after the second irradiation and the I-V characteristics are shown in Figures 4 and 5, respectively.

The calculated average pinning energies for sample #6 are inversely proportional to the critical current density. This fact is not in accordance with the usual pinning models in which J_c, being the topological and structural parameters constant, should scale as U. On the contrary, the KT parameters a(T) that depends on the vortex-antivortex coupling strength, scale as the experimental values of J_c, as reasonably predictable. We argue that a larger activation energy stems from the presence of disordered extended defects, the induced "big" defects shadow "good" pins in the irradiated samples and are implanted as well distributed point-defects due to extensive cross substitution resulting in much smaller activation energies.

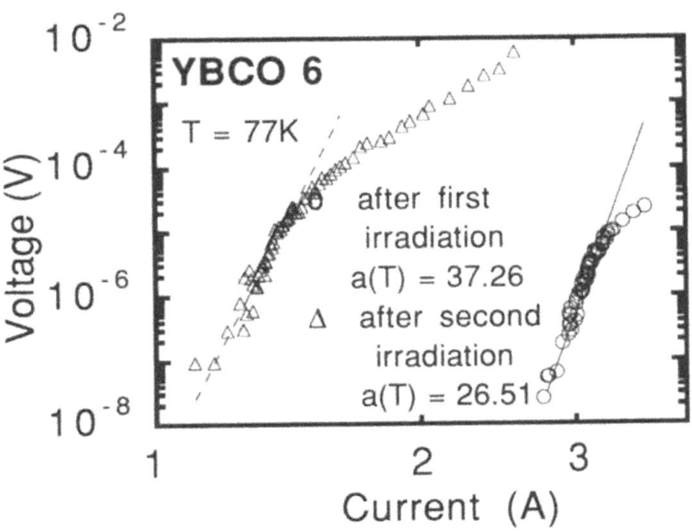

Fig.5 I-V characteristic of the sample #6 after the first and the second irradiation

However, another stimulating hypothesis, indirectly supported by the fits with the KT model, is that the enhancement is not only due to pinning effects but also to an up to now uncontrollable influence of the implanted defects on the coherence length in the sample.

Finally, we irradiated with the 6.5 MeV proton beam at the fluence of 10^{17} p/cm^2 a thin pellet of YBCO (sample #22), and we compared his properties with a reference twin pellet (sample R1), synthered in the same conditions as the first one, both having before irradiation a critical current density of 80 A/cm^2. As regards R vs T transition, a decrease of the amplitude of the transition occurs (Fig 6).

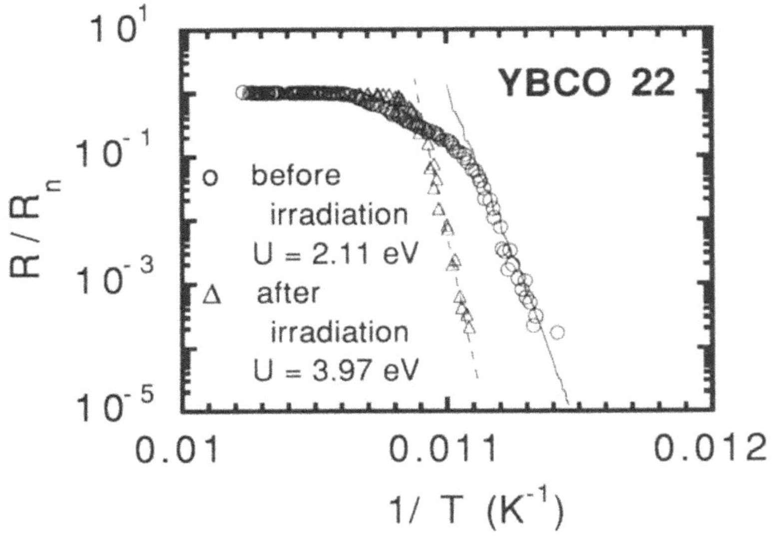

Fig.6 Arrhenius plot of the sample #22 before and after irradiation

The Arrhenius plot shows a sharp increase of the parameter U.
At 83K, the enhancement of the critical current density in the layer, taking into account the geometry of the sample, was of about 60. The two curves, before and after irradiation, are shown in Fig 7 All

superconducting properties results noticeably enhanced by the radiation.

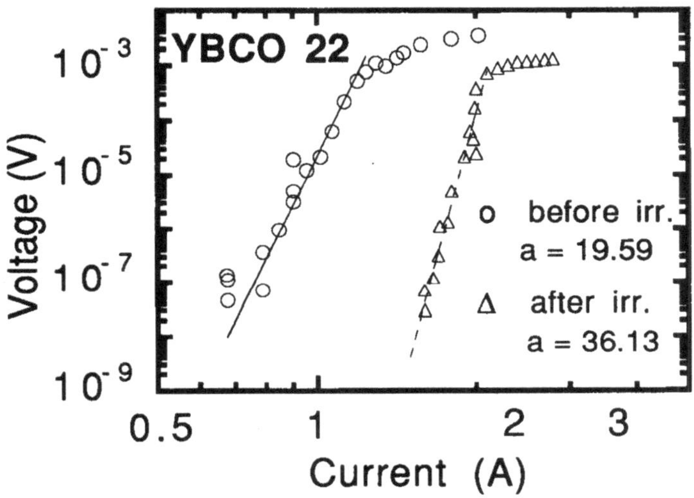

Fig.7 I-V characteristic of sample #22 before (Δ) and after (o) irradiation

4. CONCLUSIONS

In spite of the fact that these experiments on implanted layers over bulk are still at an early stage, the possibility to radiatively modulate the properties of superconducting layers by means of implanted defects is proven.

It seem likely that the enhancement is not only due to an enhanced pinning behavior: it is possible to infer that topological and structural modifications take place.

Systematic but very simple experiments, involving as few parameters as possible, will be needed; single crystal, film and layer -on-bulk behaviors must be compared.

REFERENCES

1) Y.Tokura, J.B.Torrance, T.C.Huang and A.I.Nazzal, Phys.Rev B 38 (1988), 7156
2) J.D.Jorgensen, D.G.Hinks, H.Shaded, B.Dabroiwski, B.W.Veal, A.P.Paulikas, L.J.Nowichi, L.H.Nunez, B.D.Dunlap, G.U.Segre and C.W.Kimball, Physica B 156 - 157 (1989), 877
3) D.De Fontaine, G.Ceder and M.Asta, Nature 343 (1990), 544
4) R.J.Cava, B.Batlogg, J.J.Krjewski, R.C.Farrow, A.E.White, K.T.Short, W.F.Peck and T.Y.Kometam, Nature 332 (1988), 814
5) D.G.Hinks, D.R.Richard, B.Dabrowski, A.W.Mitchell, J.D.Jorgensen and D.T.Marx, Physica C156 (1988), 477
6) J.D.Jorgensen, B.Dabrowski, S.Pei, D.G.Hinks, L.Soderholm, B.Morosin, J.E.Venturin and D.S.Ginley, Phys,Rev, B 38 (1988), 11337
7) J.Zhou, S.Sinha and J.B.Goodeniugh, Phys.Rev B 39 (1989), 12331
8) C.Chaillout, S.W.Cheong, Z.Fisk, M.S.Lehnann, B.Morosin and J.E.Schirder, Phisica C 158 (1989), 183
9) E.H.Brandt, Z.PhysiK B 80 (1990),167
10) A.Singh, N.Chandrasekhar and A.H.King, Acta Crystl. B 46 (1990),117
11) Y.Zhu, H.Zhang, H.G.Xang and M.Suenager, J.Mater.Res. 6, (1991),12
12) Z.Yi, S.Ashworth, C.Beduz and R.G.Sourlock, IEEE Trans.Magn MAG 27 (1990), 1506
13) J.Orehotsky, H.Wiesmann, A.R.Moodenbaugh, M.Suenager, H.G.Wang and H.Herman, IEEE Trans.Magn MAG 27 (1990),914
14) S.E.Babcock, X.Y.Cai, D.L.Kaiser and D.C.Larbalestier, Nature 347 (1990),167
15) G.Blatter and J.Rhyner, Proc. of the Satellite Symposium IV on High Temperature Superconductors (7th CIMTEC), Trieste, Italy, edit by P.Vincenzini (1991),21
16) J.W.Ekin, Phys.Rev. B 12 (1975),2627
17) W.J.Gallagher, T.K.Woorthington, T.R.Dinger, F.Holtzberg, D.L.Kaiser and R.L.Sanstrom, Physik 148B (1987),228
18) G.W.Crabtree, W.K.Kwok and A.Umerawa, Quantum Field Theory as an Interdisciplinary Basis, edit by F,C.Khanna,

H.Umezawa, G.Kunstatter and H.C.Lee (World Scientific, 1988)

19) D.Andreone, R.Cherubini, E.Mezzetti and B.Minetti, Il Nuovo Cimento, 12D (1990),1017

20) E.Mezzetti. D.Andreone, G.Castagno, R.Cherubini, R.Gerbaldo, B.Minetti, Proc.of the ICMC'90 Topical Conference on Materials Aspect of High-Temperature Superconductors, Garmisch-Partenkirchen, edit. by H.C.Freyhardt, R.Flukiger, M.Peuckert (1990),1079

21) E.M.Chudnovsky, Phys.Rev.Lett 65 (1990), 3060

22) H.W.Weber and G.W.Crabtree, Studies of high Temperature Superconductors, edit by A.V.Narlikar (Nova Science, NewYork, 1991). in press

23) M.A.Kirk, in Defects in Marerials, edited by P.D.Bristowe, J.E.Epperson, J.E.Griffith and Z.Liliental-Weber, (Mater.Res.Soc.Symp, Proc.209, Pittsburg, PA, 1991),743

24) E.Mezzetti, R.Gerbaldo, D.Andreone, F.Abbattista, M.Vallino, R.Cherubini, Proc. of the Satellite Symposium IV on High Temperature Superconductors (7th CIMTEC), Trieste, Italy, edit by P.Vincenzini (1991),459

25) M.Murakami, M.Morita, D.Doi and K.Miyamoto, Jap.J.Appl.Phys.28 (1989),1189

26) F.M.Sauerzopf, H.P.Wiesinger, W.Kritscha, H.W.Weber, G.W.Crabtree and J.Z.Liu, Phys.Rev. B 43 (1991), 3091

27) M.A.Kirk, M.C.Frishherz, J.Z.Liu, L.R.Greenwood and H.W.Weber, Phil.Mag.Lett. 62 (1990),41

28) F.Abbattista, D.Andreone, R.Cherubini, E.Mezzetti, M.Vallino, Proc. 3rd Nat. Meeting "HTC", Genova, Italy, (1990),287

29) R.Cherubini, G.Moschini, R.Nino, R.Policroniades and A.Varela, Nuclear Instrum. Meth A269 (1988),623

30) D.Dew-Hughes, Cryogenics 28 (1988),674

31) P.H.Kes, J.Aarts, J.van deu Berg, C.J.van der Beek and J.A.Mydosh, Supercond.Sci.Technol.,1 (1989),242

32) T.T.M.Palstra, B.Batlogg,L.F.Schneemeyer, J.V.Waszczak, Phys.Rev.Lett.61 (1988),1662; T.T.M.Palstra, B.Batlogg, R.B.van Dover, L.F.Schneemeyer, J.V.Waszczak, Appl.Phys.Lett.54 (1989),763; Phys.Rev.B 41 (1990),6621

33) J.M.Kosterlitz and Thouless, J.Phys.C 6 (1973),1181

34) R.M.Boysel, A.D.Caplin, M.N.B.Dalisius and C.N.Guy, Phys.Rev. B27 (1983),554

Magnetic Shielding by High-T_c Superconductors: Shielding Effects of YBCO Cylinders

Mineo Itoh

Department of Electronic Engineering,
Faculty of Science and Technology, Kinki University,
3-4-1, Kowakae, Higashi-Osaka, Osaka 577, JAPAN

ABSTRACT

The critical current density (J_c) of a high-T_c superconductor is an important parameter in magnetic shielding. However, the value of J_c has no close relationship to the degree of magnetic shielding. For the best performance, magnetic shielding requires special techniques related to the configuration of the shield. In this paper, the magnetic shielding is shown to be improved with the use of three special techniques. Experimental results reveal good characteristics of the shielding process. They include, among others, the effect of the excitation magnetic field on the inner magnetic field within the cylinders, the temporal change of the trapped magnetic field in the YBCO cylinders, and the distribution of the magnetic shielding field along the axial direction of the YBCO cylinders. The present paper also examines ways to improve the superconducting characteristics of the YBCO cylinder.

1. Introduction

Magnetic shielding is required in many applications to reduce the influence of the environmental magnetic field, and to protect the environment from the strong leakage of a generated magnetic field. There has been an increasing need for arrangements of magnetic shields capable of limiting regions to very low magnetic fields, less than 10^{-10} T. Examples of such needs are the measurements of magnetic fields in biomagnetic fields[1)] and Johnson noise thermometry[2)], making use of a SQUID magnetometer, which is a very sensitive, and low noise detector.

The environmental magnetic fields can be classified into two groups: (1) high-frequency fields due to electromagnetic waves, and (2) low-frequency fields due to magnetic fields. The magnetic shielding for high-frequency fields is effective and easily accomplished by using a low resistance metal, which results in a loss of eddy current. The shielding for low-frequency fields is provided by two different types. The first is passive magnetic shielding by a ferromagnetic material[3]. The second is active magnetic shielding with a superconducting material[4], which is suitable for shielding of very low frequencies such as the $1/f$ fluctuation, where f is the frequency. In fact, by using high-T_c superconducting shielding it is possible to consider measurements in a region containing a zero field at frequencies from the direct current up to 10^{11} Hz[5]. Magnetic shielding composed of both materials create a synergistic effect[2] in the measurements of very low magnetic fields by a SQUID magnetometer.

The author has been studying the fabrication of superconducting YBCOs for the purposes of applications to magnetic shielding, magnetic bearings, and electronic sensors from two different approaches. The first[6,7] is the development of coating techniques for high critical temperature (T_c) and critical current density (J_c) thick films which will enable the construction of complicated-shaped magnetic shields without the machining of brittle superconductors. The second[8-10] is improvements in the mechanical and electrical properties of superconductors which have high mechanical strength, in order to withstand machining, as well as maintaining high superconducting characteristics. This paper is focused on the latter approach.

The J_c value of high-T_c superconductors is one of the important factors in magnetic shielding. Improvement of the J_c value of YBCO has been achieved, in the past, by adding Ag powder to the pre-sintering YBCO powder[10]. The J_c value of a cylinder with 7 wt% Ag is about 9 times greater than a cylinder without Ag. However, the maximum magnetic shielding field of a cylinder with 7 wt% Ag was only determined to be approximately 20 percent higher than a cylinder without Ag[11], such as shown in Fig. 2(b). Further, the shielding effect is not closely related to the value of J_c.

For the best performance, magnetic shielding requires special techniques related to the configuration of the shield. Among these are the following: (1) the use of a superconductor (bulk, film, tape, fiber, or powder) as a supplement in the superconducting cylinder, (2) the superposition of one superconducting cylinder on another, (3) the superposition of a magnetic cylinder on a superconducting cylinder, and (4) the relationship between the diameter and the length of the superconducting cylinder. In this paper, magnetic shielding is shown to be improved by the special techniques of (1), (2), and (3).

The present paper also examines ways to improve the superconducting characteristics of the YBCO cylinder, such as the critical current density (J_c), and the critical temperature (T_c).

2. Evaluation of Magnetic Shielding Effects

The evaluation of magnetic shielding effects can be accomplished by two methods which are the evaluation of the shielding vessel in a homogeneous magnetic field[9], and the evaluation of a portion of the shielding vessel by a heterogeneous magnetic field[12]. It is considered here to use a DC magnetic field, an AC magnetic field, and pulse magnetic field for conducting the evaluations. Therefore, there are different methods used to carry out this evaluation. In order to simply the theoretical analysis, the author has been evaluating the shielding effects of a superconducting cylinder in a homogeneous DC magnetic field (B_{ex}), either perpendicular or horizontal to the axial direction of the superconducting cylinder, as follows. The effects of the excitation magnetic field (B_{ex}) on the inner magnetic field (B_{in}) within the cylinders, the temporal change of the trapped magnetic field (B_t) in YBCO cylinders, and the distribution of the magnetic shielding field (B_s) in the cylinders, were performed as shown in Fig. 1.

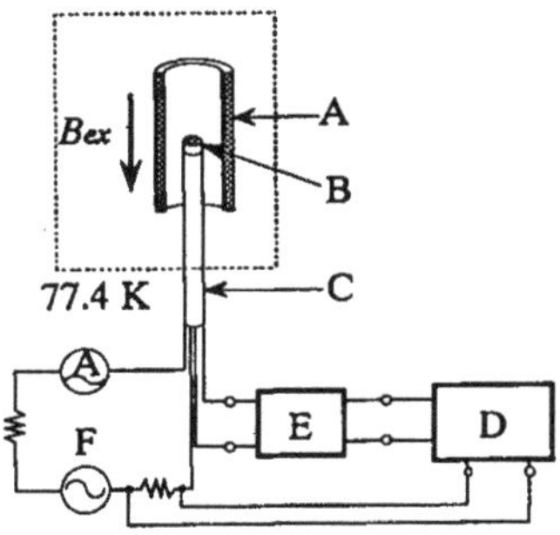

A : YBCO cylinder
B : Calibrated GaAs Hall device
C : Holder of Hall device
D : Lock-in amplifier
E : Pre-amplifier
F : Oscillator

Fig. 1. Schematic diagram showing the apparatus used to measure the degree of magnetic shielding of the superconducting YBCO cylinder.

In figure 1, a schematic diagram shows the apparatus used to measure the magnetic shielding effects of the superconducting YBCO cylinder. The values of B_{in} within the cylinder were measured using a calibrated GaAs Hall device (Toshiba Co., Ltd., THS 102A)[13] operated at the boiling point of liquid nitrogen (77.4 K). The control current in the Hall device was a constant AC current (2 kHz) of 1 mA, while the Hall device was measured by a lock-in amplifier, as shown in Fig. 1. The sensitivity of the measuring system was 100 mV/gauss.

3. Characteristics of YBCO Cylinders

3-1. Fabrication of YBCO Cylinders

The material used to form the superconducting cylinders was a mixture prepared by homogeneously mixing and grinding prescribed amounts[11] of Y_2O_3, $BaCO_3$, and CuO powders, all of which were 99.99 % pure, into an agate mortar. The mixture was calcined in a flow of very pure air (18 l/min, N_2:O_2=79:21, both of which were 99.999 % pure) to a temperature of 925 °C at the heating rate of 10 °C/min. The mixture was kept at 925 °C for 4 hours and then cooled to room temperature at the cooling rate of about 1.5 °C/min[8]. The calcined powder was again ground with either 0 or 7 wt% Ag powder added to a portion of the ground powder. With the use of a compressor (0.1 GPa) the powder was pressed into the shape of rods measuring 19 mm in length and 9 mm in diameter. The cylinder (5 mm inner diameter) was fashioned from a rod which was bored at a low speed using a carbide cutting tool. The sintering process for the cylinders was the same as that use in the calcining process[8]. The durations of the air flowing in the calcining and sintering processes were continued for 6 hours and 4 hours at temperatures above 600 °C, respectively. The values of the critical current density (J_c) and critical temperature (T_c) decrease with an increase of air impurities[8,14]. X-ray photoelectron spectroscopy (XPS) analysis has shown that low J_c and T_c materials contain much higher amounts of carbon compounds than high J_c and T_c materials[9]. For this reason, very pure air was used in the present experiment.

3-2. Superconducting Properties of YBCO Cylinders

The pellets (8 mm diameter, 1.0 mm thickness) were cut into rectangular-shaped test specimens (1.3 mm width) with the use of a diamond saw at low cutting rates to inhibit the effects of heat. The fabrication process of the specimens was the same as that described for the cylinders. Ag powder (99.99 % pure) was mixed into the calcined compound, with the Ag content ranging from 0 wt% to 28 wt%, having a grain size of about 20 μm or smaller. From X-ray diffraction analysis[11], the crystal structures of the present specimens, with and without Ag, were confirmed as $Y_1Ba_2Cu_3O_7$[15]. The four-terminal method was used for the resistance measurements of specimens. The measurements of temperature dependence of resistance for specimens were obtained using a calibrated Pt thermometer (Teijin Engineering Co., Ltd., PTF-7) at temperatures ranging between room temperature (300 K) and the boiling point of liquid nitrogen (77.4 K). The current density and resistivity were simply defined as the amount of current and resistance divided by the cross-sectional area of each specimen, respectively.

Table 1 lists the results of specimens with 0, 3, 5, 7, 10, and 14 wt% Ag. These results are for the temperature of the resistance drop (T_{cd}), the temperature of zero resistance (T_{cz0}) with 0 gauss, the temperature of zero resistance (T_{cz50}) with 50 gauss, the resistivity (ρ_{100}) at 100 K, the critical current density (J_{c0}) at 77.4 K with 0 gauss, and the critical current density (J_{c10}) at 77.4 K with 10 gauss[10]. The magnetic fields were applied perpendicular to the specimens. The compression pressure (P) of all specimens was 0.1 GPa. The J_{c0} was increased in proportion to the compression pressure (P , 0.05 - 1.0 GPa) of the specimens without Ag, and was decreased proportional to P of specimens with Ag. The T_{cz0} temperature of specimens with Ag was increased as the pressure (P, 0.05 - 0.25 GPa) decreased, but J_{c0} was improved markedly with a Ag content (3 - 14 wt%). The T_{cd} of specimens with and without Ag was insensitive to P[10]. The mechanism leading to the critical current density enhancement by the addition of Ag powder could be that the suitable amount of Ag powder (3 - 10 wt%) is the source of an oxygen supply for the specimens during sintering[16], and also acts as a binder among the grains of the specimen. However, the present experimental results reveal no direct proof of the above mechanism, and further studies to confirm the existence of such a mechanism should be performed.

Table 1. Properties of YBCO cylinders with and without silver.

silver (wt%)	0	3	5	7	10	14
T_{cd}(K)	94.8	96.8	96.8	95.3	95.2	95.1
T_{cz0} (K) at 0 gauss	91.6	94.5	94.6	93. 4	91.6	89.4
T_{cz50} (K) at 50 gauss	85.9	90.2	90.2	90.0	—	86.0
ρ_{100} (μΩm) at 100 K	15.8	7.2	7.6	5.6	7.3	8.5
J_{c0} (A/cm^2) at 0 gauss	51.0	306	357	464	350	185
J_{c10}(A/cm^2) at 10 gauss	46.0	—	—	264	—	88.0

(The compression pressure in forming the specimens was 0.1 GPa.)

3-3. Shielding Effects of YBCO Cylinders

Figure 2 (a) shows the configuration of the superconducting YBCO cylinders (5 mm inner diameter, 9 mm outer diameter, 19 mm length) with and without 7 wt% Ag.

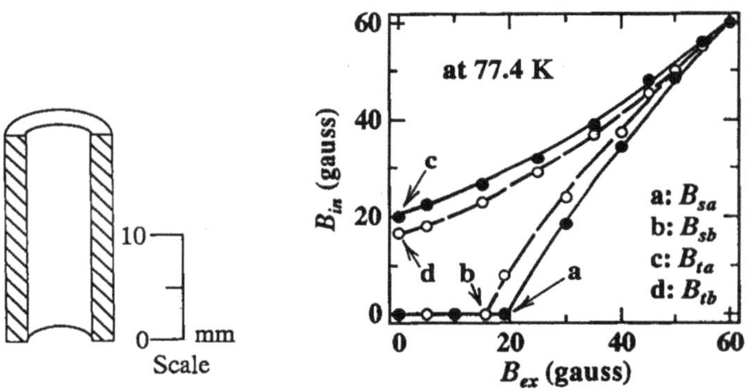

(a) (b)

Fig. 2. Typical characteristics of the magnetic shielding in the center of the cylinders (5 mm inner diameter, 9 mm outer diameter, 19 mm length) at 77.4 K. These are the characteristics of the excitation magnetic field (B_{ex}) on the inner magnetic field (B_{in}) in the center of the cylinders. (a) Typical configuration of the cylinder. (b) Solid line (solid circles) is characteristic of a cylinder with 7 wt% Ag, while the dashed line (open circles) is characteristic of a cylinder without Ag.

Figure 2 (b) shows a typical response of magnetic shielding in the center of the cylinders at 77.4 K, which are the effects of the excitation magnetic field (B_{ex}) on the inner magnetic field (B_{in}) within a center of the cylinders, when B_{ex} is applied horizontal to the axial direction of the cylinders. The solid line is characteristic of the cylinder with 7 wt% Ag (solid circles), while the dashed line is characteristic of the cylinder without Ag (open circles). The B_{sa} and B_{sb} of cylinders with and without 7 wt% Ag, the maximum magnetic shielding fields, are 19 gauss and 16 gauss, respectively. The J_c of a cylinder with 7 wt% Ag is about 9 times greater than a cylinder without Ag such as shown in Table 1. However, the B_{sa} within the center of the cylinder with 7 wt% Ag was approximately 20 percent higher than the B_{sb} within a cylinder without Ag[11], such as shown in Fig. 2 (b). Further, the maximum magnetic shielding field was not closely related to the value of J_c.

The damping ratio of the magnetic field in the center of the cylinders with and without 7 wt% Ag was found to be below 10^{-4} using the measuring system shown in Fig. 1, when the excitation magnetic field (B_{ex}) is kept constant at 10 gauss. The value and fluctuation of magnetic fields, less than 10^{-3} gauss, in the cylinder are now being investigated by using a RF SQUID magnetometer.

Figure 3 shows the temporal changes of B_{ta} and B_{tb}, trapped magnetic fields, as shown in Fig. 2 (b), in the center of YBCO cylinders with and without 7 wt% Ag at 77.4 K. The solid line (solid circles) and dashed line (open circles) are B_{ta} and B_{tb}, respectively. B_{ta} is very stable over the 100 hours, but for B_{tb} several values show a decrease with time. These characteristics will be important criteria in evaluating the magnetic shielding effect, and the properties of superconductors.

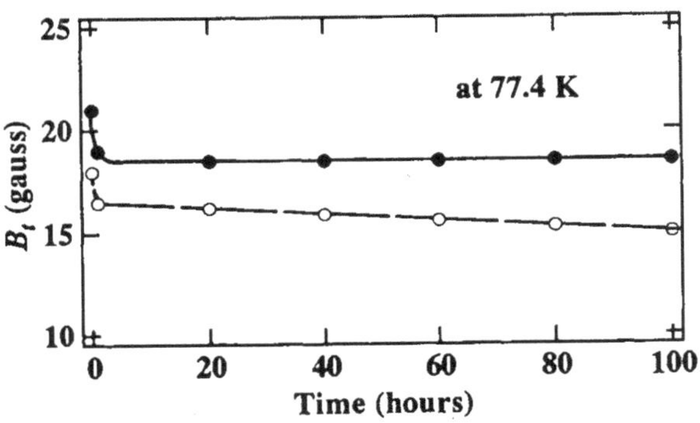

Fig. 3. B_{ta} and B_{tb} in the center of YBCO cylinders, with and without 7 wt% Ag trapped magnetic fields, versus time (hours) at 77.4 K. The solid line (solid circles) and dashed line (open circles) are B_{ta} and B_{tb}, respectively.

Figure 4 shows the distribution of maximum magnetic shielding fields (B_s) along the axial direction of the YBCO cylinder at 77.4 K.The solid line (solid circles), dashed line (open circles), and dash-dotted line (open squares) are characteristic of B_{sa}, B_{sb}, and B_{sc}, respectively. B_{sa} and B_{sb} in the cylinder with and without 7 wt% Ag are the maximum magnetic shielding fields when the excitation magnetic field (B_{ex}) is applied horizontal to the cylinder axis. B_{sc} in the cylinder with 7 wt% Ag is the result when the B_{ex} is applied perpendicular to the cylinder axis. The outside of the YBCO cylinder was wrapped with several turns of a fluoroplastic (PTFE) tape.

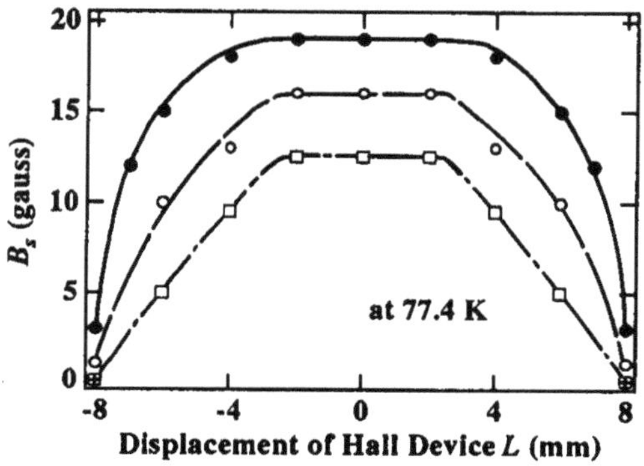

Fig. 4. Distribution of the maximum magnetic shielding fields, B_{sa}, B_{sb} and B_{sc}, along the axial direction of the YBCO cylinder at 77.4 K. The solid line (solid circles), dashed line (open circles), and dash-dotted line (open squares) are B_{sa}, B_{sb}, and B_{sc}, respectively. B_{sa} and B_{sb} are the results when the excitation magnetic field is applied horizontal to the cylinders. B_{sc} is the result when the field is applied perpendicular to the cylinder axis.

The cylinders were subjected to a thermal cycle between 300 K and 77.4 K a total of 120 times, but no remarkable changes in Fig. 2 (b), Fig. 3, and Fig. 4 were observed.

4. Improvements of Magnetic Shielding Effects

4-1. Use of Superconductor as Supplement : Supplemented Cylinder

Superconductors in the form of the bulk, film, tape, and powder can be considered as supplements for improving the magnetic shielding effect. In this paper, the use of a superconductor composed of a YBCO powder[11] as the supplement is examined in which the grain size of the YBCO particles was about 20 μm or smaller. Figure 5 (a) shows the configuration of the supplemented cylinder which consists of YBCO powder uniformly occupying the space between an innermost YBCO cylinder with 7 wt% Ag (6 mm inner diameter, 11 mm outer diameter, 36 mm inner length, 40 mm outer length) and a nylon cylinder (15 mm inner diameter, 17 mm outer diameter, 45 mm length).

Figure 5 (b) shows the distribution of the maximum magnetic shielding field (B_s) along the axial direction of the YBCO cylinder at 77.4 K when the excitation magnetic field (B_{ex}) was applied horizontal to the cylinder axis. The solid line (solid circles), dashed line (open circles), and dash-dotted line (open squares) are for the innermost cylinder having 7 wt% Ag, the supplemented cylinder such as shown in Fig. 5 (a), and a cracked cylinder as the innermost YBCO cylinder, respectively. The distribution of the shielding field of the supplemented cylinder when using a cracked YBCO cylinder exhibited the same as characteristics of B_s of the 7 wt% innermost cylinder. The shielding effect using the supplemented cylinder was improved by about 2 gauss when compared with the YBCO cylinder having 7 wt% Ag, and also a renovated cracked cylinder. The fabrication of supplemented cylinders with higher effective shielding are now being investigated by using superconductors composed of tape, fiber, and film.

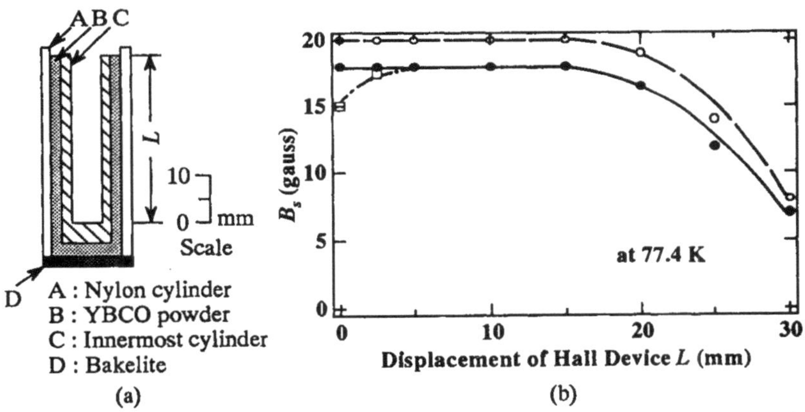

Fig. 5. Typical distribution of maximum magnetic shielding fields (B_s). (a) Typical configuration of the supplemented cylinder. (b) The solid line (solid circles), dashed line (open circles), and dash-dotted line (open squares) are the innermost YBCO cylinder with 7 wt% Ag, supplemented cylinder of Fig. 5 (a), and a cracked cylinder as the innermost cylinder, respectively.

4-2. Double Superconducting Cylinder

The double cylinder was constructed by superimposing a high-T_c Bi-Pb-Sr-Ca-Cu-O (BPSCCO) cylinder on a YBCO cylinder having 7 wt% Ag such as that shown in Fig. 6 (a). The YBCO cylinder in the double cylinder configuration is the same as that shown in Fig. 2 (a). The outside of both cylinders was wrapped with several turns of a fluoroplastic (PTFE) tape. The material for the BPSCCO cylinder was fabricated as described in references 12) and 17). The BPSCCO cylinder (17 mm inner diameter, 23 mm outer diameter, 123 mm length) was formed under pressure of 0.15 GPa using the cold isostatic pressing method[12, 17]. The critical current density (J_c) and critical temperature (T_c) of the BPSCCO cylinder were measured at approximately 150 A/cm^2 and 103 K, respectively.

Figure 6 (b) shows the distribution of the maximum magnetic shielding field (B_s) along the axial direction of the cylinders at 77.4 K when the excitation magnetic field (B_{ex}) was applied horizontal to the cylinder axis. The solid line (solid circles), dashed line (open squares), and dash-dotted line (solid squares) are the B_s for the YBCO cylinder with 7 wt% Ag content (B_{sa}), the BPSCCO cylinder (B_{sd}), and the double cylinder (B_{se}), respectively. The B_s in the double cylinder remained constant over the region of ± 4 mm from the center of the cylinder. On the other hand, the region of constant B_s for the single YBCO cylinder was ± 2 mm. The value of the maximum shielding field (B_s) for the double cylinder was about 1.8 times that of the single YBCO cylinder. The double cylinder was subjected to repeated thermal cycle between 300 K and 77.4 K 20 times, but no remarkable change in Fig. 6 (b) was observed.

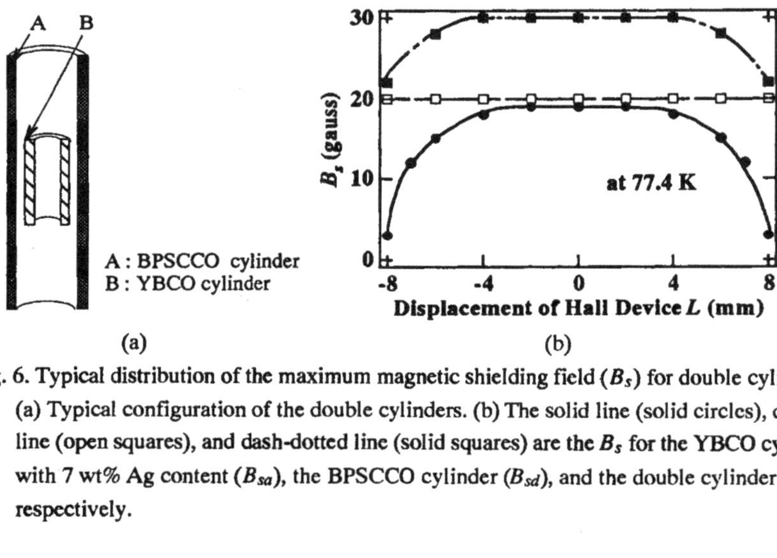

Fig. 6. Typical distribution of the maximum magnetic shielding field (B_s) for double cylinders. (a) Typical configuration of the double cylinders. (b) The solid line (solid circles), dashed line (open squares), and dash-dotted line (solid squares) are the B_s for the YBCO cylinder with 7 wt% Ag content (B_{sa}), the BPSCCO cylinder (B_{sd}), and the double cylinder (B_{se}), respectively.

4-3. Superposition of Magnetic Cylinder on Superconducting Cylinder : Superimposed Cylinder

Figure 7 (a) shows the configuration in which a magnetic cylinder is superimposed on a YBCO cylinder. The YBCO cylinder is the same as YBCO cylinder having a 7 wt% Ag shown in Fig. 2 (a). The magnetic cylinder (10 mm inner diameter, 12 mm outer diameter, 45 mm length) is made of soft iron. By using a strong AC magnetic field, the magnetic cylinder was degaussed at 77.4 K before taking measurements of the shielding effect. However, the soft iron cylinder had a residual magnetism of about 3 gauss. The relative permeability of the soft iron was measured as approximately 10 at 1 kHz at the boiling point of liquid nitrogen (77.4 K)[5].

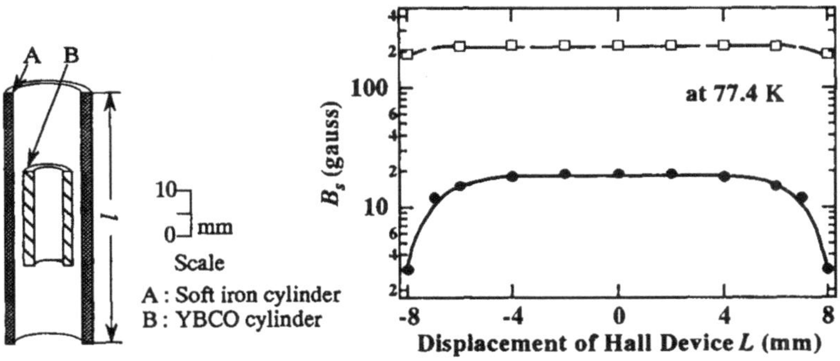

(a) (b)

Fig. 7. Typical distribution of the maximum magnetic shielding field (B_s) for the superimposed cylinders. (a) Typical configuration of the superimposed cylinders. (b) The solid line (solid circles), and dashed line (open squares), are the B_s for the YBCO cylinder with 7 wt% Ag content (B_{sa}), and the superimposed cylinder (B_{sf}), respectively.

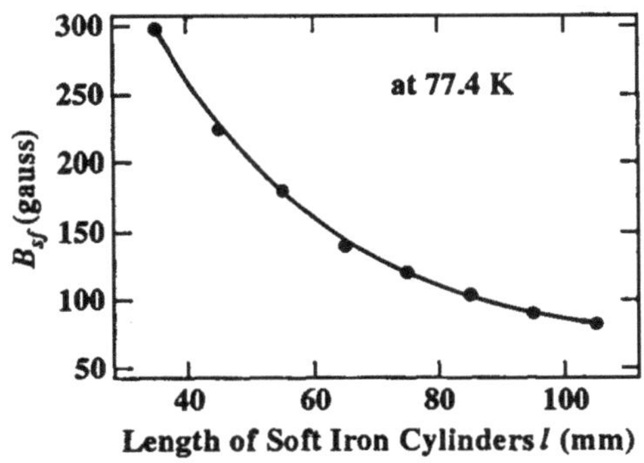

Fig. 8. Maximum magnetic shielding field (B_{sf}) in the center of the YBCO cylinder, within the superimposed cylinder as a function of the length (l) (10 mm inner diameter, 12 mm outer diameter) of the soft iron cylinder.

Figure 7 (b) shows the distribution of the maximum magnetic shielding field (B_s) along the axial direction of the cylinders at 77.4 K when the excitation magnetic field (B_{ex}) was applied horizontal to the cylinder axis. The solid line (solid circles), and dashed line (open squares) are the B_s for the YBCO cylinder with 7 wt% Ag content (B_{sa}), and the superimposed cylinders (B_{sf}), respectively. The value of the maximum shielding field for the superimposed cylinder is about 11 times greater than that of the single YBCO cylinder. The cylinders were subjected to repeated thermal cycles between 300 K and 77.4 K for 120 times, but no remarkable change in Fig. 7 (b) was observed.

Figure 8 displays the maximum magnetic shielding fields (B_{sf}) in the center of the YBCO cylinder with 7 wt % Ag as resulted to the length (l) of the outer magnetic cylinder (10 mm inner diameter, 12 mm outer diameter) shown in Fig. 7 (a) at 77.4 K. The B_{sf} decreases exponentially as l ranges between 35 mm and 110 mm. Therefore, to design a superimposed cylinder for magnetic shielding, it is necessary to determine the optimal ratio of the length to the diameter of the magnetic cylinder. The damping ratio of the magnetic field in the center of inner YBCO cylinders has been determined to be less than 10^{-5} by using the measuring system shown in Fig. 1, when the excitation magnetic field (B_{ex}) is kept constant at 150 gauss on the superimposed cylinders.

5. Summary and Conclusions

Superconducting cylinders must be uniform in their characteristics for good magnetic shielding when any position of a cylinder is sampled for inspection. In particular, if large sized superconducting cylinders are constructed an important problem will be to measure the uniformity of cylinder. Therefore, the present research group has been studying the measurements of uniformity (two dimensional) of high-T_c superconductor materials by using new equipment[18-20] to measure the repulsive force due to the Meissner effect. The method of fabrication of larger sized cylinders with greater effective characteristics of magnetic shieldings for wide applications is now under investigation. The values and fluctuations of magnetic fields of less than 10^{-3} gauss in the cylinder are now being investigated by using a RF SQUID magnetometer.

The magnetic shielding by YBCO cylinders can be summarized as follows:

(1) The J_c of a cylinder with 7 wt% Ag is about 9 times greater than a cylinder without Ag such as shown in Table 1. However, the B_{sa} in the center of the cylinder with 7 wt% Ag was approximately 20 percent higher than the B_{sb} in a cylinder without Ag[11], such as shown in Fig. 2 (b). Further, the maximum magnetic shielding field was not closely related to the value of the J_c.

(2) The temporal change of the trapped magnetic field (B_t) in the superconducting cylinder, such as in Fig. 3 is an important criterion in evaluating the magnetic shielding effect and the properties of superconductors.

(3) The cylinders were subjected to repeated thermal cycles between 300 K and 77.4 K for 120 times, but no remarkable changes in the results were observed.

(4) The shielding effect of the supplemented cylinder, using YBCO powder as the supplement, was improved by about 2 gauss when compared with the YBCO cylinder with 7 wt% Ag, and also a renovated cracked cylinder.

(5) The value of the maximum shielding field B_{se} in a double cylinder configuration was raised by about 1.8 times that of the single YBCO cylinder.

(6) The value of B_{sf} for the superimposed magnetic cylinder on a YBCO cylinder, was approximately eleven times that of the single YBCO cylinder.

6.Acknowledgements

The author would like to thank Mr. Takashi Ohyama for his cooperation in the conduction of the experiments. The author also wishes to acknowledge the valuable discussions with Dr. Hiroyuki Ishigaki, and Prof. Takumi Minemoto.

7.References

1. G. L. Romani, S. J. Williamson, and L. Kaufman, *Rev. Sci. Instrum.* **53**, (1982) 815.
2. M. Itoh, T. Mizusaki, T. Ohmi, M. Motomura, and A. Hirai, *Jpn. J. Appl. Phys.* **25**, (1986)1097.
3. L. Boesten, *Jpn . J. Soci. Instrum. Cont. Engi.* **17**, (1978) 107.
4. B. Cabrera, *Ph. D. Thesis*, Stanford University (1975).
5. T. Ohyama and M. Itoh, *IEEE Trans. on Magnetics* **28**, No.2 (1992), in press.
6. M. Itoh and H. Ishigaki, *Jpn. J. Appl. Phys.* **27**, (1988) L420.
7. M. Itoh and H. Ishigaki, *Proc. 12th Int. Cryogenic Engineering Conf.*, 1988, (Butterworths, Guildford, 1988) p.908.
8. M. Itoh, H. Ishigaki, and K. Demizu, *Jpn. J. Appl. Phys.* **28**, (1989) L1527.
9. For example, M. Itoh, H. Ishigaki, and T. Ohyama, *Cryogenics* **30**, Sept. Suppl. (1990) 863.
10. M. Itoh, H. Ishigaki, T. Ohyama, T. Minemoto, H. Nojiri, and M. Motokawa, *J. Mater. Res.* **6**, No. 11 (1991), in press.
11. M. Itoh, *OYO BUTSURI* (J. Jpn. Soci. Appl. Phys.) **60**, (1991) 478.
12. For example, K. Hoshino, H. Ohta, E. Sudoh, K. Katoh, S. Yamazaki, H. Takahara, and M. Aono, *Jpn. J. Appl. Phys.* **29**, (1990) L1435.
13. M. Itoh, *J. Fac. Sci. and Technol. Kinki Univ.* **25**, (1989) 355.
14. M. Itoh, H. Ishigaki, T. Ohyama, T. Minemoto, M. Motokawa, and H. Nojiri, *Proc. 11th Inter. Conf. on Magnet Technology*, 1989, (Elsevier, Essex, 1990) p.1476.
15. H. Eickenvusch, W. Paulus, and R. Schrüllhorn, *Mat. Res. Bull.* **22**, (1987) 1505.
16. For example, J. Moreland and L. F. Goodrich, *IEEE Tran. Magnetics* **25**, (1989) 2056.
17. T. Nakayama, H. Ohta, H. Takayama, K. Hoshino, K. Shigemitsu, E. Sudoh, S. Yamazaki, K. Katoh, H. Takahara, and M. Aono, *Proc. 2nd Int. Sym. Superconductivity*, (Springer-Verlag, Tokyo, 1990) p.709.
18. M. Itoh, H. Ishigaki, and A. Hida, *IEEE Tran. Magnetics* **25**, (1989) 2518.
19. H. Ishigaki, A. Hida, M. Itoh, and T. Minemoto, *Proc. 11th Inter. Conf. on Magnet Technology*, 1989, (Elsevier, Essex, 1990) p.1476.
20. H. Ishigaki, M. Itoh, A. Hida, H. Endo, T. Oya, T. Ohyama, and T. Minemoto, *IEEE Tran. Magnetics* **27**, (1991) 2427.

MAGNETIC SHIELDING OF LARGE VOLUMES AT ROOM TEMPERATURE BY MEANS OF HIGH T_C SUPERCONDUCTING MATERIALS *

Franco Pavese
Istituto di Metrologia "G.Colonnetti"
Consiglio Nazionale delle Ricerche
strada delle Cacce 73, 10135 Torino, ITALY

Domenico Andreone
Istituto Elettrotecnico Nazionale "G.Ferraris"
strada delle Cacce 91, 10135 Torino, ITALY

Marco Bianco
Istituto per le Ricerche di Tecnologia Meccanica e per l'Automazione SpA
10080 Vico Canavese, ITALY

ABSTRACT

The use of high Tc superconducting materials in magnetic shielding applications places specific problems, the number of which greatly increases when the shielded volume is large, and, additionally, when it must be kept at room temperature, as it is required in biomedical applications. Many solutions well suited for small shields fail when scaling up the shield size, for both technological and economical reasons. This paper briefly reviewes the available techniques for making coatings of superconducting ceramics on surfaces from the point of view of their suitability to solve the specific problem. Technological and structural problems in order to fabricate large shields are outlined. The preliminary results obtained with $YBa_2Cu_3O_{7-\delta}$ using the "Continuous Detonation Spray" technique are reported, which promise to lead to a very convenient solution of large shield fabrication.

* *Program supported by the National Research Council of Italy, under the Progetto Finalizzato "Superconducting and Cryogenic Technology" (SuCryTech)*

1. INTRODUCTION

The fabrication of superconducting shields faces quite different problems depending on whether the volume to be shielded is small or medium/large. Small volumes can easily and directly be made by using the bulk superconducting ceramic without the need of a supporting material. Moreover, the whole shield can easily be heat-treated, even at high temperature, without resorting to large furnaces, which are a big investment. The state-of-the-art of most of the available techniques already makes it possible to fabricate small shields.

Quite different the situation is when medium/large sizes are required, since the direct use of the bulk material is prevented by economical and practical reasons. The solution to be adopted critically depends on whether or not a thermal treatment at high temperature is necessary. In fact, the latter limits the maximum shield size or requires a big investment for a large high-temperature high-quality furnace. Unfortunately, the few coating methods which are able to produce directly a superconducting layer are only suitable for small-size surfaces, or their cost for producing medium-size surfaces —and thick layers— is very high.

Non self-supporting high-T_c materials need a substrate. The maximum heat-treatment temperature places constraints on the choice of the substrate material. Especially when treatment temperature is high, the substrate must accomodate for the superconducting ceramic thermal expansion. In all cases, it must accomodate for the thermal contraction of the superconducting ceramic when it is cooled from room temperature down to liquid nitrogen temperature. In addition, the substrate material must itself be compatible with the eavily oxidizing atmosphere required to the heat treatment. Finally, the substrate can poison the superconducting ceramic, affecting its properties, either during the high-temperature treatment or at room temperature.

For the fabrication of large shields, also a suitable mechanical structure has to be developed, taking into account all the cryogenic problems involved (thermal contraction of different materials, refrigeration technique to be adopted, ...) in addition to the electromagnetic problems common to the traditional shield design. At the end of this process, an overall shield project competitive from the economic point of view with the traditional ones should be obtained.

2. FABRICATION OF HIGH-T_C SUPERCONDUCTING SHIELDS

A flow chart of the decisional process underlying the design of a crogenic shield is shown in Fig.1. The first decision to be taken depends on whether a small or medium/large shield is needed. In fact, if the material preparation requires a heat treatment, this decision sets the dimensions required to the furnace, unless, after a coating process, the film can be thermally processed with a laser beam. Early tests performed by the authors with the laser technique showed that it is almost impossible to control the local temperature to a reasonable degree of accuracy, since the temperature value is completely determined by the *local* film-to-substrate heat-transfer coefficient, which can change from spot to spot by orders of magnitude depending on local irregularities of such parameters such as the film thickness, its morphology or adhesion to the substrate, especially when the aim of the thermal treatment is sintering. The heated surface temperature has been observed with a thermocamera and found to show point-to-point variations of hundreds of degrees.

If an overall thermal treatment must thus be performed on the shield, and a medium/large shield is needed, the only way to avoid the use of a large furnace is to build the shield in modules. Immediately, the need arises for the availability of a superconducting sealing media for joining hermetically the modules from external electromagnetic interferences.

Concerning the fabrication of the shield (module) itself, there is a variety of solutions possible, ranging from the use of unsupported bulk materials (in form of cylinders, cups, tiles) to the use of films (generally thick for this application, as the penetration field is thickness dependent[1]) coating a variety of substrates.

The use of the bulk material is limited to small shields, by economical reasons, though it has been proposed also for making one-piece large cylindrical enclosures, or it could be considered for making 1-3 mm thick tiles.

Every substrate material places some specific problems, which will be analyzed in the following. A variety of coating techniques are available; the main types are listed in Table I.

All the decisions about the shield technology must be taken before the refrigeration method and the mechanical structure for the enclosure fabrication can be selected and adopted, since the latter must match the specific problems arising from the former.

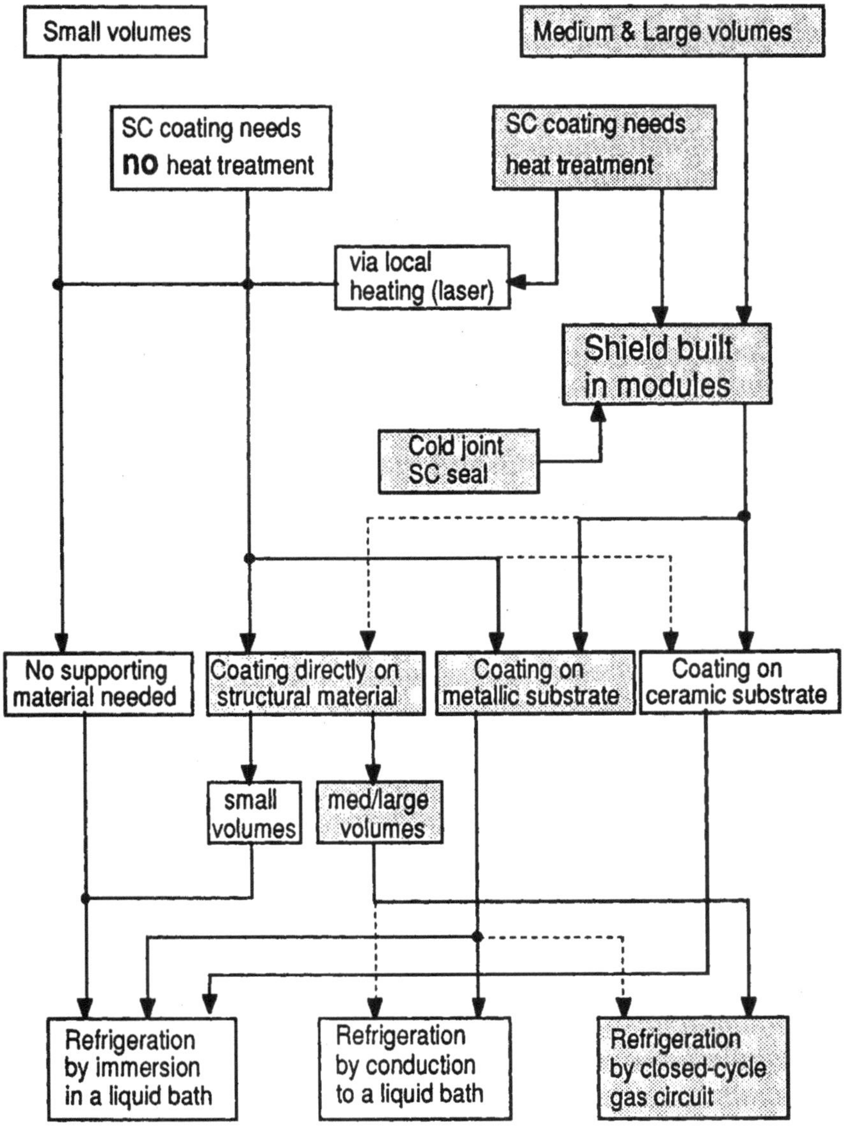

Fig.1 - Flow chart of the decisional process in designing a superconducting shield using high-T_C superconducting materials. Shaded boxes indicate authors' research path.

TABLE I

Selection of possible techniques for fabricating superconducting shields

Technique	Requirements and limitations
Bulk ceramic parts	Self-supporting. Cylinders: very limited size. Plates: very thick, very costly.
Electrophoretic coating	Low density of shielding material. High temperature sintering required. Small/medium modules. Silver thin sheet substrate.
Laser ablation	Thin layers. No subsequent heat treatment required. Very small targets.
Sputtering	Thin layers. No subsequent heat treatment required. Small modules only. Very costly
Laser deposition	Extremely critical parameters. Critical high temperature heat treatment required. Small modules.
Plasma spray	Very high coating temperature. Large-area coating possible. High temperature heat treatment required.
Ultrasonic spray (CDS)	Coating of large areas possible with little grain melting. High layer density. High temperature sintering required. Structural materials can be used as substrate.
Ink coating	Low density of shielding material. High temperature contamination possible from dispersing media. High temperature sintering required, limiting module size.
Filled resin, paint	Very low density of the shielding material. No subsequent heat treatment required. No constraints on module size.

3. DESIGN OF SUPERCONDUCTING CRYOSHIELDS

Figure 2 shows some possible designs of superconducting cryoshields, depending on their size.

3.1 Small-size Shields

An elongated cylinder or a cup of bulk material can be used in this cases. This is also the size of a small-scale prototype of coated-substrate shield intended for initial shielding-factor characterization.

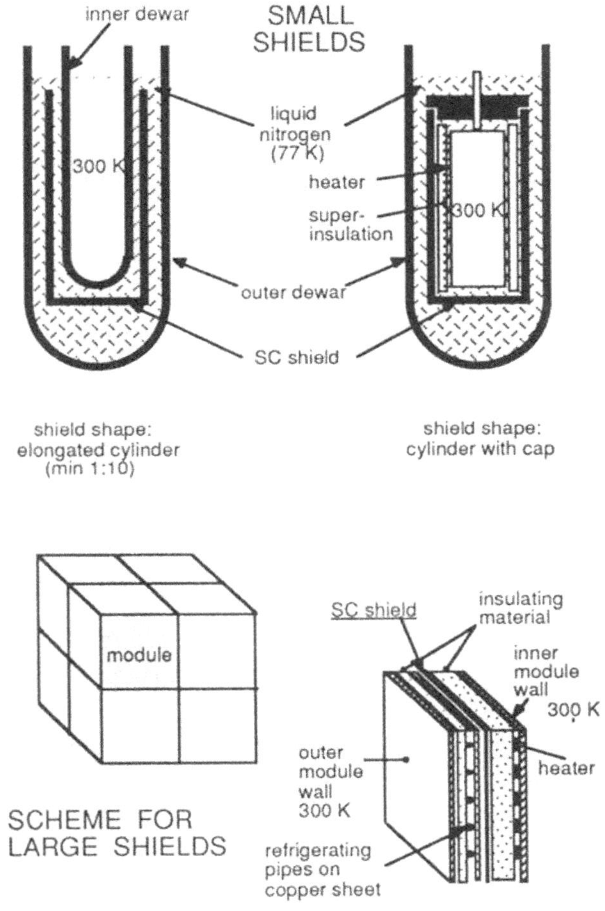

Figure 2 - Layouts of possible mechanical designs for the fabrication of superconducting cryoshields, depending on the size.

For example, a cylinder with a length-to-diameter ratio of at least 10 can be used for this purpose, made up of a layer of superconducting material on a silver or copper tube. At a later stage, a cup-type shield with a cap can be used, which also allows to study the shielding effectiveness of the cap, and, in general, of the technique for making joints electromagnetically tight between two pieces of the shield. In both cases, the shield assembly can directly be immersed in the refrigerant (most commonly, boiling liquid nitrogen at 77 K). The internal room-temperature volume can be obtained by simply using an inner dewar or, more generally, by using a heated inner box, separated from the refrigerant by means of some kind of passive insulation. This solution is certainly the simplest for a volume of the shields not exceeding ≈ 1 L.

3.2 Large Shields

When the shield size needs be increased, or be designed in order to make scaling up of dimensions possible, the shield must be brocken in modules, which will form the full shield after assembling, unless an *in-situ* technique is found for making the ceramic layer or for performing a low-temperature heat-treatment.

Each module forms a package made of several functional layers. The two outer walls at 300 K contain the low temperature plate needed by the superconducting ceramic to form the actual shielding surface. This plate is separated from the outer walls by a thermal insulation, made, for instance, of some kind of plastic foam. The assembly also includes a heating system for the inner wall and a refrigerating system for the superconducting plate. The latter can act also as a structural part of the mechanical assembly, or can only perform the shielding function (e.g., in the case of a silver foil). The difficulty of the mechanical design comes from the fact that all the portions of the superconducting surface must be designed so that they are strictly contiguous or overlapping, since each gap would lower the shield effectiveness. Also, the thermal paths from room to low temperature must be air-tight, to prevent air convection and air or moisture condensation. These requirements may induce topological difficulties in the mechanical design. The size of a single module depends on the maximum size allowed by the heat-treatment furnace, but can hardly be less that 10×10 cm.

A wide choice of shielding plate types can be envisaged, whose relative merits are still under scrutiny. A first, though expensive, possibility is the use of tiles made straight on of superconducting ceramics, 1-3 mm thick. Another similar solution is to use a ceramic substrate coated with the superconducting ceramic, via a possible buffer barrier with thermal-matching and contamination-stopping purposes. Their efficient refrigeration is probably difficult, since most ceramics have a low thermal conductivity. A metallic substrate is a much better choice from the thermal point of view, but this solution is weaker from the point of view of matching the superconducting ceramic thermal expansion. When only a metal thin foil can be used, it is again difficult to achieve a simple and efficient refrigeration. The best solution would be to coat with a superconducting layer directly a thick copper plate.

For refrigeration, it is difficult to consider other than pipes where a cold or liquified gas flows, which must be thermally connected to high thermal conductivity plates in thermal contact with the superconductor *. Liquid nitrogen can circulate in them, like in a flow-cryostat. Or, a cold gas can be forced to circulate in them, obtained either from evaporation of liquid nitrogen or by means of a closed-cycle refrigerating machine. The latter solution allows a wider choice of working temperatures, since a modern single-stage cold-head can easily provide a platform temperature of 50 K, and a two-stage one can reach temperatures below 15 K.

4. RESEARCH WORK UNDER WAY IN THE FRAME OF THE "SUCRYTECH" NATIONAL PROJECT IN ITALY

The goal of the task concerning magnetic shielding is to develop the technologies aimed to implement on industrial scale and with competitive costs large-scale shielded enclosures having the internal volume kept at room temperature.

The flow mainstream of the research is shaded in Fig.1. It was mainly determined by the fact that, not considering for economical reasons the use of bulk materials and the techniques for fabricating a thick extended superconducting layer which are too expensive, in order to meet the goal of final economic competitiveness of the process, only modular shields can be developed. This choice involves paralleling the optimization of a high-productivity technique for producing extended thick layers with the subsidiary development of a superconducting media not requiring heat treatments after application, for sealing the gaps between modules.

The task can be subdivided in three main steps.

4.1 Development of Extended Films on Metallic Support

An analysis of the techniques reported in Table I indicates that laser ablation and all varieties of sputtering techniques, which only would allow to obtain directly superconducting films without requiring further thermal treatments, are nowadays extremely

* Unfortunately the working temperature of the present ceramic superconductors is still too low for Peltier-cooling devices.

expensive for extended films and not readily available. The other methods can be subdivided into two main group: techniques which lead to low-density films, such as electrophoretic deposition, and high-productivity techniques, such as the ultrasonic spray and the plasma spray. The latter being not readily available, the former has been selected, which is implemented in the laboratory of one of the authors (MB).

The ultrasonic spray techniques available on the market are now commonly referred to as High Velocity Oxygen Fuel (HVOF); the equipment used in these experiments, and forming part of the HVOF group, is the "Continuous Detonation Spray" (CDS) [2]. It consists of a gun where the combustion of a calibrated mixture of propane and oxygen is generated in the form of short detonations having a frequency of 300 s^{-1}, which cause the oxygen-rich exhaust gases and the feeding powder to escape at a supersonic velocity (~ Mach 3-4 for the gases and Mach 1.5-2 for the powder respectively) and at a relatively low temperature. Both characteristics are quite different from other "thermal spray" techniques, as shown in Fig. 3.

These characteristics are all likely to be beneficial for the aimed application. In fact, the very short transit time of the powder, dispersed in the oxygen as carrier gas, in the combustion chamber makes each grain to hardly reach the YBCO melting temperature (~1000 °C). This low temperature allows to start from fully reacted powders already treated for good superconducting characteristics, since the grain structure is likely to be mostly retained. The high impact velocity on the substrate, whose temperature can easily be controlled, ensures a strong adhesion and also makes the density of the sprayed coating extremely high.

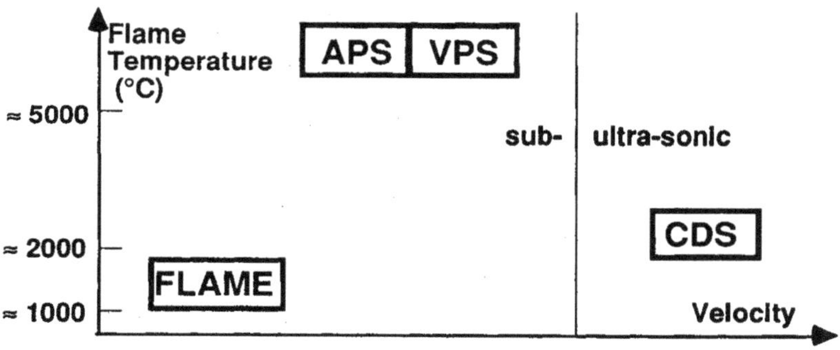

Fig.3 - Characteristics of the CDS technique compared with other "thermal spray" techniques.

Contrarily to plasma spray, the thickness of the coating made with CDS can finely be controlled and consequently made uniform, an important requirement in magnetic shielding applications. A productivity up to 0.1 m^2 min^{-1} can easily be achieved.

Metallic substrate materials will be preferred, since, for large enclosures this makes the refrigeration problems easier to solve. Silver is the obvious choice, but copper is the final goal, since it is the best structural material for the shield and, as an additional bonus, it can allow higher sintering temperatures: however, a protection from eavy oxidation in strongly oxidizing atmospheres must be developed. The use of austenitic steel has been reported [3)], but its use requires buffer layers for increased adhesion and reduced superconductor contamination, which would increase the production difficulties and costs.

4.2 Development of Subsidiary Superconducting Media for "Cold" Use

The fabrication of the shield in modules to be assembled requires the development of a joining media that cannot further be heat-treated after application. A resin, and/or a grease, will be developed eavily filled with a very fine superconducting powder. The shielding effect completely relies, in this case on the quality of the original powder and is severely limited by the absence of intergrain connections. On the other hand, its use is confined to filling of narrow gaps: it has already been shown that results reasonably good can be obtained even with unsupported mildly compressed powders [4)].

Also the development of the less efficient ink technique will be considered, since it must be expected, when making very extended film surfaces, that some repairs of cracks or pinholes will become necessary. These repairs could simply be done with the preceding "cold" technique. However, they will be more efficiently made by painting the defective parts with the ink and by performing anew the full heat treatment of the module.

4.3 Fabrication of Small-scale Prototypes for Feasibility Demostration of the Scaling-up of theTechniques Developed

From a technological point of view, there is a substantial difference in building up an efficient electromagnetic shield when scaling small targets or one-piece cylindrical samples up to modular assemblies, even when simple refrigeration by immersion in a bath of liquid nitrogen is used. Therefore, the simplest scalable prototype, which allows for an accurate measurement of the shielding factor, will be in the shape of a squared box with at least one module on each of its six faces. Non-trivial problems must be solved such as follows: geometrically-accurate flat perimeter of each module; mounting frame, allowing for thermal contraction and inter-module gap sealing; magnetometer probe penetration duct; capability to map the internal field. As an example, a too small prototype scale prevents from using a helium-temperature-cooled SQUID magnetometer, and high-T_C SQUIDs are not yet readily available, so limiting the maximum measurement sensitivity.

Finally, the thermal design of a shielded enclosure will be studied to solve the problems of refrigeration by circulation of cold gases in pipes, of avoiding frosting of the cold parts and of keeping the internal working volume safely at room temperature.

5. PRELIMINARY RESULTS

Since the first tests on silver sheets and copper plates, YBCO* coatings sprayed with the CDS technique were found impervious to stripping and cracking for thickness values up to 100-200 μm.

Fig.4 - Microphotograph of a CDS-sprayed YBCO coating on silver (white) substrate before heat treatment.

* Sometimes added with several percent of silver.

The layers were compact (Fig.4) and virtually immune from pinholes (with proper powder distribution and powder distributor well cleaned from hard impurities). Density values of 95 ± 5% the theoretical density were measured, as can generally be obtained only in pellets after isostatic compression at very high pressures.

The heat treatment, performed in a horizontal cylindrical furnace with a central uniform space controlled to within ± 2 °C and under a controlled atmosphere, consisted first in a high-temperature step of several hours to increase grain connectivity, followed by an oxygenation step in the range 450-600 °C, depending on the runs, all the operations being done in flowing pure oxygen. The influence of the highest cycle temperature on the resistive transition has been studied systematically[5]. Preliminary results show that sintering must be achieved at 920-925 °C for transition to become sharp and to stabilize above 90 K, but that some grain connectivity can be observed at temperatures as low as 650 °C, indicating that most of the original grain structure is retained even after spraying.

Resistive transition were measured on samples were the four electrodes were obtained by applying a special low-melting alloy with a commercial ultrasonic and thermal soldering iron.: ohmic contacts with a resistance of less than few tenths of an ohm, and stable down to liquid helium, were obtained. Temperature was measured with a reproducibility of ± 0.1 K and an accuracy of 1%, using a simple dip-in probe of proper thermal design. Typical resistive transitions are shown in Fig.5.

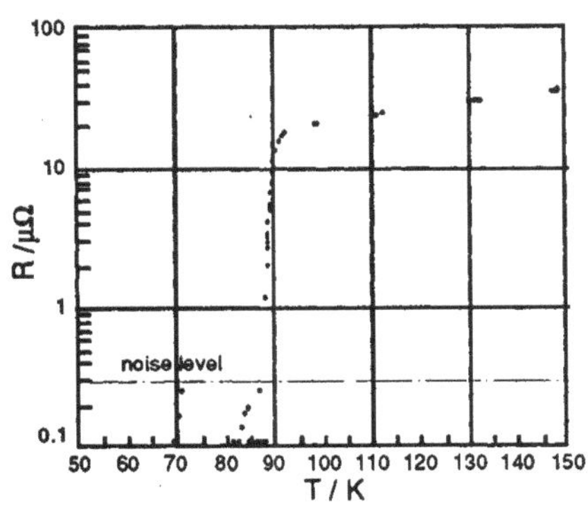

a)

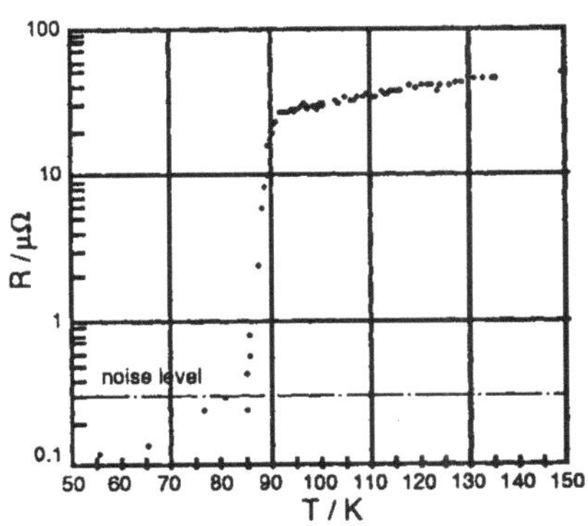

b)

Fig.5 - Typical resistive transition of a CDS-sprayed YBCO coating on silver substrate: a) without silver added to YBCO; b) witrh 8% silver added.

Preliminary magnetic measurements were performed with a Hall probe having a sensitivity of ± 10^{-7} T using cylinders 20 mm in diameter and 200 mm long, immersed in liquid nitrogen (77.3 K) or liquid helium (4.2 K). Typical results are shown in Fig.6.

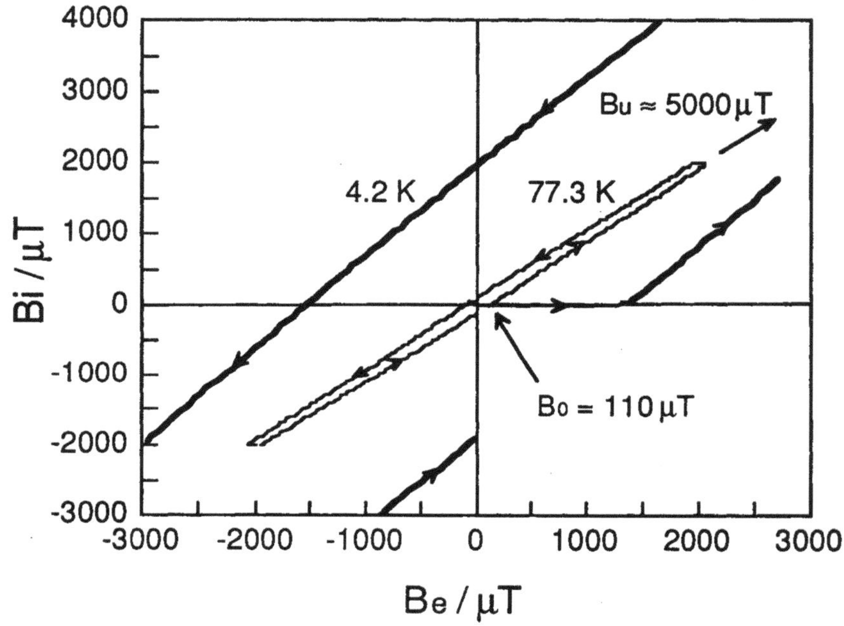

Fig.6 - Hysteretic behaviour obtained with a CDS-sprayed YBCO coating (80 μm thick) on a silver cylinder Ø 20×200 mm. a) 77.3 K; b) 4.2 K.

When scaled for thickness, values for B_0 scaled to film thickness of 1.4 mT mm^{-1} at 77.3 K and 18 mT mm^{-1} at 4.2 K were found, comparable with the avarage results in the literature. The upper field B_u defined where the material does no more show a shielding effect is about 5 mT at 77.3 K, a ratio $B_u/B_0 \approx 50$, much larger than generally observed in the literature. The shape of the hysteresis loop at 77.3 K is close to linear in most of its range.

The ink technique is still under development [6)] and tests on a filled resin compound are going to be started.

6. CONCLUSIONS

A research work is under way in Italy for the realization of magnetic shields scalable up to (very) large dimensions and with the working volume kept at room temperature, as it is necessary in biomedical applications.

The coating of metallic substrates to form extended thick films of superconducting materials is based on the CDS technique. Preliminary results of coatings on silver made with commercial YBCO powders show that the resistive transition temperature is ≈ 90 K. The maximum field (B_0 scaled to film thickness) that can be completely shielded is ≈ 1.4 mT mm^{-1} at 77.3 K and 18 mT mm^{-1} at 4.2 K.

The ink technique and a superconducting filled resin are under development as subsidiary techniques for mounting the shield modules.

7. REFERENCES

1 Bean C.P., Rev.Mod. Phys. **36** (1964) 31.

2 CDS apparatus commercially available from Plasma Technique, Switzerland.

3 Barbezat G., private communication.

4 Itoh M., see these Proceedings.

5 To be published.

6 Ohyama T., Itoh M., Isigaki H., Minemoto T., Andreone D., Pavese F., Numata K., Hoshino K., 39th Spring Meeting Japan Society Applied Physics and Related Sciences (1992).

Part 3

Weak superconductivity: junctions and devices
Tunneling spectroscopy

On the Nature of Grain Boundaries in the High Temperature Superconductors

R. GROSS and B. MAYER

Lehrstuhl Experimentalphysik II, Universität Tübingen
Morgenstelle 14, D-7400 Tübingen, Germany

ABSTRACT

The recent experimental data on the superconducting transport properties of grain boundaries in the high temperature superconductors, particularly in $YBa_2Cu_3O_{7-\delta}$ epitaxial films, are summarized. Based on the reproducible experimental data the nature of the weak link behavior and theoretical models describing the transport and noise characteristics of grain boundaries are discussed. We show that many aspects of the behavior of the current transport across grain boundaries can be explained by assuming that the grain boundary interface consists of a bad insulator containing a high density of localized defect states.

1 Introduction

The superconducting transport properties of grain boundaries in the high temperature superconductors has been studied intensively using individual grain boundaries.[1–20] Although the observed weak link nature of grain boundaries in the high temperature superconductors is disappointing with respect to the use of polycrystalline materials in applications requiring high critical current densities, the same weak link behavior allows the development of grain boundary Josephson junctions for microelectronics applications. Understanding the mechanisms responsible for the weak link nature of grain boundaries and other extended planar defects in the high-T_c oxides is crucial. For high current density applications this probably will allow to develop remedies to enhance the superconducting coupling across grain boundaries what, in turn, would significantly hasten the use of the high temperature superconductors for many applications. For those applications making use of the weak link nature of grain boundaries the knowledge of the mechanisms determining the weak link properties is required for the development of controllable high-T_c Josephson devices in which single grain boundaries are used. Up to now different types of artificially generated, so-called engineered Grain Boundary Junctions (GBJs) such as bicrystal,[1–12] bi-epitaxial,[18–20] and step edge GBJs[14] have been fabricated and shown to be useful for numerous applications.[21–30]

At present, despite of the intensive research effort we do not unequivocally understand the precise microstructural origin of the weak link behavior of grain boundaries in the cuprate superconductors. However, as we will discuss in this review, recent experiments narrow the range of possible mechanisms. At present, most valuable information is obtained from the study of well controllable, individual GBJs. Although the vast majority of the work has been carried out on the $YBa_2Cu_3O_{7-\delta}$ superconductor, bicrystals of the bismuth superconductor[10] and, more recently, on bicrystal films of the neodimium, bismuth, and thallium superconductor show similar characteristics.[11] In this article we summarize our present knowledge on the weak link nature of grain boundaries in the high temperature superconductors. Based on reproducible experimental data from a large number of samples theoretical models describing the current transport across grain boundaries in the high temperature superconductors are discussed. A summary of recent applications of GBJs is given in Ref. 12 and 21. The present status of other types of high-T_c Josephson elements is covered by another article in this volume.

2 Nature of the Weak Link Behavior

The precise origin of the weak link behavior of grain boundaries in the cuprate superconductors is not completely understood. However, the available experimental data on the structural and electrical transport properties (for a recent review see Ref. 12) can be used to narrow the range of possible mechanisms. The experimental observations can be summarized as follows:

- phenomenologically, $YBa_2Cu_3O_{7-\delta}$ GBJs can be described by the RSJ-model[31], that is, they behave as ideal overdamped Josephson junctions which are characterized by a critical current I_c and a normal resistance R_N.[4,6–9,12–18]
- the critical current $I_c(T)$ is proportional to $(1-T/T_c)^2$ close to T_c .[6,8,13]
- the normal resistance R_N is independent of temperature.[6,12–15]
- for large bias voltages, $V \geq 3-4I_cR_N$, R_N decreases with increasing bias voltage.[5,8,12,13]
- the observed I_c (B) dependences and the direct imaging of the current density distribution of GBJs by Low Temperature Scanning Electron Microscopy[5] demonstrate that the current flow across the grain boundaries, though spatially inhomogeneous, is not restricted to some narrow filaments.[5,7–9,12,15,17]
- the critical current density decreases with increasing misorientation angle in the same way for [001] tilt, [100] tilt, and [100] twist grain boundaries.[4,7,12,13,17]
- the I_cR_N products are small compared to the gap voltage Δ/e.[4,6–9,12–20]
- the I_cR_N or equivalently $J_c\rho_N$ products scale proportionally to $1/(\rho_N)^q$ with q ranging between 1 and 1.5.[7,12,13,15] Here, ρ_N is the normal resistance times the

area of the GBJ and J_c its critical current density.

- GBJs show large amounts of low frequency noise caused by fluctuations of the critical current and the normal resistance.[8,12,13,21–30]
- the structural analysis of GBJs shows that there are no second phases at the grain boundary. However, dislocations and stress effects may play an important role.[32]
- oxygen and ozon annealing experiments demonstrate that the oxygen stoichiometry at the grain boundary plays a crucial role.[9,26]

Discussing possible origins of the weak link nature of GBJs we can show quite readily that the behavior of grain boundaries cannot be attributed to the possible p- or d-wave symmetry of the superconducting order parameter. In the same way it cannot be attributed to the anisotropy of the high temperature superconductors. This follows directly from the similar angle dependence of the critical current of the tilt and twist boundaries. The critical current density J_c was found, within experimental error, to be a function only of the misorientation angle θ and not of the type of misorientation. If the symmetry of the order parameter would be important, the three types of misorientation should have yielded different angle dependences.

It has been speculated that the weak link behavior of grain boundaries is caused by the presence of second phases. However, there is no experimental evidence for large scale composition deviations at the grain boundary (see, for example, Ref. 32 for a summary of the transmission electron microscopy results). This strongly suggests that the weak link behavior is not associated with large scale composition deviations. However, stress effects and the oxygen stoichiometry seem to be important. Oxygen and ozone annealing experiments give direct evidence that deviations from the ideal oxygen stoichiometry at the grain boundary is, at least in part, responsible for the weak link behavior.[8,9,26]

We also can speculate that there is a large density of charged scattering centers at grain boundaries.[33] An obvious source of such scattering cites are dislocations, which are likely to be charged due to the strongly covalent bonding in the high-T_c material.[34] The density of dislocations increases with increasing misorientation angle and thus provides an obvious explanation for the observed decrease of the critical current density. Although dislocations were discussed early as a possible origin of the weak link nature of grain boundaries there is, to date, no satisfactory explanation of their precise role. However, some aspects of their impact can be ruled out. Firstly, GBJs with different misorientation geometries containing edge or screw dislocation show similar properties. Therefore, the type of elastic strain does not control the superconducting properties of GBJs. Secondly, the sign of elastic strain changes along, say, an edge dislocation. Therefore, also the sign of strain seems to play no important role. A likely explanation is that the order parameter is reduced in the strained region around the dislocation core. Once the density of dislocations is high enough that these regions overlap, the order parameter is reduced at the entire grain boundary and weak link behavior is expected to be observed. This is the case for a misorientation angle larger than about 10°.

With respect to the dislocation core idea one has to discuss some special grain boundaries including twin boundaries, 90° boundaries, and the set of coincidence boundaries possible for geometrical reasons. The twin boundaries, although obtained by a large angle rotation does not show weak link behavior. The situation is similar for the 90° boundaries.[35,36] Note that neither the twin nor the 90° grain boundaries require dislocations cores to be present at the interface. Hence, the absence of weak link behavior is consistent with the dislocation core idea. However, there are two additional features that have to be taken into account, namely the continuity of the CuO_2 planes across the grain boundary and the orientation of the plane of the grain boundary interface. The discontinuity of the CuO_2 planes is expected to strongly influence the superconducting properties of the boundary. Furthermore, deviations from a symmetrical grain boundary geometry can introduce dislocations resulting in turn in weak link behavior. This may be important for step edge GBJs.

The discussion of possible mechanisms responsible for the weak link behavior of grain boundaries suggests that the most probable origin is the presence of dislocations and strain at the boundaries. The dislocation core idea is in accord with the existing data. However, it must be kept in mind that we arrived at this idea by a process of elimination. A direct experimental verification of this conception is still required.

3 Models for $YBa_2Cu_3O_{7-\delta}$ Grain Boundary Junctions

Developing theoretical models for $YBa_2Cu_3O_{7-\delta}$ GBJs a crucial question is whether these junctions are SIS, SNS, SNINS, SS'IS'S, or SS'NS'S type junctions (S=superconductor, S'=superconductor with reduced order parameter, N=normal metal, I=insulator), that is, whether the interface layer at the grain boundary is a normal conducting (N) or an insulating (I) layer. Transmission electron microscopy studies have shown that the grain boundary interfaces are clean and that the lattice distortion at the grain boundary interface is confined to a few lattice constants. Oxygen and ozone annealing experiments[8,9,26] suggest that there is an oxygen deficient layer at the grain boundary interface. Oxygen deficiency or disorder is known to render YBCO insulating and not normal conducting. Furthermore, the measured values of ρ_N vary between about 10^{-7} and $10^{-8}\Omega cm^2$. Therefore, assuming an interface layer thickness of about 1nm results in a resistivity of about 0.1 .to 1Ωcm for this layer. This value is typical for a semiconductor or bad insulator but is by several orders of magnitude larger than the normal state resistivity of $YBa_2Cu_3O_{7-\delta}$. Finally, we note that oxygen disorder in the Cu-O chains removes holes from the CuO_2 planes resulting in an insulating phase. Additionally, the intrinsic relaxation of $YBa_2Cu_3O_{7-\delta}$ at internal interfaces and surfaces likely yields a high density of localized states that are electron like. These facts give evidence that there is an insulating (I) layer at an oxygen deficient and disordered grain boundary interface, that has a thickness of up to more than 1nm.

Up to now, several models[13,37–39] have been developed to describe the electrical transport characteristics of $YBa_2Cu_3O_{7-\delta}$ GBJs. The models can be classified into tunneling (SIS) and non-tunneling (SNS) type models. However, as will be discussed below, in some cases the differences between these models are quite small. In order to explain the observed transport properties of GBJs, in SNS-type models a normal metal with low conductivity close to the metal-insulator transition is assumed.[38] On the other hand, in SIS-type models an insulator with a high density of localized defect states, that is, an insulator close to the insulator-metal transition, is assumed.[13] Up to now, SNS-type models[37–39] were favored, since they naturally explain the low values of the I_cR_N products of GBJs and the temperature dependence of the critical current close to T_c . However, it was shown that also a SIS-type model[13] can explain most aspects of the observed transport and, moreover, of the noise characteristics of $YBa_2Cu_3O_{7-\delta}$ GBJs. In this model the insulator is assumed to contain a high density of localized defect states allowing resonant and intermediate state tunneling.[40]

3.1 SIS-type Models

There are well established theories describing the superconducting transport characteristics of ideal SIS-type Josephson junctions, which are formed by two ideal superconducting banks separated by a perfect insulator.[31] Using SIS-models to model high-T_c GBJs two important facts have to be considered. Firstly, it is well known that a characteristic feature of the high-T_c superconductors is their short coherence length. The implications of the short coherence length have to be taken into account in SIS-type junction models . Secondly, the chemistry of $YBa_2Cu_3O_{7-\delta}$ implies that an insulating layer at grain boundaries will most likely contain a high density of defect states. The influence of such defect states on the electrical transport characteristics has to be included in SIS-type junction models.

3.1.1 Implications of the Short Coherence Length

Deutscher and Müller[41] pointed out very early that the short coherence length ξ_0 of the high-T_c materials would modify the order parameter at a superconductor-insulator interface. In general, on the Ginzburg-Landau scale (lengths of the order $\xi_S(T) \gg \xi_0$)

the effect of the boundary between a superconducting and a non-superconducting material on the order parameter at the boundary, Δ_i, can simply be described by means of the boundary condition. Due to the small coherence length of the high-T_c cuprates the order parameter at the grain boundary interface is given by $\Delta_i/\Delta_0 \simeq b/\xi_S(T)$ in a wide temperature regime. Here, Δ_0 is the bulk value of the order parameter, $b \simeq \xi_0^2/a$ is the so-called extrapolation length,[41] and a is the interatomic distance. Since $\xi_0 \sim a$ we have $b < \xi_S(T)$ and $\Delta_i/\Delta_0 < 1$.

The reduction of Δ_i results in an unusual temperature dependence of the critical current density of SIS-type Josephson junctions close to T_c. According to the Ambegaokar-Baratoff (AB) expression[42] the critical current density is proportional to the product of the order parameter on both sides of the junction, $J_c \propto \Delta_i^+\Delta_i^-$. That is, one obtains $J_c \propto [\Delta_0(T)(b/\xi_S(T))]^2 \propto (1-T/T_c)^2$ instead of the usual result $J_c \propto [\Delta_0(T)]^2 \propto (1-T/T_c)$. This is in good agreement with the dependence observed for GBJs. Furthermore, the reduction of Δ_i results in a reduced value of the characteristic voltage $V_c = I_cR_N \simeq \Delta_i/e$ as observed for GBJs. However, at lower temperatures we have $b/\xi_S(T) \simeq b/\xi_0 \simeq 1$ and much of the strength of the order parameter at the interface is restored. This should be the case especially for the [001] tilt GBJs where the longer ab-plane coherence length, ξ_{ab}, is the important length scale. However, for [001] tilt $YBa_2Cu_3O_{7-\delta}$ GBJs low I_cR_N products are observed even at low temperatures. These small I_cR_N values at low temperatures cannot be explained solely by a reduction of the order parameter at the grain boundary due to the small coherence length.

3.1.2 Resonant and Intermediate State Tunneling

Above we have considered the case that the I-layer which separates the superconducting electrodes is formed by a perfect insulator. We will now discuss a SIS-type model[12,13] where the insulator contains a high density of defect states. Note that the AB-expression for the tunneling current of superconducting SIS tunnel junctions is derived under the assumption that both the paired and unpaired charge carriers have to tunnel through the same barrier. When localized states are present in a tunneling barrier, two different elastic tunneling processes (channels) are conceivable. Channel 1 is the usual direct tunneling process, in which an electron tunnels all the way across the barrier without any scattering. The channel 2 represents indirect tunneling, in which an electron tunnels from one electrode onto a localized state and then in turn tunnels out to the other electrode. The simplest indirect tunneling process is resonant (elastic) tunneling. According to simple one-dimensional theories of resonant tunneling, the conductivity due to this process is enhanced at biases for which the Fermi level of the electrode coincides with the energies of the localized states in the barrier.

In real materials the impurity states generally have a small radius $a_i \sim 10\AA$. This causes a strong Coulomb repulsion, $U \geq 0.1$eV, of two electrons that occupy one level. This repulsion usually does not lead to a substantial change of the normal current, since it is given by the successive tunneling of individual electrons. In the absence of Coulomb repulsion, also the paired electrons may tunnel through resonant states.[43] However, since the Josephson current is a consequence of the correlated tunneling of pairs of charge carriers it can be suppressed considerably by Coulomb repulsion. The importance of the Coulomb repulsion is given by the ratio of two time scales. The first time scale, τ_1, is determined by the decay time of a localized state into the conduction electron state and is inversely proportional to the width Γ of the resonant impurity state. The second, τ_2, is determined by the correlation time of the electrons in a Cooper pair and is inversely proportional to the critical temperature T_c . If the decay time of the localized state is long ($\tau_1/\tau_2 > 1$), the resonant Josephson current would be dominated by processes for which the two electrons of a pair are simultaneously at a center. Usually, Coulomb repulsion prohibits such processes resulting in a strong suppression of the Josephson current. In this case we have two different tunneling channels for the paired (mostly direct tunneling) and unpaired (mostly indirect tunneling) electrons and the AB-expression can no longer be applied. Due to the suppression of the Josephson current one expects an I_cR_N product which is much smaller than that given by the AB-expression. On the other hand, for $\tau_1/\tau_2 \ll 1$ the tunneling processes of the two electrons of a pair are separated in time. In this case, Coulomb repulsion is not important and the Josephson current is determined by the tunneling amplitudes of the single electrons,[44] that is, paired and unpaired electrons tunnel through the same channel.

The decay time of a localized state increases as $\tau_1 \propto \exp(\kappa x)$ with increasing distance x from the superconducting electrode. Thus the importance of Coulomb repulsion will increase strongly with increasing distance of the localized states from the electrodes. Here, $\kappa = \sqrt{2mE_b/\hbar^2}$ is the attenuation constant of the conduction electrons in the insulator. Tunneling experiments indicate a tunnel barrier height $E_b > 0.5$eV. Therefore, the attenuation is strong and $1/\kappa \approx 0.1 - 0.2$nm can be estimated. That is, the resonant or intermediate state tunneling of paired electrons via localized states is expected to be small even if the localized states are quite close to the superconducting electrodes.[43] From this we conclude that the supercurrent through the insulating layer is determined by the direct tunneling current. Assuming that the localized states are in the center of the I-layer with thickness $2d_i$ we obtain

$$J_c \;\propto\; N(0)^2 \exp\left(-2\kappa\, d_i\right) \quad . \tag{1}$$

In contrast, the normal current will be dominated by resonant or intermediate tunneling current supported by the localized states with density N_l

$$J_N \;\propto\; 1/\rho_N \;\propto\; N_l \exp\left(-\kappa\, d_i\right) \quad , \tag{2}$$

if N_l or d_i is large. As discussed above, the presence of a large density of localized defect states is very likely at grain boundaries in $YBa_2Cu_3O_{7-\delta}$. A density $N_l \approx 10^{18}cm^{-3}$ which seems to be reasonable for a disordered cuprate material,[45] would be sufficient to make resonant tunneling dominant. Combining eq. (1) and (2) at once yields[13]

$$J_c\rho_N \propto \exp(-\kappa d_i) \propto 1/\rho_N \quad \text{or equivalently} \quad J_c\rho_N \propto \sqrt{J_c}\,. \tag{3}$$

This result is close to the experimentally observed scaling behavior.[7,12–15] Within the proposed model the scaling behavior of the $J_c\rho_N$ product is simply caused by different tunneling channels of the paired and unpaired charge carriers. The strongly reduced values of the $J_c\rho_N$ products result from an increased barrier width for the supercurrent as compared to the normal current. Note that $\exp(-\kappa d_i)$ should be temperature independent. Therefore, with $\rho_N(T)$ =const., similar $J_c(T)$ dependences are expected for junctions with different absolute $J_c\rho_N$ values in agreement with the experimental observation.[8,12,13]

According to eq.(3), the $J_c\rho_N$ product decreases proportional to $\exp(-\kappa d_i)$ with increasing barrier thickness. That is, GBJs with a large intrinsic insulator thickness should have small $J_c\rho_N$ products. The observed decrease of $J_c\rho_N$ with increasing misorientation angle[13] then suggests that large angle grain boundaries have a thicker insulating barrier. This is reasonable, since disorder and strain and, hence, the thickness of the barrier layer is expected to increase with increasing misorientation angle. Furthermore, the $J_c\rho_N$ products of the GBJs can be increased by ozone annealing. Annealing and oxygen supply is expected to improve order at the grain boundary interface thereby reducing the thickness of the barrier layer and, hence, increasing the $J_c\rho_N$ product.

It is known that localized states in oxides adjacent to metals hybridize with conduction electron states forming interface states.[45] By overlap with the Cooper pairs of the bulk electrodes these surface states have a superconducting gap, $\Delta_S(x) \propto \Delta_0\exp(-\tau_1(x)/\tau_2)$, which decays rapidly with increasing distance x from the bulk electrode due to the increase of the on-site Coulomb repulsion counteracting the superconducting attraction.[45] It is likely that $\Delta_S \leq 5$meV already for $x \geq 0.2$nm.[45] Therefore, the supercurrent mediated by the surface states is expected to decrease very fast with increasing x. Qualitatively, the surface states can be considered as to mediate a continuous transition between the metal electrode and the insulating barrier layer (space-wise metal-insulator transition). By this, a narrow part of the insulating layer becomes metal-like and in turn superconducting by proximity effect. This results in a so-called SNS-like temperature dependence of the critical current, $J_c \propto (1-T/T_c)^2$, for temperatures close to T_c . For $T \ll T_c$, a $J_c(T)$ dependence close to the AB-dependence is expected. Of course, the modification of the order parameter at the S/I interface due to the small coherence length of the high-T_c superconductor has to be taken into account.

The proposed SIS-type model which includes resonant and intermediate tunneling via localized states can explain most characteristics observed for $YBa_2Cu_3O_{7-\delta}$ GBJs. Beyond the $J_c(T)$ dependence and the reduced values and the scaling behavior of the I_cR_N products it also explains the excess low frequency noise in $YBa_2Cu_3O_{7-\delta}$ GBJs. The resonant tunneling via localized states depends on the local potential of the localized states in the insulator. This potential and thus the tunneling current can easily be modulated by nearby charging. This usually yields a large amount of low frequency noise as found for $YBa_2Cu_3O_{7-\delta}$ GBJs and SQUIDs.[8,9,13,21–29] Ozone annealing is likely to decrease the density of defect states and, hence, the amount of low frequency noise.[24] Note that within this model the scaling behavior and the excess low frequency noise have the same physical origin, namely the tunneling via localized states. This is in perfect agreement with the experimental observation that both the transport (J_c and ρ_N) and noise (δI_c and δR_N) characteristics of $YBa_2Cu_3O_{7-\delta}$ GBJs show the same scaling behavior.[12] Furthermore, a white distribution of localized states explains the linear increase of conductance with increasing bias voltage, $dI(V)/dV \propto |V|$, as found for some bicrystal GBJs.[5] Finally, the model explains that similar $J_c(T)$ dependences are observed for GBJs with I_cR_N products differing by two orders of magnitude.

3.2 SNS-type Models

We now consider a SNS-type junction formed by two superconducting banks, S, with normal state resistivity ρ_S separated by a thin normal metal layer, N, with resistivity ρ_n and thickness d_n. The analysis of BCS based models for S/N interfaces shows that the proximity effect at these interfaces can be characterized by two dimensionless parameters. In the dirty limit these parameters are given by[37]

$$\gamma_S = \rho_S\xi_S \,/\, \rho_n\xi_n \qquad\qquad \gamma_B = \rho_B \,/\, \rho_n\xi_n \tag{4}$$

Here, ξ_n is the decay length of the superconducting order parameter in the normal metal and ρ_B is the boundary resistance of the S/N interface.

We first will consider the case of negligible boundary resistance, i.e. $\gamma_B = 0$. This case was discussed recently by Deutscher and Chaudhari[38] under the assumption that the N-layer consists of a low conductivity metal. The variation of the order parameter at the S/N interface can be treated in close analogy to the case of an S/I interface discussed above. The type of boundary condition that applies to the order parameter at the S/N interface is given by[38]

$$b = \xi_n \coth\left(\frac{d_n}{\xi_n}\right) \quad . \tag{5}$$

The structural analysis of the grain boundaries shows that d_n should be small. Therefore, we can assume that $d_n < \xi_n$. This yields $b \propto \xi_n^2/d_n$ in good approximation. Furthermore, since $I_cR_N \ll \Delta_0/e$, we can assume that b is small compared to ξ_S. In this limit the ratio of the order parameter at the grain boundary, Δ_i, to the bulk order parameter, Δ_0, is given by b/ξ_S. With these assumptions we obtain

$$\Delta_i \propto \xi_n^2 \,/\, d_n \tag{6}$$

In the dirty limit, $\xi_n^2 \propto 1/\rho_n$. Then, using $I_cR_N \simeq \Delta_i/e$ at small reduced temperatures and $\rho_n d_n = \rho_N = R_N A$ results in

$$J_c\rho_N = I_cR_N \propto 1/\rho_N \quad \text{or equivalently} \quad J_c\rho_N = I_cR_N \propto \sqrt{J_c} \tag{7}$$

This is in close agreement with the experimentally observed scaling behavior.

The assumption that the grain boundary layer is composed of a low conductivity metal is not essential. It also might be composed of an insulator close to the metal-insulator transition having a localization length $\xi_l > d_n$. By extension of the dirty limit to this case one obtains $\xi_n^2 \propto D(d_n) \propto 1/\rho_n$ [38] and the above result is recovered. Here, $D(d_n)$ is the diffusion coefficient on scale d_n.

Kupriyanov et al.[37] pointed out recently that the condition $\gamma_B \gg \gamma_S \geq 1$ is inherent to high-T_c superconductor - normal metal interfaces. Therefore, the case discussed above ($\gamma_B = 0$) should not apply to such interfaces. In the case of finite γ_B the order parameter undergoes a jump at the S/N interface. The jump is increasing with increasing ρ_B. For $\gamma_B \gg 1$ one can neglect the suppression of the order parameter in the superconducting electrodes due to the proximity effect. The detailed analysis shows that for $\xi_n < d_n$ one obtains $I_cR_N \propto 1/\rho_B$. With $\rho_N \simeq \rho_B$ for $\gamma_B \gg 1$, the same scaling behavior, $I_cR_N \propto 1/\rho_N$ is obtained. That is, the result $I_cR_N \propto 1/\rho_N$ is obtained quite generally no matter whether the normal resistance of the grain boundary is determined by the normal metal layer ($\gamma_B \ll 1; \rho_N \simeq \rho_n d_n$) or the boundary resistance ($\gamma_B \gg 1; \rho_N \simeq \rho_B$).

The SNS-type models can well explain both the scaling law $I_cR_N \propto 1/\rho_N$ and the temperature dependence of the critical current near T_c. However, the observation that GBJs with different I_cR_N products show very similar $I_c(T)$ dependences is difficult to explain within these models. Furthermore, these models cannot explain the relationship between the transport and noise properties of $YBa_2Cu_3O_{7-\delta}$ GBJs.

4 Acknowledgement

Financial support of this work by the Bundesminister für Forschung und Technologie (project 13N5843) is gratefully acknowledged.

5 References

1. P. Chaudhari, J. Mannhart, D. Dimos, C.C. Tsuei, C.C. Chi, M.M. Oprysko, M. Scheuermann, *Phys. Rev. Lett.* 60, 1653 (1988)
2. D. Dimos, P. Chaudhari, J. Mannhart, F.K. LeGoues, *Phys. Rev. Lett.* 61, 219 (1988)
3. J. Mannhart, P. Chaudhari, D. Dimos, C.C. Tsuei, T. R. McGuire, *Phys. Rev. Lett.* 61, 2476 (1988)
4. D. Dimos, P. Chaudhari, J. Mannhart, *Phys. Rev. B* 41, 4038 (1990)
5. J. Mannhart, R. Gross, K. Hipler, R. P. Huebener, C. C. Tsuei, D. Dimos, P. Chaudhari, *Science* 245, 839 (1989); J. Mannhart, R. Gross, R. P. Huebener, P. Chaudhari, D. Dimos, C.C. Tsuei, *Cryogenics* 30, 397 (1990)
6. R. Gross, P. Chaudhari, D. Dimos, A. Gupta, G. Koren, *Phys. Rev. Lett.* 64, 228 (1990)
7. R. Gross, P. Chaudhari, M. Kawasaki, A. Gupta, *Phys. Rev.* B 42, 10735 (1990)
8. R. Gross, P. Chaudhari, M. Kawasaki, A. Gupta, *IEEE Trans. Magn.*, MAG-27, 3227 (1991)
9. R. Gross, P. Chaudhari, M. Kawasaki, A. Gupta, *Supercond. Sci. Techn.*, 4, S253 (1991); M. Kawasaki, P. Chaudhari, R. Gross, A. Gupta, unpublished
10. N. Tomita, Y. Takahashi, Y. Ishida, *Jpn. J. Appl. Phys.* 29, L30 (1990)
11. M. Kawasaki, P. Chaudhari, A. Gupta, W. Lee, to be published
12. R. Gross, *Grain Boundary Josephson Junctions in the High Temperature Superconductors*, in Interfaces in Superconducting Systems, S. Shinde and D. Rudman eds., Springer Verlag, New York (1992)
13. R. Gross, B. Mayer, *Physica* C 180, 235 (1991)
14. K. Herrmann, Y. Zhang, H. M. Mück, J. Schubert, W. Zander, A. I. Braginski,
15. S. E. Russek, D. K. Lathrop, B. H. Moeckly, R. A. Buhrmann, D. H. Shin, *Appl. Phys. Lett.*, 57, 1155 (1990)
16. B. H. Moeckly, S. E. Russek, D. K. Lathrop, R. A. Buhrmann, J. Li, J. W. Mayer, *Appl. Phys. Lett.*, 57, 1687 (1990); *Appl. Phys. Lett.*, 57, 2951 (1990)
17. Z. G. Ivanov, P. A. Nilsson, D. Winkler, J. A. Alarco, T. Claeson, E. A. Stepantsov, A. Ya. Tzalenchuk, *Appl. Phys. Lett.*, to be published
18. K. Char, M. S. Coclough, S. M. Garrison, N. Newman, G. Zaharchuk, *Appl. Phys. Lett.*, 59, 733 (1991)
19. K. Char, M. S. Coclough, L. P. Lee, G. Zaharchuk, *Appl. Phys. Lett.*, (1991), in press; see also *Proceedings of SQUID'91 Conference*, Berlin, Germany (1991)
20. L. P. Lee, K. Char, M. S. Coclough, G. Zaharchuk, *Appl. Phys. Lett.*, to be published
21. R. Gross, P. Chaudhari, *Status of DC-SQUIDs in the High Temperature Superconductors*, in "Principles and Applications of Superconducting Quantum Interference Devices", A. Barone ed., World Scientific (1991)
22. R. Gross, P. Chaudhari, M. Kawasaki, M. B. Ketchen, A. Gupta, *Appl. Phys. Lett.* 57, 727 (1990)
23. R. Gross, P. Chaudhari, M. Kawasaki, M. B. Ketchen, A. Gupta, *Physica* C 170, 315 (1990)

24. R. Gross, P. Chaudhari, M. Kawasaki, M. B. Ketchen, A. Gupta, *IEEE Trans. Magn.* MAG-27, 2565 (1991)
25. M. Kawasaki, P. Chaudhari, T. Newman, A. Gupta, *Appl. Phys. Lett.*, 58, 2555 (1991)
26. M. Kawasaki, P. Chaudhari, A. Gupta, *Phys. Rev. Lett.*, to be published
27. R. Simon, J. B. Bulman, J. F. Burch, S. B. Coons, K. P. Daly, W. D. Dozier, R. Hu, A. E. Lee, J. A. Luine, C. E. Platt, M. J. Zani, *IEEE Trans. Magn.* , 27, 3209 (1991); *Appl. Phys. Lett.* , 58, 543 (1991)
28. M. Siegel, F. Schmidl, K. Zach, E. Heinz, J. Borck, W. Michalke, P. Seidel, *Physica C* , (1991), in print
29. G. Cui, Y. Zhang, K. Hermann, Ch. Buchal, J. Schubert, W. Zander, A. I. Braginski, Ch. Heiden, in *Nonlinear Electronics and Josephson Devices*, N. F. Pedersen, M. Russo, A. Davidson, G. Costabile, and S. Pagano eds., Plenum Press, New York (1991); see also *Supercond. Sci. Tech.*, 4, S130 (1991); *Physica* , C 175, 545 (1991) *Supercond. Sci. Tech.*, to be published
30. A. H. Micklich, J. J. Kingston, F. C. Wellstood, J. Clarke, K. Char, M. S. Coclough, G. Zaharchuk, *Appl. Phys. Lett.*, 59, 988 (1991)
31. A. Barone, G. Paterno, *Physics and Applications of the Josephson Effect*, Wiley&Sons, New York (1982)
32. M. S. Chisholm, S. J. Penneycook, *Nature* 351, 47 (1991)
33. P. Chaudhari, Proceedings of the *M^2HTSC Conference*, Kanazawa, Japan (1991), to be published in Physica B
34. K. Jagannadham, J. Narayan, *Mat. Science and Engin.* B8, 5 (1991);
35. S. E. Babcock, X. Y. Cai, D. L. Kaiser, D. C. Larbalestier, *Nature* 347, 178 (1990);
36. C. B. Eom, A. F. Marshall, Y. Suzuki, B. Boyer, R. W. Pease, T. H. Geballe, to be published
37. M. Yu. Kupriyanov, K. K. Likharev, *Sov. Phys. Uspekhi* 160, 49 (1990); *IEEE Trans. Magn.* MAG 27, N2 (1991); *Sov. Phys. JETP* 96, 1420 (1989); Extended Abstracts ISEC'89, Tokyo (1989), p. 534; *Proceedings of SQUID'91 Conference*, Berlin, Germany (1991)
38. G. Deutscher, P. Chaudhari, *Phys. Rev.* B , (1991), in press
39. J. Mannhart, P. Martinoli, *Appl. Phys. Lett.* 58, 643 (1991)
40. E. L. Wolf, *Principles of Electron Tunneling Spectroscopy*, Oxford University Press, New York (1985)
41. G. Deutscher, K. A. Müller, *Phys. Rev. Lett.* 59, 1745 (1987)
42. V. Ambegaokar, A. Baratoff, *Phys. Rev. Lett.* 10, 486 (1963)
43. A. G. Aslamasov, M. V. Fistul, *Sov. Phys. JETP* 55, 681 (1982)
44. L. I. Glazman, K. A. Mateevul, *JETP Lett.* 49, 659 (1989)
45. J. Halbritter, *Surface Science* 122, 80 (1982); *J. Appl. Phys.* 58, 1320 (1985); *IEEE Trans. Mag.* MAG-19, 799 (1983); *Sol. State Com.* 18, 1447 (1976); *Sol. State Com.* 34, 675 (1980); *Phys. Rev. B*, submitted for publication

MULTILAYER-TYPE JOSEPHSON JUNCTIONS AND DEVICES FABRICATED THEREOF

G.J. GERRITSMA
Faculty of Applied Physics, University of Twente,
P.O.Box 217, 7500 AE Enschede, The Netherlands

ABSTRACT

A short overview will be given of the various ways thin film high temperature Josephson junctions may be fabricated and which results have been obtained so far world wide. Examples are for instance $YBa_2Cu_3O_x/PrBa_2Cu_3O_x/YBa_2Cu_3O_x$ sandwiches developed at Bellcore or Si-implanted weak links from TRW.
In order to make multilayer-type junctions, the fabrication of such layers of high quality is a necessity. Not only a high transition temperature or a high current density are of importance, but also such factors as homogeneity, surface rougness etc. play a role. Furthermore not only the junction fabrication itself is of importance but also that of the circuit wiring.This holds especially for the cross-over of wires or the superconducting connection to a top electrode of a sandwich junction. Fabrication and extensive measurement data on ramp-type junctions of Twente will be presented in order to gain insight both in the controllability of the fabrication process as well as in the device physics. Various theoretical junction models like SIS, SNS, SINS etc. will be reviewed and there predictions will be compared to the measurement data. Good agreement may be obtained using a SNINS model taking magnetic pair breaking in the $PrBa_2Cu_3O_x$ barrier into account.
Finally some future developments will be discussed.

1. Introduction

Since the discovery of high temperature ceramic superconductivity a lot of effort has been spent on the preparation and study of high-T_c thin film Josephson junctions and SQUIDs. The first dc-SQUID was made at IBM using natural grain boundaries that may occur in $YBa_2Cu_3O_x$ (YBCO) thin films.[1] It was observed that the critical current is of the form $(1-T/T_c)^{1.5}$, just as in granular Al_2O_3 films with Josephson tunnel junctions between the grains. By applying a magnetic field perpendicular to the SQUID loop a clear voltage modulation could be observed up to temperatures of 68 K. This was rapidly being followed by several other groups.[2-5] Comparable results were obtained on some other materials, i.e. $Bi_2Sr_2Ca_1Cu_2O_x$ (BSCCO)[6] and $Tl_2Ba_2Ca_2Cu_3O_y$ (TBCCO)[7].

All junctions mentioned so far belong to a wider class of single layer junctions as opposed to multilayer sandwich junctions or multilayer edge-type junctions. In the next sections we will give a brief description of the various junction types, except for grain boundary junctions which are covered by another article in this volume. This is followed by the description of two fabrication methods of high quality multilayer thin films and their properties with respect to device behaviour. Next a section on high-T_c device physics is included followed by a discussion of the properties of some multilayer junctions and finally some future trends and developments will be briefly outlined.

2. Various junction types

As was already mentioned in the introduction the Josephson junctions described in this article will be divided into three major classes. Single layer junctions consist of just one layer of super conducting material that covers a substrate that may either be flat or into which a step has been etched. In the latter case a weak link is formed at the step. Such junctions have been fabricated using c-axis oriented YBCO on (100) $SrTiO_3$ substrates.[8,9] Using High-resolution transmission electron microscopy (HREM) it was shown that grain boundaries are formed at the step edge.[9] Using flat substrates weak links may be formed by natural grain boundaries in polycrystalline thin films[1], highly textured YBCO thin films on MgO[10] or epitaxial films on $SrTiO_3$ bicrystals.[11] A variant on the latter theme avoiding the fabrication of bicrystals was developed at Conductus Inc.[12] There an r-plane sapphire substrate was covered epitaxially with an MgO seed layer. Part of this seed layer was subsequently etched away. After removing the photo resist mask the whole substrate was covered by an epitaxial $SrTiO_3$ buffer layer. This buffer layer grows in two different orientations separated by a 45° boundary. Next the substrate is covered by a YBCO layer into which also a 45° grain boundary is formed. Starting from epitaxial thin films weak links may also be created by locally reducing the dimensions, i.e. by forming a Dayem bridge or by locally destroying the super conducting properties. A Dayem bridge may for instance be created by pulsed Xe laser patterning[13] whereas destruction may be caused by 200 keV Si^{++} focused ion beam implantation[14]. The latter technique has recently been used to fabricate a series of digital logic circuits like AND, EXOR, EXNOR an Inverter, based on series arrays of interferometers (SAIL logic)[15]. Input coupling lines and control lines were made of silver separated from the superconductor by an insulating layer of SiO_x. A wide variation in device modulation characteristics was observed, giving rise to a low circuit yield of 1/20, but nevertheless working logic could be demonstrated.

Multilayer sandwich junctions are based on the epitaxial growth of a trilayer of a super conducting ground electrode, a non-super conducting barrier and a super conducting top electrode. As YBCO and related materials are highly anisotropic two kinds of sandwiches may be distinguished, those with the c-axis perpendicular to the substrate or those with the ab-plane perpendicular. The barrier may be normal conducting, semi conducting or insulating. The first working sandwich junctions of the c-axis type were fabricated by a team at Bellcore[16]. The trilayer consisted of YBCO/$PrBa_2Cu_3O_x$(PBCO)/YBCO followed by a gold contact layer. The layers were fabricated using pulsed laser deposition. A barrier layer thickness of 50 nm was reported. The junction critical current scaled linearly with the surface area whereas the junction resistance was inversely proportional to it, with I_cR_n products ranging up to 3.5 mV. Weak Shapiro steps were observed in response to 84 GHz mm-wave radiation. Furthermore a weak modulation of I_c with transversally applied magnetic field was observed. It was concluded that the current transport across the barrier was inhomogeneous. The same type of junction was also fabricated by a team at Matsushita[17]. In this case a trilayer of 2212 BSCCO/2201 BSCCO/2212 BSCCO with a platinum contact layer was used. The barrier layer thickness was 60 nm, the I_cR_n-product at 4.2 K was determined to be 0.1 mV. Recently the same group changed the barrier material by substituting Nd for Ca[18].

We repeated the Bellcore experiment in our group using a modified off-axis rf magnetron sputtering technique[19]. Using a PBCO barrier layer thickness of 50 nm none of our sandwich junctions carried a super current. In cooling down below the transition temperature of the electrodes the device resistance first decreased and upon further cooling increased again. Only when we reduced the barrier layer thickness to about 9 nm some of the junctions showed a super current. All measured IV-characteristics were different. From this experiment and other knowledge about the layer morphology, i.e. the occurrence of all kinds of outgrowths, we concluded that micro shorts act as weak links in these junctions. It is doubtful if in highly anisotropic super conducting materials like YBCO c-axis Josephson junctions may be fabricated. Using a high quality c-axis oriented YBCO base electrode, a Au barrier and a Nb counter electrode no super current can be observed[20]. This was only possible if a (110) $SrTiO_3$ substrate was used giving rise to a (103) orientation of the YBCO films. As YBCO and related materials may also be grown with the a-axis perpendicular to the substrate a sandwich may be fabricated that may carry a super current without having to rely on micro shorts. Such sandwich Josephson junctions were recently fabricated at Bellcore[21]. Barrier layer thicknesses of 30, 100 and 150 nm were used giving rise to an I_cR_n product at 10K of 2.5, 0.5 and 0.1 mV, respectively. From the thickness dependence of the critical current density as a function of the barrier layer thickness a normal state coherence length of the order of 10 nm at 10K was inferred. Although a-axis oriented films are reported to be smoother than c-axis oriented films[22], i.e. they are free of outgrowths, it cannot be excluded that again micro shorts play a role. Usually thin films do have some particulates that are rich in copper, especially if the T_c is high, i.e. around 90 K. A variant on the sandwich scheme is the use of a normal conducting ground layer with two separate super conducting electrodes on top of it. This method of junction fabrication was used by a team at Hitachi[23]. The ground layer was $La_{1.5}Ba_{1.5}Cu_3O_x$ which exhibits semi conducting behaviour while $HoBa_2Cu_3O_x$ super conducting electrodes were being used. The separation between the two electrodes was achieved by etching with a reactive ion beam of Ar + 50% O_2 gas using a resist pattern that was formed by electron beam lithography. From a plot of the logarithm of the critical current versus the electrode spacing a characteristic length of 65 nm at 4.2 K was inferred. Electrode spacings up to 0.5 μm have been used in this experiment. This characteristic length is much to high for a normal state coherence length and it was assumed that the ground layer was a mixture of super conducting and other phases.

We now turn to the third class of junctions, that of the multilayer edge-type. They are all based on c-axis oriented YBCO-like super conducting films. By etching away part of the film an edge is formed and the ab-plane is exposed. If the substrate is again covered by a c-axis super conducting layer the two electrodes join at the edge with their ab-planes parallel. Instead of directly joining at the edge also extra barrier layers may be incorporated. The first junctions of this type did not use an extra layer[24]. The base electrode was covered by an thick insulating layer of BaF_2 or PBCO in order to avoid a direct contact with the counter electrode. The exposed edge of the super conducting base electrode was furthermore treated by either an oxyfluoride process or a plasma oxidation. Super currents of these junctions and voltage modulation in response to an applied magnetic field in the case of dc-SQUIDs could be observed starting from 4.2 K to about 60 K. In their paper the authors assumed that the weak link between the two electrodes consists of many metallic shorts. Given the HREM investigations of step-edge junctions in Jülich[25] another explanation may be that grains are formed at the edge. At Twente we devised a variant on this technique[26,27]. However instead of a step edge a ramp edge was made using an Ar^+ ion beam. The slope of this ramp edge is usually around 20^0 and from the work done in Jülich[25] we infer that subsequent layers may be grown epitaxially on this ramp edge hence avoiding the formation of any grain boundary and overall parallel ab-planes do occur. As a barrier material we used PBCO. The highest I_cR_n product we obtained at 4.2 K was 8 mV for a 6 nm thick barrier (estimated from sputter time as c-axis growth rate) but it cannot be excluded that in this particular case micro shorts are responsible for the weak link behaviour. A more or less similar approach using a shadow mask to define the ramp edge was followed in Berkeley[28]. In this case Nb doped $SrTiO_3$ was used as barrier material with thicknesses of 13, 25 and 50 nm. From a plot of the logarithm of the critical current against temperature a coherence length in the barrier of 5 nm at 4.2 K was inferred which is not to far off the calculated value of 11.6 nm. The I_cR_n products of these devices lie between 1 and 3 mV. Instead of just making one edge at a time also two edges can be made which are than connected by a normal metal barrier. This can be done by etching a narrow trench in a film either by focused ion beam etching[29] or using electron beam lithography[30]. Either Ag or Au may be used as the normal metal to fill the trench. Another way to fabricate a similar kind of normal metal layer bridge uses a stepped substrate. This substrate is covered with a thin YBCO layer and the remaining gap at the step is filled by Ag, Au or a mixture of both[31,32]. Using an Ag-Au alloy an I_cR_n product at 4.2 K of 1 mV may be obtained[32] due to a relatively high barrier resistivity and correspondingly lower suppression of the order parameter at the interface as compared to a pure Ag or Au barrier.

3. Fabrication of high-quality (multi) layers

The growth of c-axis oriented epitaxial ceramic super conducting thin films in situ by using elevated substrate temperatures is by now a well known technique. Nevertheless the fabrication of high quality epitaxial layers requires great care. First of all single crystal substrates should be used having a good lattice match with that of the thin film, have a smooth surface without contaminants and have a low diffusivity and reactivity at the deposition temperature. Furthermore the substrates should not have a phase transition in the temperature range involved, i.e. from room temperature to the deposition temperature. Examples of good substrate materials are for instance MgO, Yttrium stabilized ZrO_2 (YSZ),$SrTiO_3$ and $NdGaO_3$. Furthermore the as deposited thin film should be of stoichiometric composition, although for the Thallium compound this seems to be impossible and a post deposition anneal treatment is needed[33]. What may happen if the composition is not stoichiometric is nicely demonstrated in a paper by Carlson et al. where YBCO was coevaporated at room temperature on $LaAlO_3$ followed by an oxidation and anneal step[34]. A variation of 1% of one of the elements Y, Ba or Cu reduces T_c by 1 K and furthermore the surface roughness increases.

So far most of the available data on the fabrication of epitaxial layers for device fabrication relate to two deposition methods, i.e. sputtering and laser ablation. Starting from a stoichiometric target great care has to be taken to avoid so called back sputtering effects. Negative oxygen species are accelerated in the cathode dark space and may bombard the substrate. Due to the different sputter yields of the elements involved the as deposited thin film will not have the right composition. This effect may be avoided by using elevated sputter gas pressures[35] or large target to substrate distances or off-axis geometries[19,36,37]. Apart from an optimum deposition temperature, which is for instance illustrated by a minimum in the RBS yield of laser ablated films[22], also the partial oxygen pressure is of importance. If it becomes too low the film will become thermodynamically unstable and may decompose[38]. Furthermore there is also an upper limit as all in situ deposited thin films are having the tetragonal phase during the deposition[38]. In our own work using a modified off-axis rf magnetron sputtering technique we found a pronounced optimum partial oxygen pressure for YSZ whereas for $SrTiO_3$, $LaAlO_3$ and MgO this is less so. Using laser ablation as a deposition technique great care has to be taken to avoid the presence of boulders. First of all in laser ablation a minimum power density is required to get good films anyhow, but a careful study shows that above this threshold the number of boulders may rapidly increase with laser power density[39]. Furthermore the number of boulders depend on the number of shots on the target. By regularly grinding the target surface this problem may be reduced. From our own measurement data on sputtered films we know that high quality multilayers may be grown with a super conducting transition temperature of 90 K[40]. These films are relatively free of defects, exhibit a sharp layer-to-layer transition, but nevertheless small protrusions of height 20 to 40 nm occur at the surface. This probably explains why some of our sandwich junctions with a 9 nm PBCO barrier layer showed a super current, but most did not.

Recently is has also become possible to fabricate a-axis oriented thin films. Successful synthesis of pure a-axis oriented films has been reported by several groups using various growth techniques like off-axis rf magnetron sputtering on $SrTiO_3$[41] and laser ablation on PBCO buffered $SrTiO_3$[22]. Very recently a-axis oriented YBCO was grown on (110)$NdGaO_3$ substrates with the c-axes aligned in the plane of the substrate[42].

4. High temperature device physics

What is evident from the I_cR_n-products, that have been obtained in high temperature super conducting Josephson junctions, is that it is substantially smaller than the value predicted by the BCS theory for tunnel junctions. For $T \ll T_c$ the theoretical value is given by[43]

$$I_cR_n = \frac{\pi}{2e}\,\Delta(0) \qquad (1)$$

where $\Delta(0)$ is the energy gap at T = 0 which is in the order of 20 mV. For weak links the results depend on the particular model used but the I_cR_n products for $T \ll T_c$ are of the same order of magnitude[44]. Now one might argue that as long as the fundamental mechanism behind high temperature superconductivity is not known we cannot make any sound predictions of device properties, but we will not follow this line of reasoning as also other explanations are possible. It was pointed out by Deutscher and Müller[45] that the depression of the order parameter at the SI-boundary may be attributed to the very short coherence length of the high-T_c oxides. Inside the superconductor, the order parameter is given by the solution of the Ginzburg-Landau equation

$$\Delta(x) = \Delta_0 \tanh[(-x+x_0)/\sqrt{2}\,\xi(T)] \qquad (2)$$

where Δ_0 is the undisturbed order parameter (energy gap), x the distance from the interface into the isolator, $\xi(T)$ the temperature dependent coherence length an x_0 is determined by the De Gennes' boundary conditions[46]. As $\xi(0)$ is of the order of the lattice parameter a, $\Delta(0) << \Delta_0$ in a wide temperature range. This leads to an anomalous temperature dependence of the critical current for SIS junctions

$$I_c \propto \Delta(0)^2 \propto (1-T/T_c)^2 \tag{3}$$

This temperature dependence is identical to that of SNS junctions[47]. As T approaches zero, the order parameter at the interface is restored and hence it cannot explain why the I_cR_n product at 4.2 K is also too low.
Another possible reason for the suppression of I_cR_n lies in the nature of the proximity effect in the vicinity of an SN interface. If the metals forming an SNS junction obey the dirty limit condition, i.e. the electron mean free path l is much smaller than the normal metal coherence length ξ_n, and if the barrier thickness (junction length) $L >> \xi_n$, then I_cR_n depends on two suppression parameters[48]

$$\gamma = \frac{\rho_s\ \xi_s}{\rho_n\ \xi_n} \text{ and } \gamma_B = \frac{R_B}{\rho_n\ \xi_n} \tag{4}$$

where $\rho_{n,s}$ are the normal-state resistivities of the junction materials and $\xi_{n,s}$ their coherence lengths, while R_B is a specific resistance of the SN boundary. The quantity γ (if $\gamma_B = 0$) is a measure of the diffusion of quasi particles into the superconductor that suppress the order parameter. The second parameter γ_B differs from zero if an insulating barrier is present at the interface or if it is abrupt on the atomic scale. This parameter further reduces the order parameter at the interface. Using some typical values $\xi_s = 0.5$ nm, $\rho_s = 2 \times 10^{-3}\ \Omega.cm^2$, $\xi_n = 10^2$ nm (Ag or Au), $\rho_n = 2 \times 10^{-6}\ \Omega.cm^2$ and $R_B = 10^{-9}\ \Omega.cm^2$ it is found that $\gamma = 10$ and $\gamma_B = 5 \times 10^3$ and hence an appreciable suppression of the order parameter occurs. Details of the calculations, various modifications to include SINS and SNINS junctions and graphical results are given by Golubov, Kupriyanov and Likharev[48-51]. In the last reference also magnetic pair breaking was taken into account, a phenomenon that might for instance occur in PBCO.

Furthermore it should be noticed that the device physics also may depend on the width or the area of the junctions as a result of self shielding effects. The characteristic dimension is given by the Josephson penetration depth λ_J which is equal to [52,53]

$$\lambda_J = (\hbar/2e\ \mu_0\ J_c\ t)^{1/2} \tag{5}$$

where $t = d + 2\lambda_L$, d = barrier layer thickness, λ_L = London penetration depth and J_c is the critical current density.
For the width of the junction $w \le 4\ \lambda_J$ we obtain so called narrow, or small, junctions in which no flux flow phenomena occur whereas for $w > 4\ \lambda_J$ this is indeed the case. Note that even of $w > 4\lambda_J$ for so called cross-line geometries the critical current does not saturate with increasing width but is proportional to it[52]. For $t \approx 2\lambda_L \approx 300$ nm one gets $\lambda_J \approx 300\ \mu m/\sqrt{J_c}$ taking J_c in A/cm^2.

5. Properties of some multilayer junctions

Apart from grain boundary junctions not much systematic data is available on other junction types. As already mentioned in section 2 for a-axis oriented YBCO/PBCO/YBCO sandwich junctions the critical current and normal-state conductivity scale linearly with the junction area[21]. The variation on a given 1 x 1 cm wafer was about ± 50%. Furthermore the critical current density scales exponentially with the barrier layer thickness as expected for a long barrier. A characteristic length scale of 10 nm at 10 K was found. Furthermore the I_cR_n product decreases with increasing barrier layer thickness.

For multilayer edge junctions width a Nb-doped $SrTiO_3$ barrier it was also found that the critical current scales exponentially with the barrier layer thickness[28]. In this case a characteristic length scale of 5 nm at 4.2 K was found whereas a value of 11.6 nm was calculated. The I_cR_n product decreases with increasing temperature, but no specific data are given.

From our own data we will present some results that are relevant to this overview. In fig. 1 the critical current I_c and the inverse normal state resistance R_n^{-1} are plotted as a function of the junction area and it may be observed that they scale linearly, as one would expect. The variation is about ± 100% but in this case various 0.5 x 1 cm wafers were used. In fig. 2 the logarithm of the critical current density J_c at 4.2 K is given as a function of the barrier layer thickness from which a characteristic length scale of 5 to 8 nm is found. This is reasonably close to the value found for a-axis oriented sandwich junctions[21].

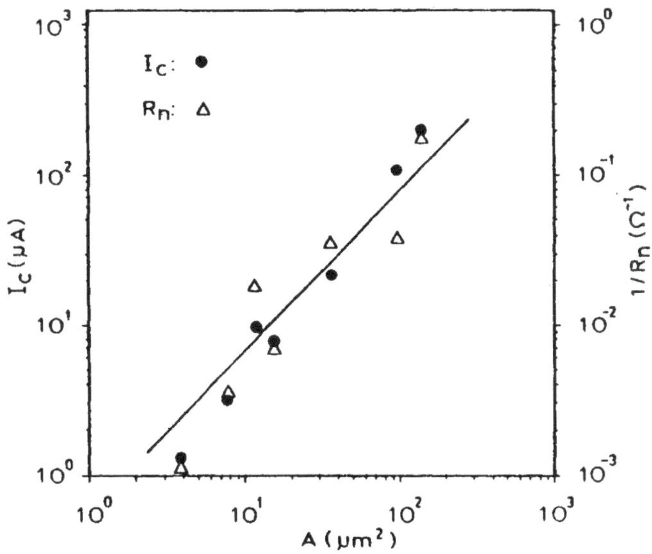

Fig. 1. Scaling of the critical current I_c and resistance R_n with the junction area A for junctions with the same barrier thickness.

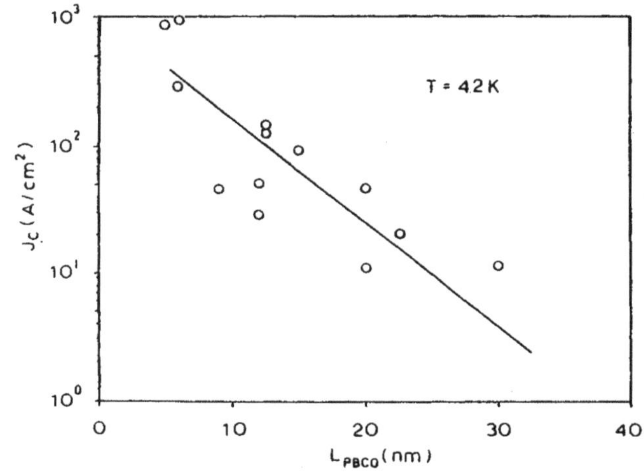

Fig. 2. Dependence of the criticalcurrent density J_c on the PBCO barrier thickness, L.

In fig. 3 the critical current density J_c is plotted as a function of the product of the normal-state resistance and the junction area. The data points are distributed along the line $J_c \propto (R_nA)^{-m}$ with m = 1.2, while for SINS structures with L << ξ_n one would expect a value of m = 1.0[44,30].

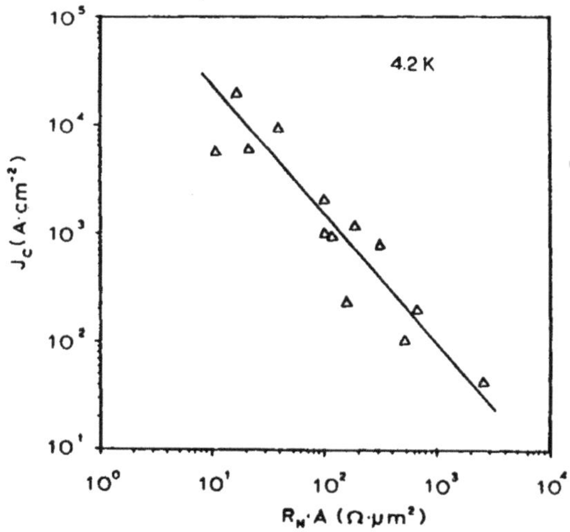

Fig. 3. Critical current density J_c as a function of the product R_nA at T = 4.2 K.

In fig. 4 we present the amplitude of the zero and first order Shapiro steps at 10 GHz as a function of the microwave power. The solid curve is calculated from the RSJ-model. The agreement is reasonable. Furthermore a clear Fraunhofer pattern was observed.

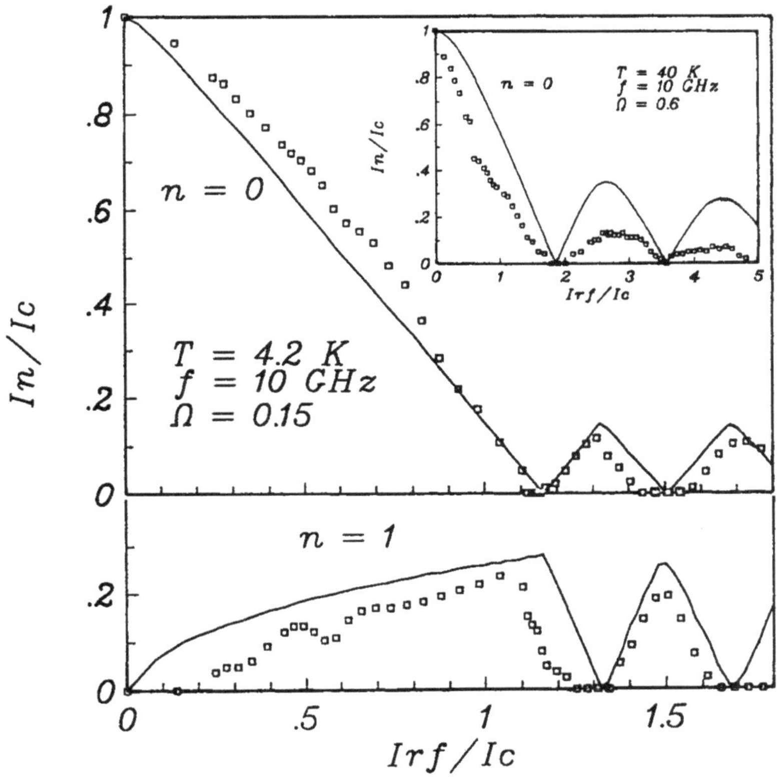

Fig. 4. Magnitude of the current steps vs the normalized microwave current.

In fig.5 the I_cR_n product of two junctions with different barrier thicknesses is plotted as a function of temperature together with a series of calculated results for various models. We considered an SNS junction and an SNINS junction with or without a temperature independent magnetic pair breaking parameter $\tau = 1/\pi T_c$. It may be observed that the assumption of pair breaking leads to reasonable agreement with the experimental data for d = 20 nm. From this we may conclude that in the PBCO barrier a temperature independent pair breaking mechanism might occur. It is however not obvious why the junction should obey an SNINS model.

Using these kind of junctions also dc-SQUIDs were fabricated, a typical set of voltage modulation curves obtained at 4.2 K is presented in fig. 6. Details have already been published elsewhere[27].

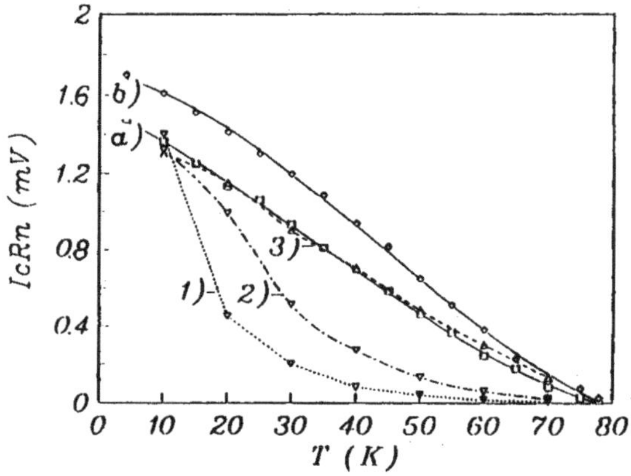

Fig. 5. Characteristic voltage $V_c = I_cR_n$ vs the temperature for junctions with barrier thicknesses d = 20 nm (curve a) and d = 8 nm (curve b). Curves 1-3 are calculated values of V_c: (1) SNS junction with $d/\xi_n = 2$ and large γ_B, (2) SNINS junction with $d/\xi_n = 2$ and small γ_B and (3) SNINS junction with $d/\xi_n = 1.5$, $\gamma_B << 1$ and magnetic pair breaking parameter $\tau^* = 1/\pi T_c$.

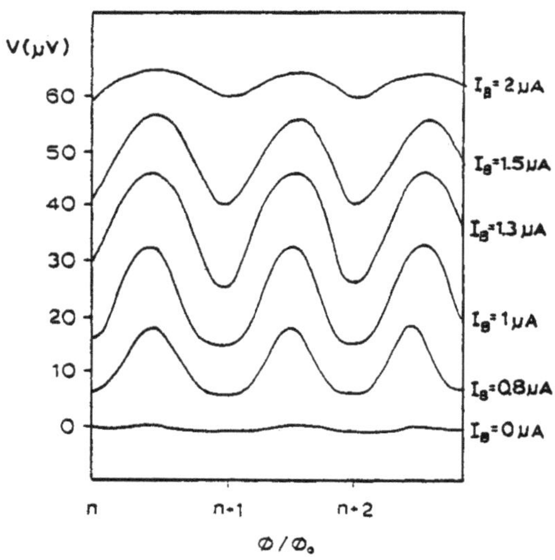

Fig. 6. A set of typical V_{mod} vs ϕ characteristics of a dc SQUID for various bias currents measured at 4.2 K.

6. Some future trends and developments

So far substantial progress has been made in the fabrication of dc SQUIDs with integrated pick-up coils, using natural grain boundary junctions and it may be expected that its application is not too far off, especially if signal-to-noise ratio is not of utmost importance. It may be expected that the same holds for various edge-type junctions. For sandwich junctions the situation is much more complex as a way has to be found to make a reliable high-T_c connection between top electrodes that are separated by a gap. It requires a much more sophisticated process than the fabrication of cross-overs for multiturn input coupling coils of dc-SQUIDs, a problem that has been solved satisfactorily[54,55].

With respect to more complicated circuits, i.e. those circuits requiring various Josephson junctions, it is necessary that the variation in the critical currents of the junction on one chip at least should be reduced, otherwise very low device yields occur[15]. In our own laboratory we have set up a small scale production unit to churn out chips on a more or less regular basis, about once a week. In this way we hope to be able to tighten the production constraints. This starts with substrate source selection, incoming inspection of substrates, surface cleaning procedures etc. We expect that within a couple of years some simple logic circuits may be fabricated with an acceptable yield and repeatable device characteristics.

7. References

1. R.H. Koch, C.P. Umbach, G.J. Clark, P. Chaudhari and R.B. Laibowitz, *Appl. Phys. Lett.* **51** (1987) 200.
2. H. Nakane, Y. Tarutani, T. Nishino, H. Yamada and U. Kawabe, *Jpn. J. Appl. Phys.* **26** (1987) L1925.
3. B. Häuser, M. Diegel and H. Rogalla, *Appl. Phys. Lett.* **52** (1988) 844.
4. R. Yusa, M. Nakao, S. Fujiwara, K. Kaneda, S. Suziki and A. Mizukami, in *Proceedings 5th International Workshop on Future Electron Devices High Temperature Superconducting Devices* (R. and D. Assoc. for Future Electronic Devices, Tokyo, 1988), p.225.
5. I. Takeuchi, J.S. Tsai, H. Tsjuge and N. Matsukura, *Jpn. J. Appl. Phys.* **27** (1988) 2265.
6. I.S. Gergis, J.A. Titus, P.H. Kobrin and A.B. Harker, *Appl. Phys. Lett.* **53** (1988) 2226.
7. R.H. Koch, W.J. Gallagher, B. Bumble and W.Y. Lee, *Appl. Phys. Lett.* **54** (1989) 951.
8. R.W. Simon, J.B. Bulman, J.F. Burch, S.B. Coons, K.P. Daly, W.D. Dozier, R. Hu, A.E. Lee, J.A. Luine, C.E. Platt, S.M. Schwarzbek, M.S. Wire and M.J. Zani, *IEEE Trans. Magn.* **27** (1991) 3209.
9. G.C. Cui, K. Hermann, Y. Zang, C.L. Jia, Ch. Buchal, J. Schubert, W. Zander, A.I. Braginski and Ch. Heiden, in *Nonlinear Superconductive Electronics and Josephson Devices* (ed. N.F. Pedersen et al., Plenum Press, New York, 1991).
10. S.E. Russek, D.K. Lathrop, B.H. Moeckly, R.A. Buhrman, D.H. Shin and J. Silcox, *Appl. Phys. Lett.* **57** (1990) 1155.
11. P. Chaudhari, J. Mannhart, D. Dimos, C.C. Tsuei, J. Chi, M.M. Oprysko and M. Scheuermann, *Phys. Rev. Lett.* **60** (1988) 1653.

12. K. Char, M.S. Colclough, S.M. Garrison, N. Newman and G. Zaharchuk, *Appl. Phys. Lett.* **59** (1991) 733.
13. M.A.M. Gijs and R.J.E. Jansen, *Appl. Phys. Lett.* **56** (1990) 1484.
14. M.J. Zani, J.A. Luine, G.S. Lee, J.M. Murduck, R. Hu, M.J. Lewis, R.A. Davidheiser and L.R. Eaton, *IEEE Trans. Magn.* **27** (1990) 2557.
15. S.M. Schwarzbek, R.A. Davidheiser, G.R. Fischer, J.A. Luine and N.J. Schneier, *Appl. Phys. Lett.* **59** (1991) 866.
16. C.T. Rogers, A. Inam, M.S. Hegde, B. Dutta, X.D. Wu and T. Venkatesan, *Appl. Phys. Lett.* **55** (1989) 2032.
17. K. Mizuno, K. Higashino, K. Setsune and K. Wasa, *Appl. Phys. Lett.* **56** (1990) 1469.
18. K. Mizuno and K. Setsune, *Jpn. J. Appl. Phys.* **30** (1991) L1559.
19. J. Gao, B. Häuser and H. Rogalla, *J. Appl. Phys.* **67** (1990) 2512.
20. H. Akoh, C. Camerlingo and S. Takada, *Appl. Phys. Lett.* **56** (1990) 1487.
21. J.B. Barner, C.T. Rogers, A. Inam, R. Ramesh and S. Bersey, *Appl. Phys. Lett.* **59** (1991) 742.
22. A. Inam, C.T. Rogers, R. Ramesh, K. Remschnig, L. Farrow, D. Hart, T. Venkatesan and B. Wilkens, *Appl. Phys. Lett.* **57** (1990) 2484.
23. Y. Tarutani, T. Fukazawa, U. Kabasawa, A. Tsukamoto, M. Hiratani and K. Takagi, *Appl. Phys. Lett.* **58 (1991) 2707.**
24. R.B. Laibowitz, R.H. Koch, A. Gupta, G. Koren, W.J. Gallagher, V. Foglietti, B. Oh and J.M. Viggiano, *Appl. Phys.Lett.* **56** (1990) 686.
25. C.L. Jia, B. Kabius, K. Urban, K. Hermann, G.C. Cui, J. Schubert, W. Zander, A.I. Braginski and C. Heiden, *Physica C* **175** (1991) 545.
26. J. Gao, W.A.M. Aarnink, G.J. Gerritsma and H. Rogalla, *Physica C* **171** (1990) 126.
27. J. Gao, W.A.M. Aarnink, G.J. Gerritsma, D. Veldhuis and H. Rogalla, *IEEE Trans. Magn.* **27** (1991) 3062.
28. D.K. Chin and T. van Duzer, *Appl. Phys. Lett.* **58** (1991) 753.
29. M.G. Forrester, J. Talvacchio, J.R. Gavaler, M. Rooks and J. Lindquist, *IEEE Trans. Magn.* **27** (1991) 3098.
30. D.B. Schwartz, P.M. Mankiewich, R.E. Howard, L.D. Jackel, B.L. Straughn, E.G. Burkhart and A.H. Dayem, *Proc. SPIE* **948** (1988) 37.
31. M.S. DiIorio, S. Yoshizumi, K-Y. Yang, J. Zhang and M. Maung, *Appl. Phys. Lett.* **58** (1991) 2552.
32. R.H. Ono, J.A. Beal, M.W. Cromar, T.E. Harvey, M.E. Johansson, C.D. Reintsema and D.A. Rudman, *Appl. Phys. Lett.* **59** (1991) 1126.
33 W.Y. Lee, V.Y. Lee, J. Salem, T.C. Huang, R. Savoy, D.C. Bullock and S.S.P. Parkin, *Appl. Phys. Lett.* **53** (1988) 329.
34 D.J. Carlson, M.P. Siegal, J.M. Philips, T.H. Tiefel and J.H. Marshall, *J. Mater. Res.* **5** (1990) 2797.
35. U. Poppe, J. Schubert, R.R. Arons, W. Evers, C.H. Freiburg, W. Reichert, K. Schmidt, W. Sybertz and K. Urban, *Solid State Comm.* **66** (1988) 661.
36. C.B. Eom, J.Z. Sun, K. Yamamoto, A.F. Marshall, K.E. Luther, T.H. Geballe and S.S. Laderman, *Appl. Phys. Lett.* **55** (1989) 595.
37. J. Geerk, G. Linker and O. Meyer, *Mat. Sc. Rep.* **4** (1989) 193.
38. R.H. Hammond and R. Bormann, *Physica C* **162-164** (1989) 703.
39. D.H.A. Blank, *PhD Thesis, University of Twente* (1991).
40. J. Gao, W.A.M. Aarnink, G.J. Gerritsma, A.J.H.M. Rijnders, H. Rogalla, F. Hakkens, W. Coene and M.A.M. Gijs, *Physica C* **177** (1991) 384.

41. C.B. Eom, A.F. Marshall, S.S. Laderman, R.D. Jacowitz and T.H. Geballe, *Science* **249** (1990) 1549.
42. K.H. Young and J.Z. Sun, *Appl. Phys. Lett.* **59** (1991) 2448.
43. V. Ambegaokar and A. Baratoff, *Phys. Rev. Lett.* **10** (1963) 486 and **11** (1963) 104 (erratum).
44. K.K. Likharev, *Rev. Mod. Phys.* **51** (1979) 101.
45. G. Deutscher and K.A. Müller, *Phys. Rev. Lett.* **59** (1987) 1745.
46. P.G. De Gennes, *Rev. Mod. Phys.* **36** (1964) 225.
47. J. Clarke, *Proc. Roy. Soc.* **A208** (1969) 447.
48. M.Yu. Kupriyanov and K.K. Likharev, *Sov. Phys. Usp.* **33** (1990) 340 and references therein.
49. A.A. Golubov and M.Yu. Kupriyanov, *Sov. Phys. JETP* **69** (1989) 805.
50. M.Yu. Kupriyanov and K.K. Likharev, *IEEE Trans. Magn.* **27** (1991) 2460.
51. A.A. Golubov and M.Yu. Kupriyanov, *Physics Letters* **A154** (1991) 181.
52. A. Barone and G. Paterno, *Physics and Applications of the Josephson Effect* (Wiley & Sons, New York, 1982).
53. K.K. Likharev, *Dynamics of Josephson Junctions and Circuits* (Gordon and Beach, New York, 1986).
54. F.C. Wellstood, J.J. Kingston and J. Clarke, *Appl. Phys. Lett.* **56** (1990) 2336.
55. J.J. Kingston, F.C. Wellstood, P. Lerch, A.H. Miklich and J. Clarke, *Appl. Phys. Lett.* **56** (1990) 189.

YBCO BASED SNS JOSEPHSON JUNCTIONS

C.Camerlingo, B.Ruggiero, M.Russo
Istituto di Cibernetica del Consiglio Nazionale delle Ricerche
Arco Felice (Naples), Italy

H.Akoh and S.Takada
Electrotechnical Laboratory, Tsukuba, Japan

ABSTRACT

Recent progress in the fabrication and characterization of SNS junction based on superconducting YBCO films are considered. YBCO-Au-Nb Josephson junctions based on epitaxially grown films have been fabricated. Measurements show a relevant anisotropic behavior in respect to the YBCO lattice orientation. The effect of a depression of the YBCO energy gap at the barrier interface is discussed in the framework of the S/N proximity effect.

1. *Introduction*

Well known technological difficulties have hampered all the attempts of fabricating Josephson tunnel junctions involving high T_c superconductors (HTCS). Problems arise mainly from the short coherence length ξ_S and the anisotropy of these materials. Up to day, great efforts have been devoted to overcome such a technological impasse, but current-voltage characteristics exhibiting evidences for both tunnel and Josephson effects have been reported only for junctions on $Ba_{1-x}K_xBiO_3$, the most "conventional" among the non-conventional superconductors[1].

Advances in understanding the physics of weakly coupled HTCS[2,3] have been achieved by novel structures based on grain boundary properties which use bicrystals and step-like shaped substrates. Furthermore they constitute the base of a viable technology to fabricate devices such as SQUIDs operating at high temperature. However, in spite of a claimed versatility, such structures defect of both control and reproducibility in comparison with film sandwich structures, this being a fundamental requirement in all fabrication processes for realistic devices.

SNS junctions, in which a "normal" metal layer is the coupling between two superconducting electrodes, provide an alternative structure to the artificial insulating barrier junctions, with the advantage of an easier fabrication.

SNS junctions clearly featuring Josephson effect have been fabricated on YBCO films by using Au and Au/Ag films[4,5] as normal layer. Gold and silver provide a good low resistance contact to the YBCO surface, avoiding any oxygen desorbtion and impurity diffusion. Moreover the large value of the coherence length in these metals ($\xi_N \approx 100$ nm in Au) allows to employ "thick" normal layers with obvious advantages for uniformity and fabrication parameter control. By gold and silver, YBCO films have been weakly coupled to low T_c superconductors such as lead and niobium. Logic gate circuits[6] have been successfully realized by using YBCO-Ag-Pb:In junctions to prove the feasibility of electronic devices involving HTC superconductors.

In spite of a satisfactory general behavior of such junctions, experimental evaluations of the critical current and the resistance product I_cR_N are quite low, being in the range of a few hundreds of microVolts. This parameter gives an indication of the junction quality[7]; in HTCS based junctions it is expected to be considerably large. In fact, in the limit of a very thin normal layer, the I_cR_N value is proportional to $(\Delta_1\Delta_2)^{1/2}$ and a value as high as 5 mV is expected, if energy gap values $\Delta_1 \approx 21$ meV and $\Delta_2 \approx 1$ meV are assumed for the YBCO and the conventional superconductor respectively. Proximity effect could be the main responsible of such a discrepancy. Taking in account the mutual influence between interfaced superconducting and normal layers, this effect is often employed in modeling the HTCS phenomenology related to both intrinsic mechanisms (surface "dead" layer and boundary resistance effect between copper planes) and incidental ones (surface degradations and oxygen removing occurring in the fabrication process). As recently pointed out by Deutscher and Simon,[8] in the framework of the proximity effect, the low value of I_cR_N is related to the material chosen as normal layer and a low density of states material, such as a degenerate semiconductor, is expected to give a higher value of the parameter.

In the present paper such aspects are discussed because they deserve interesting implications for the future of HTCS electronic devices. Results concerning epitaxially grown YBCO-Au-Nb sandwich junctions are reported referring to different orientations of the YBCO crystal lattice. The proximity effect in these structures is considered in the "dirty limit" (electron mean free path shorter than coherence length) following the generalized model by Barone and Ovchinnikov[9] which provides an appropriate scenario for HTCS-based SNS' structures.

2. *Anisotropic behavior in YBCO-Au-Nb junctions*

YBCO-Au-Nb Josephson junctions have been fabricated, with different crystalline orientation of the YBCO base-electrode.

Superconducting YBCO films are epitaxially grown on (100) MgO and (110) $SrTiO_3$ substrates by an oxygen rf plasma assisted coevaporation process[10]. Copper is evaporated by a current heated crucible, yttrium and barium by two e-guns. The film deposition is performed at a substrate temperature of 600-700 °C and it is followed by a cooling process in an oxygen rf plasma at low pressure (4 10^{-4} Torr). The films are about 100 nm thick and T_c is in the range 70-80 K. After the YBCO film deposition, the samples are transferred into another vacuum system where they are cleaned by a oxygen rf plasma process and a gold film (5-50 nm thick) is deposited. Subsequently, in different deposition stations, a SiO insulating layer and a Nb counterelectrode are deposited to complete the structure. The geometry of each film is patterned by depositing through stencil masks to obtain SNS' junctions[4,11,12] with an area of about (0.8 x 0.8) mm^2. A schematic junction is sketched in Fig.1.

Junctions with a c-axis oriented YBCO base-electrode are obtained on (100) MgO substrate while a (103) orientation of the copper planes is observed in samples on (110) $SrTiO_3$. In the latter case the copper planes are slightly tilted in respect to the normal direction to the film surface allowing the fabrication of junctions with a large current contribution directed along the ab-axis. An anisotropic junction behavior is observed depending on the different YBCO base-layer orientation.

Looking at the current-voltage (I-V) characteristics, in (100) YBCO based junctions, no dc Josephson current can be detected

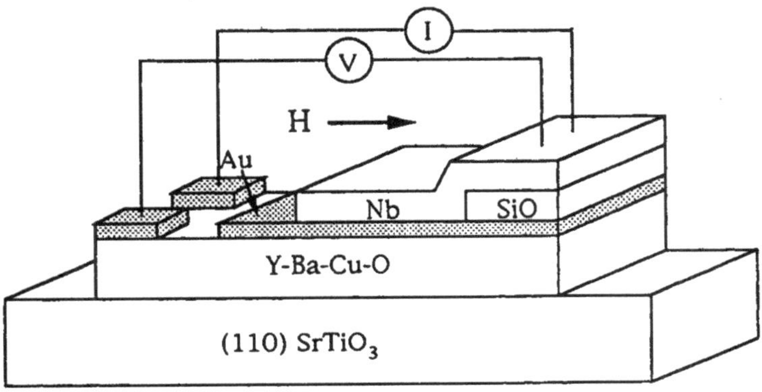

Fig.1 Schematic drawn of a YBCO-Au-Nb junction (not in scale).

whatever the gold normal layer thickness is. On the contrary, in (103) YBCO based structures, the dc Josephson effect is clearly observed together with current steps induced by a microwave irradiation.

Another aspect of the influence of the YBCO orientation on the junction behavior is related to the external magnetic field dependence of the maximum dc Josephson current, I_c-H. The magnetic field modulates the critical current maximum value in a "regular" way but, depending on the copper plane direction in respect to the field, different values of the $I_c(H)$ periodicity are observed. Such an effect is shown in Fig.2 for two junctions having different configurations; a similar anisotropic behavior has been observed in the same junction by changing the magnetic field direction[13].

To justify the different values of the magnetic field which reduces to zero the critical current, a model based on a large junction behavior has been proposed in Ref.13. In that framework the observed anisotropy is ascribed to the anisotropic value of the London penetration depth λ_Y in YBCO films as measured on trasmission striplines by Kohjiro et al.[14]. In fact in a large junction the critical field is:

$$B_{COi} = 2\lambda_{ji}\, J_c\, D_i/d_i \qquad (1)$$

with

$$\lambda_{ji} = (c/4\pi)\ (h/e\ J_c\ D_i)^{1/2} \tag{2}$$

$$D_i = \lambda_{Yi}\coth(t_Y/\lambda_{Yi}) + t_{Au} + \lambda_{Nb}\coth(t_{Nb}/\lambda_{Nb}) \tag{3}$$

$$d_i = \lambda_{Yi}\ \tanh(t_Y/2\lambda_{Yi}) + t_{Au} + \lambda_{Nb}\ \tanh(t_{Nb}/\lambda_{Nb}) \tag{4}$$

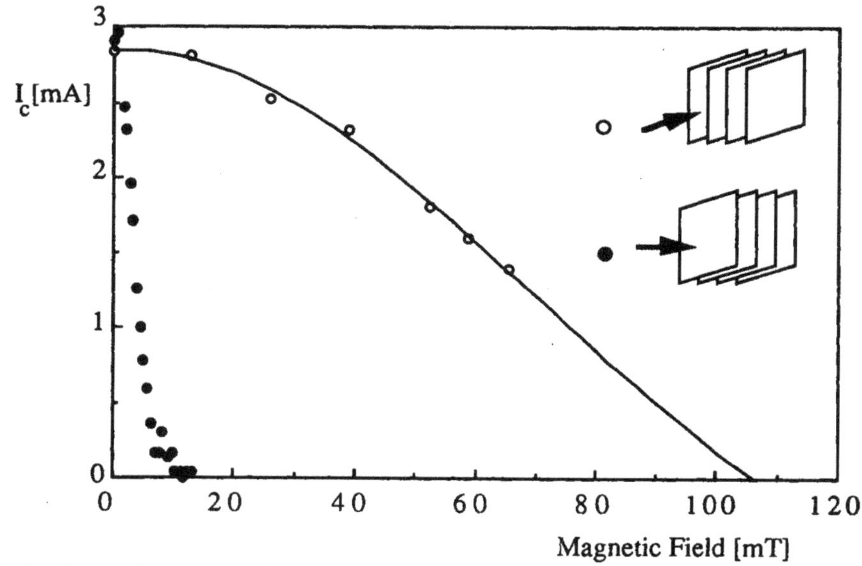

Fig.2 Dependence of the maximum dc Josephson current on the applied magnetic field in two YBCO-Au-Nb junctions. The cases of a magnetic field directed perpendicularly to the Cu planes (dots) and along the Cu planes (circles) of the YBCO base electrode are shown. The two junctions have the same dimensions. All the measurements have been performed at 4.2 K.

where λ and t are the London penetration depth and the film thickness respectively. Since in the experiments $t_Y << \lambda_{Yi} > t_{Au}$ and $\lambda_{Yi} > \lambda_{Nb} << t_{Nb}$, Eq.3 becomes:

$$D_i \approx \lambda_{Yi}^2 / t_Y \tag{5}$$

On the other hand, being $t_Y \ll \lambda_{Yi}$ and $\lambda_{Nb} \ll t_{Nb}$,Eq.4 becomes:

$$d_i \approx t_Y/2 + t_{Au} + \lambda_{Nb} \tag{6}$$

As a consequence, the ratio between the critical field in the two directions considered is:

$$B_{COA}/B_{COB} \approx \lambda_{YA}/\lambda_{YB} \tag{7}$$

However the observed anisotropy can be only qualitatively accounted because if the λ_Y values measured in Ref.14 are used, the expected value of the ratio B_{COA}/B_{COB} is about 3.5. In the present case a value as large as 10 is measured. A similar result is also reported in Ref.13 and it is generally considered as compatible with the discussion reported above.

3. *Proximity effect in a S/N system*

The occurrence of a low I_cR_N value in YBCO-Au-S junctions has been discussed[8] in the framework of the de Gennes approach to the proximity effect[15]. In the dirty limit approximation, a depression of the order parameter Ψ_S in the superconductor is prevised. Referring to Fig.3, a measure of the depression is provided by the ratio b/ξ_S, where b is the extrapolation length defined as:

$$b^{-1} = \frac{1}{\Psi} \left.\frac{d\Psi}{dx}\right|_{x=0} \tag{8}$$

and $\xi_S = (hD_S/2\pi k_B T)^{1/2}$ is the coherence length, D_S is the diffusion coefficient and the other symbols have the usual meaning.

In Ref.8 b/ξ_S has been calculated:

$$\frac{b}{\xi_S} = \frac{N_S(E_F)}{N_N(E_F)} \frac{\xi_S}{d_N} \tag{9}$$

where d_N is the thickness of the normal layer, $N_S(E_F)$ and $N_N(E_F)$ are the densities of states at the Fermi energy in the superconductor and in the normal metal respectively .

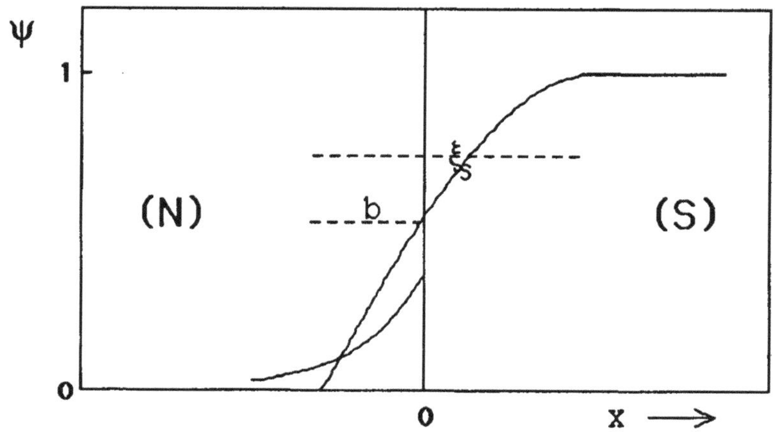

Fig.3 Schematic view of the spatial dependence of the superconducting order parameter near the interface of a S/N system.

In a YBCO/Au bilayer, due to the small value of ξ_S, a large depression of the order parameter and of the energy gap Δ is expected. Being $N_S(E_F) \approx N_N(E_F)$ and assuming $d_N \approx 10$ nm, for c-axis oriented YBCO films, such a situation is probably so critical to avoid any superconducting coupling in a junction based on it. In a-axis oriented films, a larger value of $\xi_S \approx 1$ nm leads to $b/\xi_S \approx 0.1$. These unfavorable conditions are also due to a $N_N(E_F)$ value[16] larger than expected from the carrier density n as measured by Hall effect. This is not surprising if an anisotropy of the Fermi surface is considered. It invalidates the $N_N(E_F)$ proportionality to n. For instance, it is easy to verify that $N_N(E_F)$ is proportional to the electron effective mass m* rather than to n for a cylindrical Fermi surface[17].

In order to limit the interface depression of Δ, only a suitable choice of the normal layer material can compensate a short ξ_S by a more favorable $N_S(E_F)/N_N(E_F)$ ratio. So low $N_N(E_F)$ materials (for instance degenerate semiconductors) could be more suitable for fabricating HTCS-N-S devices exhibiting high I_cR_N values.

To discuss such a possibility, it is worth noting that the approach of Ref.8 is not properly adopted. It is strictly valid only at temperatures $T_{cS} \approx T_{cN}$ and $T >> |T_{cS} - T|$, where $T_{cS(N)}$ is the transition temperature of the superconducting (normal) layer. Experiments on HTCS-N-S structures do not meet such conditions. Barone and Ovchinnikov[9] have extended the de Gennes approach by a description in terms of Green functions for $T_{cN} << T_{cS}$.

In this framework, the continuity of the logarithmic derivative of Ψ at the S/N interface cannot be still assumed. It is substituted by the following relation for Δ:

$$\frac{d\Delta}{dx} = \Delta\left(\frac{8T}{\pi D_S}\right)^{1/2}\frac{1}{\gamma} \tag{10}$$

where γ is a parameter depending on the value of the ratio:

$$\beta = \frac{N_N(E_F)}{N_S(E_F)}\left(\frac{D_N}{D_S}\right)^{1/2} \tag{11}$$

The dependence $\gamma(\beta)$ is reported in Fig.4 for $T_{cN} = 0$.

Depending on the $N_N(E_F)/N_S(E_F)$ ratio, the two asymptotic limits of Eq.10 give:

$$\begin{aligned} &\Delta(x=0)=0 && \text{for} \quad N_N(E_F)/N_S(E_F) \to \infty \\ &\left.\frac{d\Delta}{dx}\right|_{x=0} = 0 && \text{for} \quad N_N(E_F)/N_S(E_F) \to 0 \end{aligned} \tag{12}$$

The second limit suggests that a quite small value of N_N leads to an undepressed Δ at the S/N interface.

The considerations previously reported can be extended to SIS junctions and to point contact systems. Due to degradation effects during the junction fabrication processes, the YBCO surface layer becomes not superconducting. The actual junction results in a S/NIN/S structure.where the superconducting gap is depressed leading to a reduction of the critical current density. In such a situation the I_cR_N value has been calculated, in the framework of an elastic tunneling process, by Mannhart and Martinoli[18] for an arbitrary depression of the order parameter.

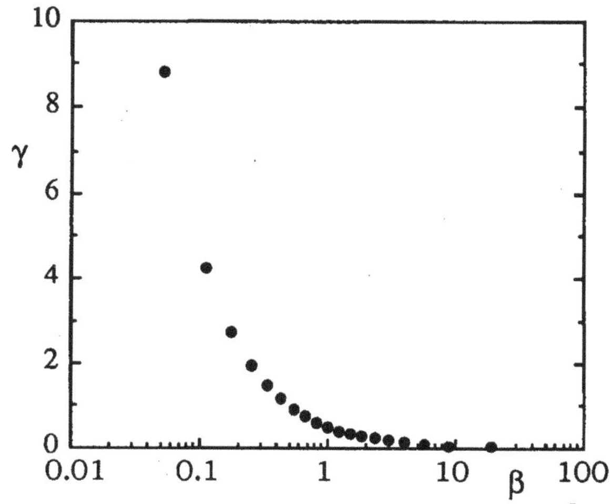

Fig.4 Dependence $\gamma(\beta)$ as calculated by Barone and Ovchinnikov[9], for $T_{cN}=0$.

In literature I_cR_N product values ranging from some to some hundreds of μV are reported. As discussed above, small values are related to the use of a normal layer by either gold or silver. A higher value $I_cR_N \cong 1$ mV at T= 4.2 K has been measured in YBCO/YBCO step-like junctions using an Ag-Au alloy (50:50 atomic percent) as normal layer[20]. This alloy exhibits a resistivity five times larger than that of pure silver. YBCO/YBCO edge junctions with a degenerate semiconductor ($SrTiO_3$:Nb) as normal layer, Chin and Van Duzer[20] have measured $I_cR_N \cong$ 1-3 mV. In such structures at a niobium concentration 10^{17}-10^{20} cm^{-3} the $SrTiO_3$ exhibits a metallic behavior with a carrier density roughly equal to the doping concentration and an estimated $\xi_N \cong 11.6$ nm. Due to an excellent lattice matching to the a-axis YBCO, $PrBa_2Cu_3O_{7-y}$ also provides good quality N-like barriers[21,22]. The I_cR_N values reported are noticeably large. In Fig.4 data concerning YBCO-$PrBa_2Cu_3O_{7-y}$-YBCO as reported by Barners et al. are compared with the results obtained by Yoshida et al.[5] for Au/Ag barrier junctions. The I_cR_N values are reported as a function of the critical current density at T=4.2 K. In agreement with the density of states in $PrBa_2Cu_3O_{7-y}$ (6×10^{19} $cm^{-3}eV^{-1}$)[23], two order of magnitude lower than that of Ag/Au, larger I_cR_N have been obtained.

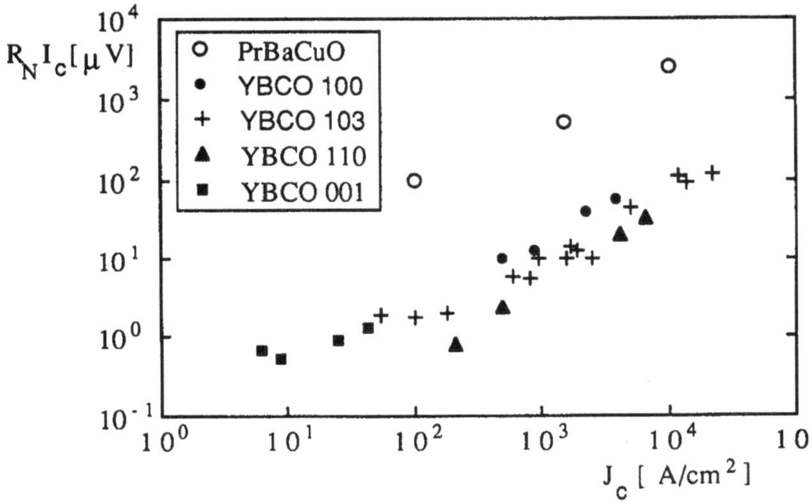

Fig.5 Experimental dependence of I_cR_N on the critical current density for YBCO-Au/Ag-Pb/In junctions at 4.2K (after Yoshida et al.) and for YBCO-$PrBa_2Cu_3O_{7-y}$-YBCO junctions at 10K (after Barner et al.). The data of Yoshida refers to different base electrode orientations, as specified in the inset.

However the simple model previously discussed should be carefully applied in the case of semiconductors, due to a possible occurrence of interface potentials, as Schottky barriers, and conductivity mechanisms other than diffusion ones in the limit of low density of carriers, like Mott variable range hopping[25].

4. *Acknowledgements*

This work is partially supported by CNR under the Progetto Finalizzato *Tecnologie Superconduttive e Criogeniche.*

5. *References*

1) A.N.Pargellis, F.Sharifi, R.C.Dynes, B.Miller, E.S.Hellman, J.M. Rosamilia, E.H.Hartford Jr, Appl. Phys. Lett. **58** (1991), 95.
2) R.Gross, P.Chaudari, M.Kawasaki, A.Gupte, IEEE Trans. Magn. **MAG 27** (1991), 3227.
3) A.I.Braginsky , Physica **C 185-189** (2091), 391.
4) H.Akoh, F. Shinoki, M.Takahashi, S.Takada, Jap. J. Appl. Phys. **27** (1988), L519.
5) J.Yoshida, T.Hashimoto, S.Inoue, M.Sagoi, K.Mizushima, in IC-SQUID Proc. (1991), in press
6) T.Nishino, Y.Tarutani, H.Hasegawa, T.Fukazawa, S.Saitoh, H.Nagane, U.Kawabe, Appl. Phys. Lett. **57** (1990), 828.
7) A.Barone, G.Paternò, "Physics and Application of Josephson Effect " J. Wiley and Sons (New York 1982), Chap. 11.
8) G.Deutscher, R.W.Simon, J.Appl. Phys **69** (1991), 4137.
9) A.Barone, Yu.N.Ovchinnikov, Sov. Phys. JETP **50** (1979), 735.
10) M.Matsumoto, H.Akoh, S.Takada, J. Appl. Phys. **66** (1989), 3907.
11) H.Akoh, F.Shinoki, M.Takahashi, S.Takada, IEEE Trans. Magn. **25** (1989), 795.
12) H.Akoh, C.Camerlingo, S.Takada, Appl. Phys. Lett. **56** (1990) 1487.
13) H.Akoh, S.Kohjiro, C.Camerlingo, R.Yuasa, S.Takada, Physica **C 180** (1991), 227.
14) S.Kohjiro, T.Tachino, H.Akoh, S.Takada, Jap. J. Appl. Phys. **30** (1991), L1471.
15) D.G. de Gennes : Rev. Mod. Phys. **36** (1964), 225.
16) The experimental value is obtained by direct measurements of heat capacity.
19) V.Z.Kresin, S.A.Wolf, Phys. Rev. **B41** (1990), 4278.
18) J.M.Mannhart, P.Martinoli, Appl. Phys. Lett. **58** (1991), 643.
19) R.H.Ono,J.A.Beall, M.W.Cromar, T.E.Harvey, M.E.Johansson, C.D.Reintsema, D.A.Rudman, Appl. Phys. Lett. **59** (1991), 1126.
20) D.K.Chin, T.Van Duzer, Appl. Phys. Lett. **58** (1991), 753.
20) J.Gao, W.A.M.Aarnink, G.J.Gerritsma, D.Veldhuis, H.Rogalla, IEEE Trans. on Magn. **MAG 27** (1991), 3062.
22) J.B.Barner, C.T.Rogers, A.Inam, S.Bersey, Appl. Phys. Lett. **59**, (1991), 742.
23) D.Lew, Y.Suzuki, C.B.Eom, M.Lee, J.M.Triscone, T.H.Geballe, M.R.Beasley, Physica **C 185-189** (1991), 2553.

MICROWAVE PROPERTIES OF HIGH–T_c THIN FILMS

C. Attanasio, L. Maritato, S. Marra, and R. Vaglio

Dipartimento di Fisica, Università degli Studi di Salerno
Baronissi (Sa), I-84081, Italy

ABSTRACT

Microwave properties of high–Tc superconductors have been extensively investigated in the last years. HTCS film microwave circuits are indeed already used in practical devices and more extensive applications can be foreseen with further improvement of the surface *rf* properties of the films. Moreover *rf* measurements can provide basic informations concerning the nature of the superconducting pairing state and of the excitation spectrum. In the present paper the state of the art of the field is briefly reviewed and new experimental data are presented to show how the actual limitation in microwave performances of HTSC films are basically due to grain–boundary or twinning problems in the films. A modified version of a weak-link Josephson model is presented and is proved to account for most of the observed results. The analysis of recent data in terms of this model leads to values of the film parameters that agree well with results obtained in different contexts.

1. Generalities

Microwave components are expected to be between the first practical devices to be produced out of high T_c superconductors. After some initial insuccess, infact, the rapid development of deposition technologies has allowed the realization of high quality textured or epitaxial thin films that can present losses that are much lower than copper (Cu) at 77K, so that the realization of commercial resonators, filters, antennas and delay lines, operating at liquid nitrogen temperatures, can be easily foreseen. The main parameter that characterizes the losses in a metal is the surface resistance R_s. In Fig.1 the dependence of R_s versus frequency (f) is reported for the high T_c compound YBCO at 77K and it is compared with Cu. The behavior of niobium (Nb) at the typicall operating temperature of 4.2K is also reported[1]. From the figure many important conclusions can be drawn :

1–The surface resistance of YBCO at 77K can be substantially lower than Cu for high quality films and $f < 10GHz$.

2–The surface resistance of high T_c superconductors is still much higher than ordinary metallic superconductors.

3–The frequency dependence of R_s is well described by a quadratic law for a given sample, but there is a a large scattering of the data from sample to sample.

Though the first conclusion is fairly encouraging, and the second is not, it is clear that a lot can still be done to improve the microwave performances of high T_c thin films, since, as clear from point 3, the values of R_s are still limited by non intrinsic

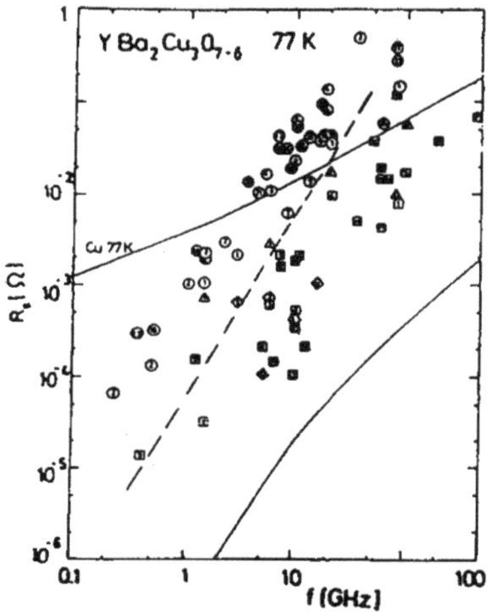

Fig.1 Surface resistance versus frequency for different YBCO samples[1]. The continous lines, for Cu at 77K and for Nb at 4.2K, are given for comparison. The broken line is a guide to the eye refering to the sample number 9 and corresponding to an f^2 behavior.

properties. In Fig.2a the temperature dependence of the surface resistance at fixed f is reported for an high quality YBCO film and for a polycrystalline sample. This type of behavior is typicall of all high T_c materials, with higher quality samples showing lower losses. The continous curve is the predicted behavior if one assumes a BCS temperature dependence of R_s and reasonable film parameters. The data are generally consistent with a BCS behavior (with $2\Delta/kT_c$=3.5–5), only very close to T_c[2], whereas at lower temperatures an anomalously high, sample dependent, residual term is present and dominates the losses. Many studies have pointed out how this anomalous term is strongly related to the film morfology. Moreover this term is strongly dependent on both dc and rf magnetic field amplitude, the dependence being high for polycrystalline samples and much lower for epitaxial films (Fig.2b).

2. Models

In the BCS framework, the surface resistance of a superconductor (if local electrodynamics holds) can be generally written as:

$$R_{BCS} = \sqrt{\frac{\mu_0\omega}{2}} \frac{\left[(\sigma_1^2+\sigma_2^2)^{1/2}-\sigma_2\right]^{1/2}}{(\sigma_1^2+\sigma_2^2)^{1/2}} \tag{1}$$

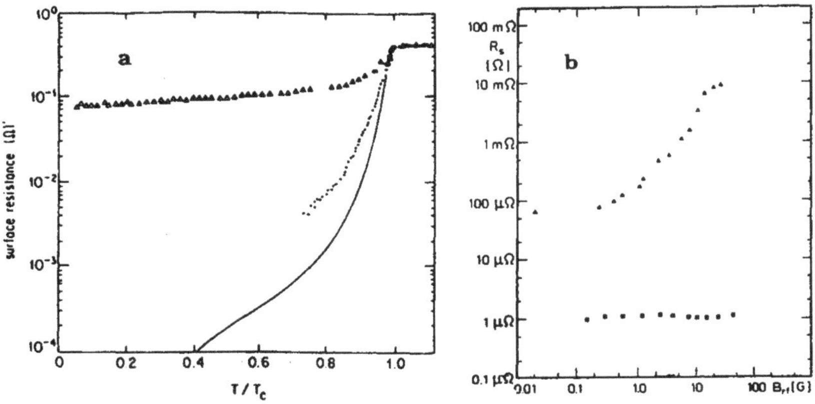

Fig.2 a) Surface resistance versus reduced temperature at the frequency of $100 GHz$ for an high-quality YBCO thin film (points) and for a polycrystalline sample (open triangles). The continous line is the BCS expression. b) YBCO surface resistance versus rf magnetic field amplitude at T=4.2K and f=821MHz for a polycrystalline (open triangles) and for a large epitaxial sample (full squares).

where σ_1 and σ_2 are, respectively, the real and the imaginary part of the superconductor complex conductivity. This expression simplifies, for $T \leq T_c/2$, to:

$$R_{BCS} = \frac{1}{2\lambda_1}\frac{\sigma_1}{\sigma_2^2} \tag{2}$$

where $\lambda_1 = (\mu_0\omega\sigma_2)^{-1/2}$ is the magnetic field penetration depth.
In a real superconductor, however, the surface resistance can be written as

$$R_s = R_{BCS} + R_r \tag{3}$$

where R_r is a temperature independent "residual" term that generally exhibits an f^2 dependence. Since in the BCS theory σ_1 it is almost frequency independent and it is $\sigma_2 \sim 1/f$ also R_{BCS} shows a quadratic frequency dependence. This implies that we can formally write R_s as in (2) adding to σ_1 a "residual conductivity" σ_r that is temperature and frequency independent.

Since it is $\sigma_1 \sim n_n \sim exp(-\Delta/kT_c)$, where n_n is the number of "normal electrons", the presence of a residual conductivity is in all equivalent to the assumption of a "residual" number of normal electrons. In this respect the anomalously

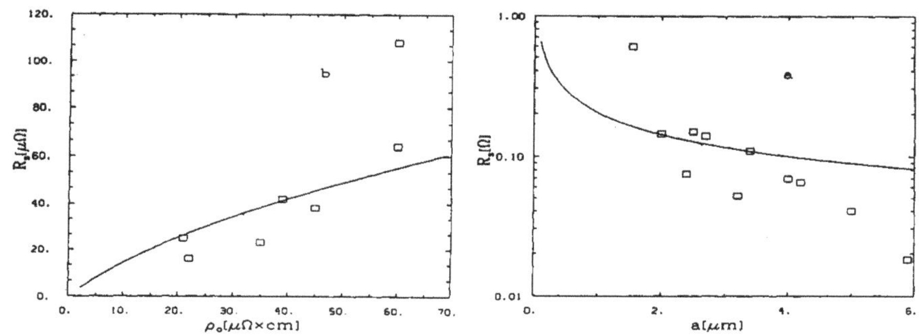

Fig.3 a) Surface resistance versus the average grain dimension for YBCO samples[1]. The continous line is the best fit curve obtained by using equation (4)(see text). b) Surface resistance as a function of the residual normal resistivity[5]. The continous line is the best fit curve.

large surface resistance observed in high T_c compounds can be simply related to an anomalously large number of unpaired carriers even at low temperatures.

However there are also other possible explanations. As an example, following previous works[3], we have recently proposed a model where an high T_c superconductor is schematically treated as a network of superconducting grains coupled via Josephson junctions[4]. The grain boundaries are here modeled by a tunnel barrier of resistance R_j and critical current density J_c. In these approximations, for $T \leq 1/2\ T_c$ and at low frequencies, we get:

$$R_{sj} = \frac{1}{2\lambda_{eff}} \frac{\sigma_1 + \sigma_r}{\sigma_2^2} \tag{4}$$

where $\lambda_{eff} = \sqrt{\lambda_1^2 + \lambda_j^2 F^{-1}}$ is the effective field penetration in the film, $\lambda_j = (\hbar/a2eJ_c\mu_0)^{1/2}$ (a is the average grain dimension), $F(I) = [1-(I/I_c)^2]^{1/2}$ (I is a current proportional to the dc or rf applied field) and $\sigma_r = \mu_0\omega\lambda_j^2(\hbar\omega/2eI_cR_j)\sigma_2^2F^{-2}$ is temperature and frequency independent. The "grain boundaries model" is clearly equivalent to the "unpaired electrons model", but in this case the magnetic field dependence of R_r is also naturally explained since σ_r (through $F(I)$) increases as a function of an applied dc or rf field. This effect is clearly more pronounced for more granular samples.

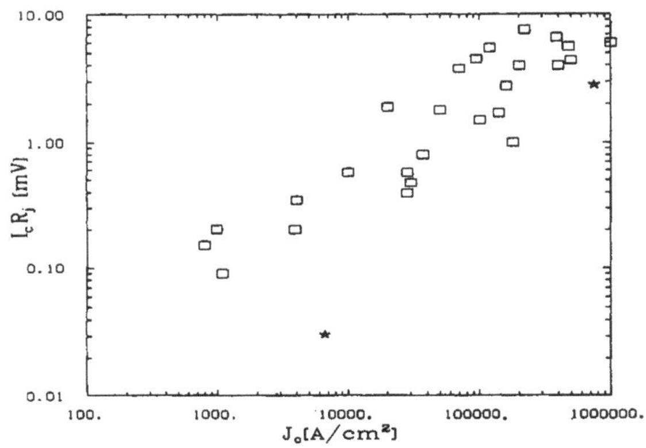

Fig.4 I_cR_j product as a function of the critical current density. The data have been measured by $I-V$ characteristics of YBCO grain-boundary junctions[6]. The two stars are the values obtained by fitting the surface resistance data, on YBCO samples, by the Josephson model.

3. Analysis of recent experimental results

The grain boundary model can be checked by careful measurements on polycrystalline films of different morfology. We are presently conducting experiments in this direction in collaboration with the Physics Department of the University of Naples, by measuring the surface resistance by means of a ring stripline resonator on well characterized high T_c superconducting films. In this paper, however, we have used the model to fit some data already reported in literature by other groups.

In Fig.3a surface resistance values as a function of the average grain dimension, for several polycrystalline YBCO samples, are shown[1]. Measurements have been performed at f=21.5GHz at the temperature of 77K. By using equation (4) a two parameter fit has been performed on this set of data. The best fit value for the critical current density is $J_c = 6.6 \times 10^3\ A/cm^2$ and for the I_cR_j product is $I_cR_j = 31\mu V$. The fitting curve is the solid line in figure 3a. A similar procedure has been adopted to fit data of surface resistance versus the residual resistivity ρ_0 for YBCO thin films[5]. The measurements were done at 10GHz and 4.2K. We have assumed, because of the very high film quality, that the extrapolated zero temperature normal state (residual) resistivity was the same as the grain boundary resistivity, $\rho_G = \frac{I_cR_j}{J_ca}$. We have found, as best fit parameters, $I_cR_j = 2.83\ mV$ and $J_c = 7.44 \times 10^5\ A/cm^2$, choosing as the average grain dimension $a = 1\mu m$. The data and the best fit curve are reported in Fig.3b. In the analysis of these two groups of data the YBCO grains have been treated as having a normal BCS behavior. The σ_1 and σ_2 values have been computed by using the Mattis–Bardeen integrals with a normal state resistivity at the transition $\rho_n = 65\mu\Omega \times cm$, $\lambda_1 = 1500$ Å, $2\Delta/kT_c = 3.52$, and, as far as concerned the data of reference 1, $T_c = 90K$. The, results of the fitting procedure, give, qualitatively, the right behavior for the surface resistance as function of the grain dimension and of the grain boundary resistivity. In addition the values obtained for J_c and for I_cR_j seem to be consistent with values for YBCO thin films[6] measured with a different technique (Fig.4). Our two points (stars in Fig.4) well agree with the others, showing a I_cR_j product almost proportional to the critical current density.

4. References

1. H. Piel and G. Müller, *IEEE Trans. on Magn.* **27** (1991), 854.
2. L. Drabeck, K. Holczer, G. Grüner, J.J. Chang, D.J. Scalapino, A. Inam, X.D. Wu, L. Nazar, and T. Venkatesan, *Phys. Rev.* **B42** (1990), 10020.
3. T.L. Hylton, A. Kapitulnik, M.R. Beasley, J.P. Carini, L. Drabeck, and G.Grüner, *Appl.Phys.Lett.* **53** (1988), 1343.
4. C. Attanasio, L. Maritato, and R. Vaglio, *Phys. Rev.* **B43** (1991), 6128.
5. S.S. Landerman, R.C. Taber, R.D. Jacowitz, J.L. Moll, C.B. Eom, T.L. Hylton, A.F. Marshall, T.H. Geballe, and M.R. Beasley, *Phys. Rev.* **B43** (1991), 2922.
6. R. Gross, P. Chaudhari, M. Kawasaki, and A. Gupta *Phys. Rev.* **B42** (1990), 10735.

SIN AND SIS QUASIPARTICLE CHARACTERISTICS OF HIGH T_c TUNNEL JUNCTIONS

J.F. ZASADZINSKI AND N. TRALSHAWALA
Illinois Institute of Technology
Chicago, Illinois 60616 USA

Q. HUANG, JUN CHEN AND K.E. GRAY
Science and Technology Center for Superconductivity
Argonne National Laboratory
Argonne, Illinois 60439 USA

ABSTRACT

SIN and SIS tunnel junction characteristics in the energy gap region are reported for several superconducting oxides. A pattern has emerged which suggests that the tunneling density of states of cuprates has a broadened BCS shape, but with distinctive features which cannot be explained by the addition of an imaginary term, Γ, to the BCS quasiparticle energy. We also report measurements of proximity effect tunneling into Au overlayers on $Bi_2Sr_2CaCu_2O_8$. A reduced energy gap is observed through Au layers as large as 60 nm thick.

1. Introduction

A number of tunnel junction devices, including switches and mixers, require quasiparticle current-voltage, I(V), characteristics which have nearly ideal, BCS behavior. These requirements include low sub-gap conductance and a sharp current rise at the gap voltage, $eV=\Delta$. However, cuprate superconductors all exhibit non-ideal characteristics[1], typified by a normalized zero-bias conductance $\sigma_s/\sigma_n(0)$ ~ 15-60%. The question is whether these tunneling characteristics are intrinsic to the cuprates (e.g. gaplessness or anisotropy) or are a manifestation of poor junction quality arising from damaged surfaces, proximity effects, bad tunnel barriers, etc. We have examined the tunneling conductances of several cuprate superconductors in both the SIN and SIS geometries and compared our results to those in the literature. A pattern has emerged which suggests that the tunneling density of states (dos) in cuprates has a non-BCS shape but with distinctive, reproducible characteristics. While this shape can be loosely described as "broadened BCS", we demonstrate that the introduction of an imaginary term, Γ, to the BCS quasiparticle energy does not describe accurately the tunneling density of states.

The tunnel current in an SIN or SIS junction can be written,[2]

$$I(V) = Ct^2 \int_{-\infty}^{\infty} \rho_1(E)\rho_2(E+eV)[f(E)-f(E+eV)]dE \quad . \qquad \text{Eq. 1}$$

Here, $\rho_1(E)$ and $\rho_2(E)$ are the quasiparticle densities of states in the two electrodes and C is a constant which contains the average band structure dos so that $\rho(E) = 1$ for a normal metal. The tunneling matrix element, t^2, has an energy dependence which is assumed to be weak over the voltage range of the energy gap, and thus has been taken out of the integral. In the limit, T=0 K, the tunneling conductance, $\sigma_S = dI/dV$, for an SIN junction becomes, for E=eV,

$$\sigma_S = Ct^2\rho(E) \qquad \text{Eq. 2}$$

Eq. 2 simply states that a measurement of the tunneling conductance at low temperatures should reveal the quasiparticle density of states, which for a BCS superconductor is given by,

$$\rho(E)_{BCS} = E/(E^2-\Delta^2)^{1/2} \qquad \text{Eq. 3}$$

An exact determination of $\rho(E)$ requires a measurement of the weak voltage-dependent background (or normal state) conductance to divide out the prefactors in Eq. 2, but such a measurement can be difficult due to the high Tc values of some cuprates. Because of the obvious additional broadening observed in cuprate junctions, the normalized conductance has generally been fit to a smeared BCS expression[1] first introduced by Dynes[3] to describe inelastic scattering of quasiparticles, and includes Γ as a smearing parameter.

$$\rho(E)_{smear} = \mathrm{Re}\ E-i\Gamma/[(E-i\Gamma)^2-\Delta^2]^{1/2} \qquad \text{Eq. 4}$$

We should note here that Eq. 4 has served merely as a convenient expression to determine the gap parameter, Δ, and that there is no underlying theoretical basis for its use to describe cuprate superconductors. In fact, as we will show, Eq. 4 does not really provide an accurate fit of the tunneling dos in cuprates.

2. Experimental SIN and SIS data

We report here representative tunneling data on bulk, polycrystalline $Ba_{1-x}K_xBiO_3$ (BKBO) and single crystals of $Nd_{2-x}Ce_xCuO_4$ (NCCO), and $Bi_2Sr_2CaCu_2O_8$ (BSCCO). Tunnel junctions were formed using the point-contact method[4,5] which has been shown to reproduce well the quasiparticle dos in Nb observed in planar junctions[6], including the subtle phonon structure which represents a deviation of ~ 1% in the conductance. The advantage of the point contact method for cuprates is that the tip can be used to scrape, clean and cleave the sample surface. Typically a Au tip was used to make SIN junctions, but in the case of BSCCO, the cleaving of the surface sometimes resulted in SIS junctions[5], presumably between a BSCCO flake and the underlying bulk material. The SIS junctions were readily identified by their unusually symmetric $\sigma_S(V)$, large gap values and prominent dip features as will be shown.

The only oxide superconductor with Tc>25K that displays a BCS energy gap is BKBO[4] as shown by the point contact junction in Fig.1. However, planar junctions can exhibit considerable broadening[7] of the BCS features. Fig. 1 (2 through 4) shows such junctions and fitting parameters from Eq. 4.

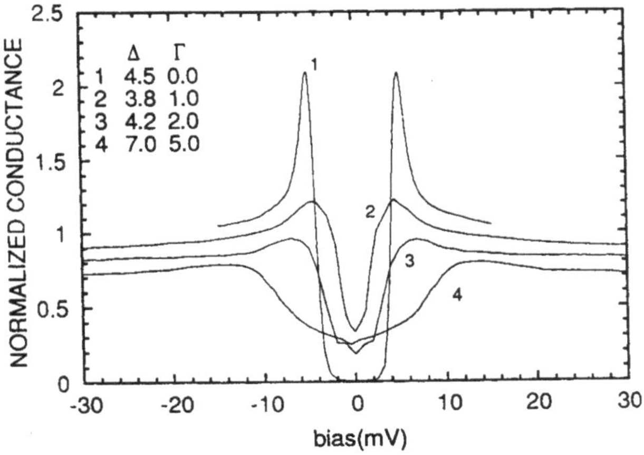

Figure 1. Normalized tunneling conductances for 1) point contact junction and 2) through 4) planar junctions on bulk pellets of BKBO. The point contact junction exhibits ideal, BCS behavior. Δ and Γ values are shown in the inset for the smeared BCS fits.

The point contact junction on BKBO in Fig. 1.1 can be fit using the BCS dos (Δ=4.5 meV) and Eq. 1 with thermal smearing only[4]. BKBO thus displays ideal characteristics and has no fundamental limitation for devices. However, planar junctions grown on the air exposed surfaces of the BKBO pellets using a thin film In counterelectrode[7] result in the types of curves (2 through4) shown in Fig. 1. These prove that poor surfaces can result in significantly broadened conductance curves. An important point to note is that the broadened junctions exhibit a conductance peak at a voltage (eV) which is much larger than the gap parameter, Δ= 4.5 meV, in BKBO. It is for this reason that Eq. 4 is used to extract the gap value in those cases where broadening is significant. Using the tunneling conductance peak to determine the gap energy may result in large overestimates. In the case of severe broadening as in Fig 1.4, the use of Eq. 4 still gives Δ values that are too large, indicating that it does not properly model the broadening.

The tunneling characteristics of cuprate superconductors generally resemble the smeared conductances of the planar BKBO junctions in Fig. 1 Some of the best tunnel junctions on cuprates have been obtained for NCCO (Tc=22K) and BSCCO (Tc=86-96K) and typical conductance curves are shown in Figs. 2 and 3 respectively. What is most notable is the similarity in shape of the superconducting conductances. Except for the difference in energy scale of the gap features arising from the widely different Tc values , it would be difficult to distinguish between BSCCO and NCCO. Thus the non-BCS-like features seem to be reproduced in these two cuprates. These features include a zero-bias conductance of 10-15% of the high bias value and a monotonic increase of the conductance about zero bias. This is in contrast to the near-zero and flat-shaped conductance at zero bias which is characteristic of a BCS superconductor as seen in Fig. 1.1 for BKBO.

Another peculiar feature is the dip in conductance for $eV>\Delta$ seen most clearly in the positive bias data for BSCCO in Fig. 3. We have noted this dip and its asymmetric character previously,[5] and it is a common feature of BSCCO data when the tunneling conductance is sharp, i.e., $\sigma_s/\sigma_n(0)$ <25%. Conductance asymmetry is intrinsic to the SIN geometry due to the barrier term t^2 in Eq. 2, but as we will show, SIS junctions on BSCCO display prominent, symmetric dips in conductance.

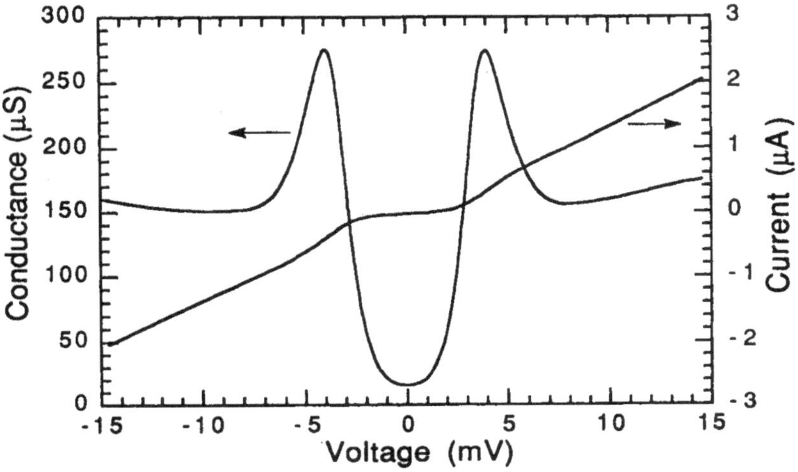

Figure 2. I(V) and dI/dV for point-contact tunnel junction on a single-crystal of NCCO using a Au tip.

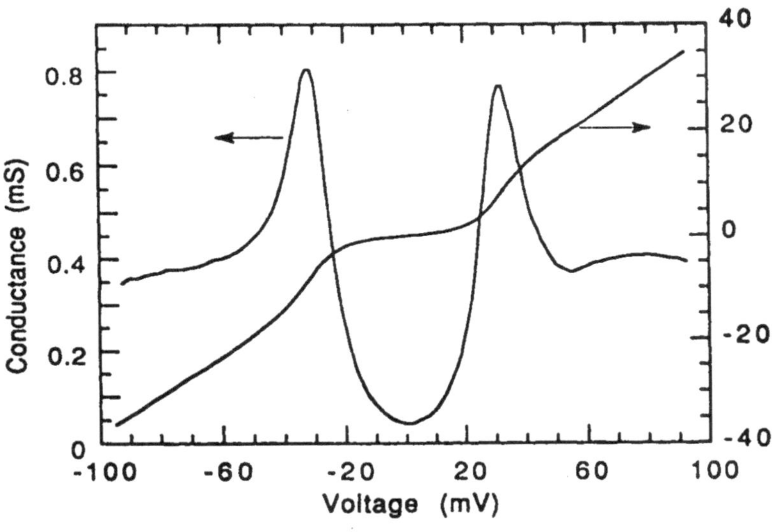

Figure 3. I(V) and dI/dV for point-contact tunnel junction on a single-crystal of BSCCO using a Au tip.

The fit of the normalized NCCO data of Fig.2 using Eq. 4 is shown in Fig. 4a (dashed line). The fit parameters are Δ=3.5 meV and Γ=0.35 meV. Several important differences with the data (solid line) are noteworthy. The BCS dos even with smearing is still flat near zero bias whereas the data exhibit the monotonic, parabolic-like increase. Also, the smeared BCS dos smoothly approaches unity at high energy, whereas the NCCO data display a more abrupt change of slope and a negative deviation from the BCS dos. This subtle effect is much more pronounced in the generated SIS curve of Fig. 4b, where noticeable dips in conductance are evident. The generated SIS curve

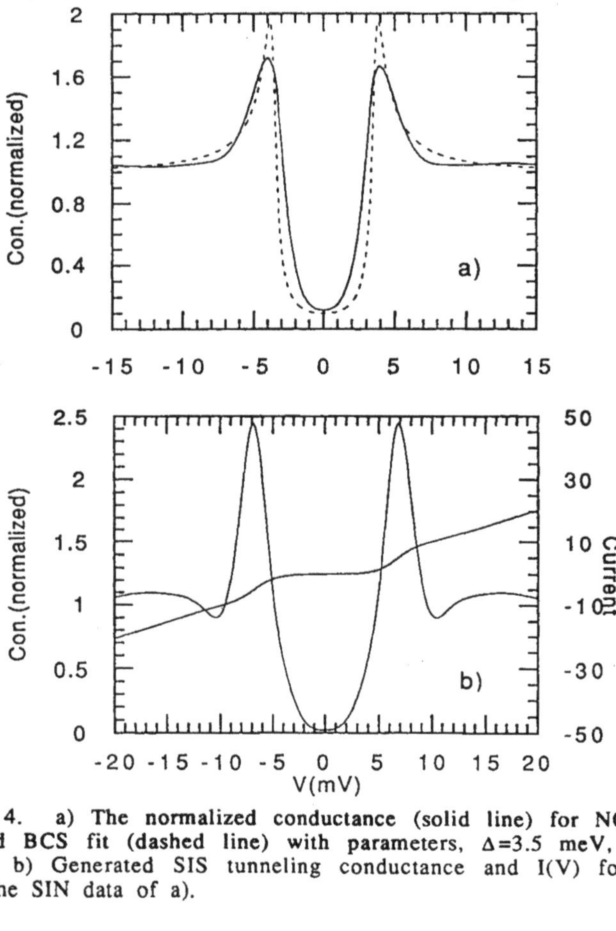

Figure 4. a) The normalized conductance (solid line) for NCCO and smeared BCS fit (dashed line) with parameters, Δ=3.5 meV, Γ=0.35 meV. b) Generated SIS tunneling conductance and I(V) for NCCO using the SIN data of a).

is obtained using Eq. 1 and the normalized NCCO conductance data for both $\rho_1(E)$ and $\rho_2(E+eV)$. While we have not obtained measured SIS curves for NCCO, we will show that the generated I(V) curves resemble those measured on SIS junctions of BSCCO.

There are a number of tunneling studies of BSCCO in the SIS configuration using break junctions[8] or mechanical contacts[9,10] and all show similar results. We compare two of our SIS point-contact junctions on BSCCO, including a Pb-doped sample, with the break junction work of Forro et al[8] and the single-crystal edge junction[9] study on Pb-doped samples. Fig. 5 a) through d) shows these two published results along with our point-contact SIS junctions. The data have been normalized by a constant (σ(187mV)) and shifted by 1 for clarity. There is a strong similarity in the shapes of the curves, indicating that the observed features are reproducible from sample to sample and among quite different tunneling methods. These features include a weakly decreasing background conductance and a conductance at zero bias which is low (~1% of the high bias value) but which increases monotonically about zero bias. In particular, we note the sharp dips observed near ~90 mV in each junction.

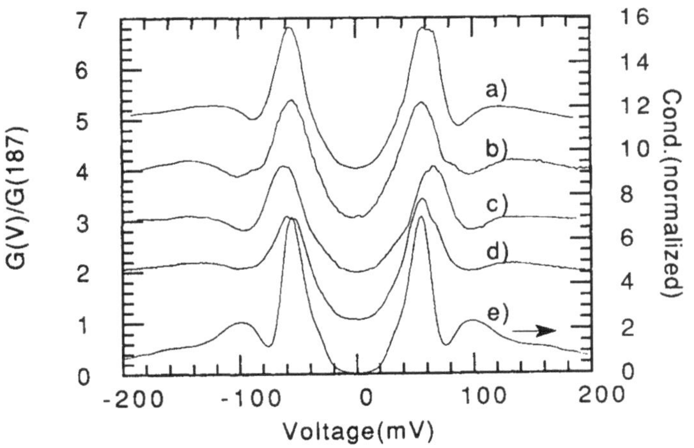

Figure 5. Comparison of SIS tunnel junctions of BSCCO from our point-contact study and published work. a) single-crystal edge junction, Pb-doped sample (ref. 9), b) break junction (ref. 8), c) point-contact junction, Pb-doped sample, d) point-contact junction. Curve e) is an SIS conductance generated from the SIN junction shown in Fig. 3.

Using the SIN data on BSCCO of Fig. 3 we have generated an SIS curve using Eq. 1 and this is shown in Fig. 5e. The SIN data have been normalized by a constant and thus the voltage dependent background has not been normalized out and shows up in the generated SIS curve, matching the observed experimental background conductance seen in curves a) through d). Most of the important features of the measured SIS junctions have been reproduced by the generated SIS curve implying a consistency between measured SIN and SIS junctions. In particular, the dip feature which is a subtle effect in the SIN data is much more prominent in the generated SIS curve. This enhancement effect explains why the measured SIS junctions always seem to show this conspicuous dip feature.

Just as the SIN data of BSCCO and NCCO have similar shapes except for the energy scale, the generated SIS curves of Figs. 5e) and 4b) are similar as well. Thus the tunneling dos in these two cuprates appear to have common, non-BCS-like features, and in the case of BSCCO these features are reproduced in many different tunneling experiments. We have also shown that these characteristics cannot be modeled by a simple, smeared BCS dos as in Eq. 4. The observation in BKBO junctions of both ideal and broadened BCS characteristics always leaves open the question of whether the non-BCS-like features observed in cuprates are due to surface problems arising from disorder, proximity effects, poor tunnel barriers, etc. The similarity between two cuprates with widely different Tc values may have an important bearing on this question. It suggests that the tunneling dos is revealing an intrinsic property of cuprates, either a bulk effect originating from the nature of the pairing interaction or a surface effect which in some manner is reproduced among widely different tunneling methods.

3. Proximity Effect Studies

Proximity effect tunneling studies of conventional superconductors with normal metal overlayers have provided a wealth of fundamental information[2] and have led to technological advances, e.g. Nb/Al Josephson junctions. For these reasons, we have investigated junctions on single crystals of BSCCO which have been cleaved in vacuum and immediately coated with a thin layer of Au to protect the surface. An important question is whether the underlying energy gap of the BSCCO is observable through the Au layer.

In Fig. 6 is shown a selected set of tunneling data for various Au thicknesses in the range 200-600 A. Here a soft In tip was used so as not to perforate the Au layer and also the In-oxide surface of the tip provides a reasonably good tunnel barrier.

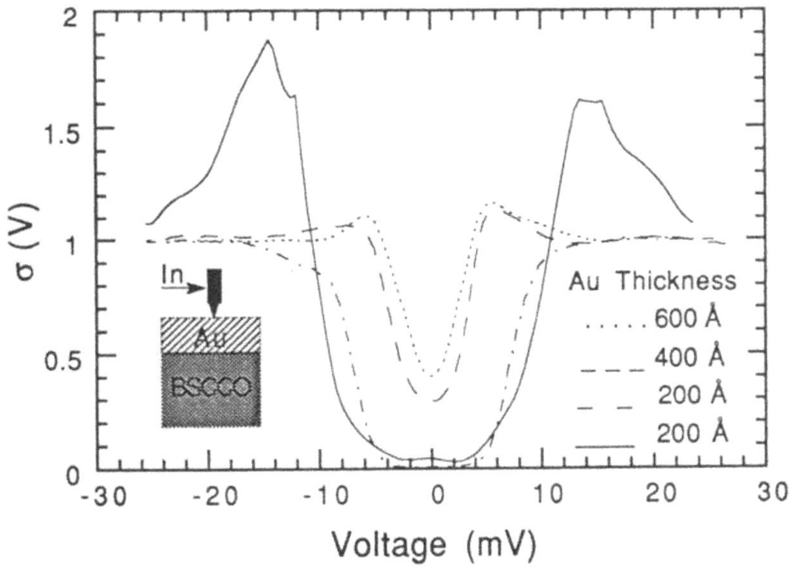

Figure 6. Normalized tunneling conductance for Au/BSCCO bilayers with various Au thicknesses. The tunneling geometry is shown in the inset.

The observed gap parameter of the Au/BSCCO bilayers decreases in a systematic fashion with Au thickness, and even the 600 A layers typically show a gap parameter of ~ 3 meV. This reduction in apparent gap with layer thickness can be explained using the proximity effect theory of Arnold[11] by assuming the conductance peak is a measure of the bound-state energy in that theory. However, we have not yet fit the individual conductance curves with the Arnold theory. Two junctions with Au thicknesses of 200 A showed a zero-bias dos close to zero, in contrast to point contact junctions formed directly on the BSCCO surface which always displayed at least 10-15% zero-bias dos. This indicates that the proximity effect approach has potential as a technique for fabricating thin-film junctions with the low sub-gap conductance necessary for device applications.

4. Conclusion

To date, all tunneling data on cuprate superconductors exhibit features which may be described loosely as "broadened BCS" and the degree of broadening may differ from study to study. The observance of both broadened and ideal characteristics in the cubic oxide, BKBO, points out the importance of surface effects in tunneling measurements. However, the best tunneling data on cuprates exhibit non-BCS features that are reproducible from sample to sample and among a variety of tunneling techniques. These features include a monotonic increase in conductance about zero-bias and conductance dips at $eV>\Delta$. Also, we have demonstrated that there is a consistency among SIN and SIS junction data by showing that the pronounced dips observed in SIS junctions are arising from a more subtle, but observable feature in the SIN data.

Most importantly, there appears to be a similarity in the tunneling data of BSCCO and NCCO, two cuprates with widely different Tc values and different signs of charge carriers. This implies that the tunneling density of states is measuring an intrinsic property of cuprates, either a bulk effect arising from the nature of the pairing interaction or a surface effect which somehow is nearly identical in these two compounds and among different tunneling methods.

The question remains whether these non-BCS features can be eliminated and ideal behavior achieved as in BKBO. Our preliminary proximity effect study of Au/BSCCO bilayers indicates that sizeable gap structures can be observed through Au layers as thick as 600 Å. Two junctions with 200 Å layers exhibit reduced zero-bias conductance values compared to junctions made directly on BSCCO. This indicates that a proximity effect approach may be useful for achieving junctions with low sub-gap conductance.

5. Acknowledgements

Work supported by U.S. Department of Energy, Division of Basic Energy Sciences, Contract No. W-31-109-ENG-38 and NSF Office of Science and Technology Centers, Contract No. STC-8809854 (Q.H., J.C., and J.F.Z.)

REFERENCES

1. See review article by, Tetsuya Hasegawa, Hiroshi Ikuta and Koichi Kitazawa in Physical Properties of High Temperature Superconductors III, ed. D.M. Ginsberg (World Scientific Publishing, 1992)

2. E.L. Wolf, Principles of Electron Tunneling Spectroscopy Ch. 3 (Oxford Univ. Press, New York, 1985).

3. R.C. Dynes, V. Narayanamurti and J.P. Garno, Phys. Rev. Lett. **41**, 1509 (1978)

4. Q. Huang et al, Nature **347**, 369 (1990)

5. Qiang Huang, J.F. Zasadzinski, K.E. Gray, J.Z. Liu, and H. Claus, Phys. Rev. **B40**, 9366-9369 (1989).

6. Qiang Huang, J.F. Zasadzinski, and K.E. Gray Phys. Rev. **B42**, 7953 (1990).

7. J. F. Zasadzinski, N. Tralshawala, J. Timpf, D.G. Hinks, B. Dabrowski, A.W. Mitchell and D.R. Richards, in High Tc Thin Films and Single Crystals, Eds. W. Gorzkowski, M. Gutowski, A. Reich and H. Szymczak, (World Scientific, 1990)

8. D. Mandrus, L. Forro, D. Koller and L. Mihaly, Nature **351**, 460 (1991)

9. J.J. Wnuk, R.T.M. Smokers, F.W. Nolden, L.W.M. Schreurs, Y.S. Wang and H. van Kempen, Supercond. Sci. Technol. **4**, 412 (1991)

10. Qun Chen and K.W. Ng, Phys. Rev. **B 45**, 2569 (1992)

11. G.B. Arnold, Phys. Rev. **B18**, 1076 (1978)

Electron Tunneling Studies of Y-Ba-Cu-O Systems

A.M. Cucolo, R. Di Leo, P. Romano
Dip. di Fisica, Università di Salerno,
84081 Baronissi, Italy

Abstract

We have studied the superconducting and normal characteristics of planar junctions made on Y-Ba-Cu-O single crystals with different oxygen stoichiometries. Oxygen losses have been induced by heating the crystals with T_C=90K at 500 °C in vacuum. As the oxygen deficiency increases, the superconducting conductance curves show gap-like structures reduced in amplitude. The normal state conductances, measured at $T>T_C$, are highly linear in voltage and have deeper slopes on the thermally treated samples.

1. Introduction

Since the discovery of the high T_C superconductors, many attempts have been made to achieve good-quality tunnel junctions on these materials in an effort to understand the nature of the superconducting state. In fact for conventional superconductors tunneling measurements probe the quasiparticle density of states from which the value of the energy gap at the Fermi level as well as information about the pairing mechanism responsible for superconductivity can be achieved. Unfortunately, making planar junctions on high-T_C superconductors has proved non-trivial. Tunneling probes superconductivity within the material's coherence length ξ that in these compounds is extremely short and anisotropic with the highest values in the range of a few tens of angstroms. In the 1:2:3 materials more complications arise due to the easy oxygen losses from the surface. The role of the oxygen is in fact very critical and affect the superconducting transition temperature as well as other important physical properties. In the $YBa_2Cu_3O_{7-\delta}$ (Y-Ba-Cu-O) compound, as the oxygen content is varied from 7 to 6.5, the oxygen can assume a number of possible orderings, so that structural transitions occur corresponding to superconducting phases with T_C=90K and T_C=60K [1].
In this paper we study the tunneling characteristics of Y-Ba-Cu-O single-crystals with different oxygen stoichiometry. We have altered the oxygen

content of our samples by simply heating them at 500°C in vacuum since heating treatments are known to reduce theY-Ba-Cu-O compound[1]. Reproducibility of the data as well as the quality of the tunnel junctions have been carefully tested. We have divided the paper in two main sections devoted to the superconducting and the normal state properties since we think that the normal state tunneling reveals some unique features of this compound that can help in understanding their superconducting properties.

2. Junction preparation

20 Y-Ba-Cu-O single crystals have been studied in these experiments. They were grown in the same bacth by the flux method[2]. According to ac and dc susceptibility measurements, they had superconducting transition temperatures of 90K. We have always processed two samples together. Each couple of crystals was etched for 30 min in 1% Br solution in methanol. At this point in a vacuum system, equipped with a cryogenic pump, one of the crystals was kept at 500°C for 1 hour while the other one was not processed. Subsequently the crystals were masked with epoxy glue to define the junction geometry. The tunnel barrier were naturally formed by exposing the samples to the ambient air for about 30 min. As counterelectrode a Pb film was thermally evaporated through a metallic mask. Details of this fabrication procedure have been reported elsewhere[3,4].
We have investigated the effect of the heating treatments on the T_c of the single crystals by near zero-field microwave absorption. In treated samples, we have found a degradation of a surface region consistent with a formation of a small fraction of 60K phase.
Junction dimensions were about 1.0x0.1 mm^2. The resistances of the junctions at room temperature, at 100 mV, ranged between a hundred and a few thousand ohm. Four terminal measurements of the differential resistances were performed using a standard low frequency lock-in technique.

3. Superconducting tunneling

In a conventional superconductor/insulator/normal-metal (S-I-N) junction, in the limit of T=0K, the conductance as function of the voltage $G_S(V)$ is given by:

$$G_S(V) = N_0 \, P_0 \, N_S(E) \qquad (1)$$

where N_0 is the normal counterelectrode density of state, P_0 is the tunneling probability, both assumed constant in the considerate energy range (few meV), and $N_S(E)$ is the quasiparticle density of states of the superconducting material.

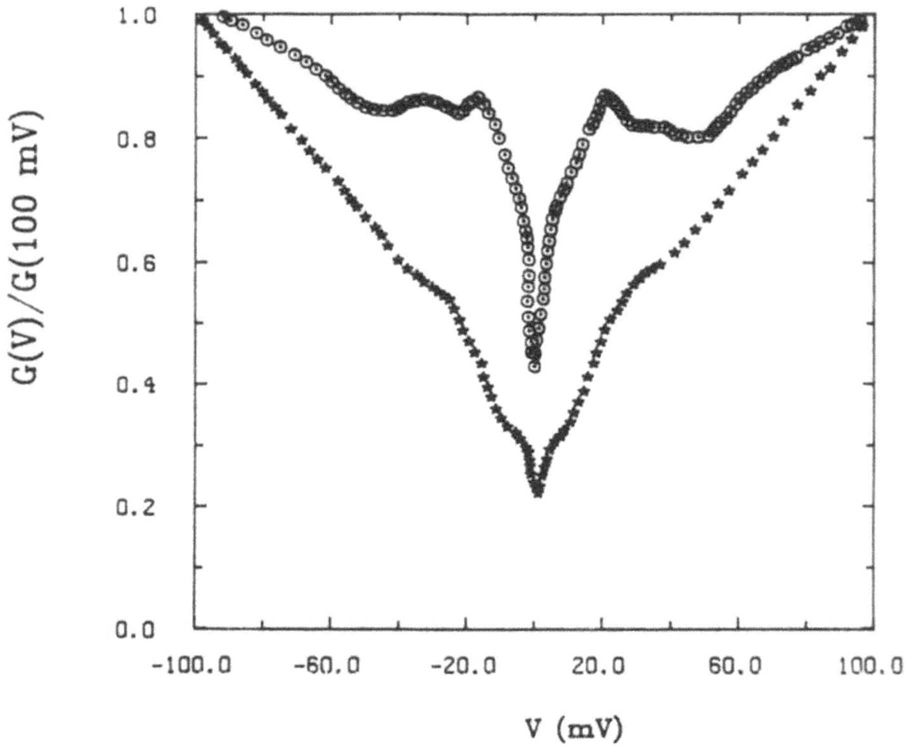

Fig. 1 Normalized conductance vs. voltage at T=8K for an untreated junction (circles) and for a thermally treated junction (stars).

In the weak-coupled BCS approximation

$$N_S(E) = N_n\,[E/(E^2-\Delta^2)^{1/2}] \qquad \text{for } E>\Delta$$
$$N_S(E) = 0 \qquad \text{for } E<\Delta \qquad (2)$$

where Nn is the normal density of state of the superconducting material.
Thus the $G_S(V)$ of a S-I-N junction measured at $T<<T_C$ is proportional to the renormalized quasiparticle density of states of the superconductor.
In Fig. 1 we report the $G_S(V)$ at T=8K normalized to the $G_S(100mV)$ value for two Y-Ba-Cu-O/Pb junctions: open circles refer to an untreated crystal and stars refer to a thermally reduced sample. Differences of these conductance curves with the BCS superconductors have been extensively discussed[5,6]; here we want to focus only on their similarities and relative changes.
We observe that:
a) at ±50 mV both conductances appear to have a striking linear dependence on the applied voltage, with the "degraded" junction (stars) exhibiting a steeper variation in voltage
b) both junctions show finite values of the zero-bias conductance and features at about ±4mV
c) the gap-like structures at about ±20 mV exhibited by untreated sampled are shifted to higher energies and strongly reduced in amplitude for "degraded" samples
d) the secondary maxima at ±36mV are completely lost for thermally treated junctions.
We have performed a number of checks in order to verify that we are free of spurious effects in our junctions. At 1.2K, in zero magnetic field, both junctions showed the Pb gap and phonon structures at the correct energies and of the correct amplitudes, indicating a single step tunneling process. Sub-gap leakage currents were less than a few percent, indicating good quality tunnel barriers, continuous and pin-hole free.
To further investigate the superconducting properties of our junctions, in Fig. 2 we show the normalized zero-bias conductance G(0)/G(100mV) as a function of temperature for untreated (circles) and treated (stars) junctions. The jump in the curve relative to the completely oxygenated crystal, is due to the contribution of the single crystals resistance going to zero at $T=T_C$, and the different behaviour above and below T_C is similar, though less pronounced, to what one would expect due to the opening of the energy gap in a metal-insulator-BCS superconductor junction. The conductance of the thermally treated sample show no discontinuity in the range 50-130K . These data

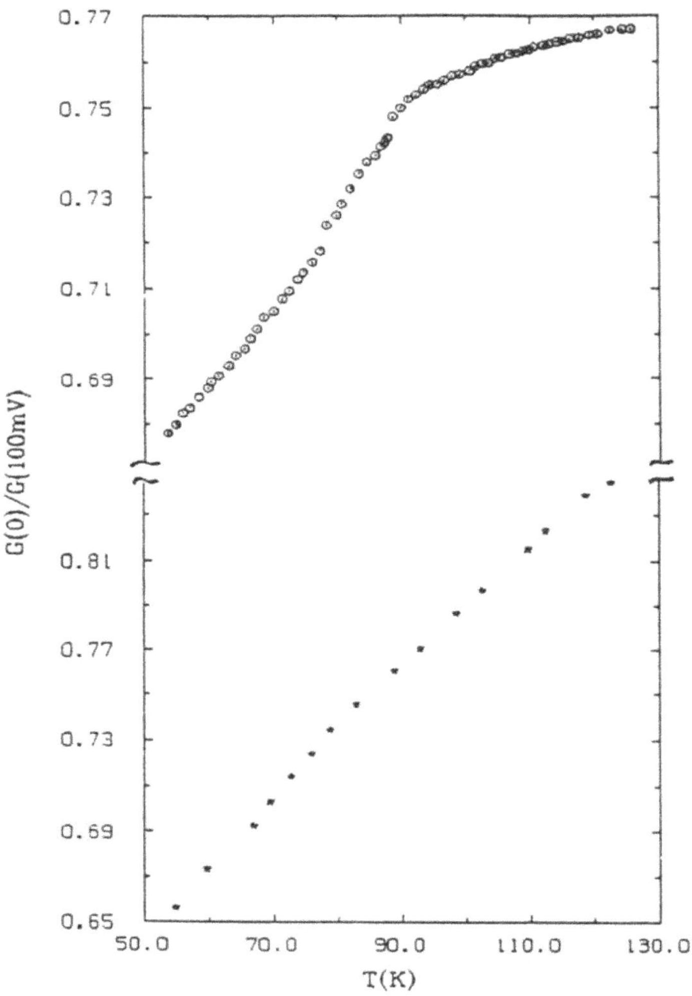

Fig. 2 Normalized conductance vs. temperature for the same junctions of Fig.1 .

demonstrate that the T_c, measured by tunneling spectroscopy, is the same as the bulk in fully oxygenated samples, while reduced samples show now evidence of a local T_c even if they are constituted by a mixture of 90K and 60K superconducting phases.
These results appear quite intriguing and seem to indicate that the simple BCS result relating smaller energy gaps to lower-T_c materials is not observed in high-T_c superconductors. If one assumes that "local" T_c can be measured only in the untreated junction, considering a gap-like structure at ±20 mV, a ratio $2\Delta/KT_c = 5.0$ is found for this sample.

4. Normal tunneling

In conventional junctions, the normal state tunneling is studied at low temperatures by means of an applied magnetic field that drives normal the superconduting electrode. With ordinary metals, the densities of states of both materials vary slowly with the energy and the normal conductance, $G_n(V)$, are analyzed, on a wider energy range (few hundred meV), in terms of an energy-dependent tunnel probability.
The P(E) depends on the barrier height, thickness and symmetry, and according to this model, the $G_n(V)$ are found to have a V^2 dependence. Due to the high H_{c2} value of theY-Ba-Cu-O compound, the normal conductance $G_n(V)$ of this kind of junctions cannot be measured at low temperatures. Therefore, inorder to analyze the normal tunneling one has to work above the critical temperature of the high-T_c material.
In Fig. 3 we show the normal tunneling data of the same junctions as in Fig.1 It is clearly seen that except for thermal rounding at voltages less than or equal to a few K_BT/e the conductance shows a linear behaviour on both samples that can be parametized by:

$$G_n(V) = G_{n0}(1+\alpha|V|)$$

where G_{n0} is the extrapolated zero-bias conductance and α is the linear slope. We have found $\alpha = 0.009\ mV^{-1}$ for the reduced junction (stars) and $\alpha = 0.035\ mV^{-1}$ for the fully oxygenated sample.
These α values are highly reproducible with variations of less than 20% on the same kind of junctions. They do not depend on the junction resistance and are quite insensitive to temperature variations. In fact the same α values can be deduced by extrapolating the linear background conductance of Fig. 1.

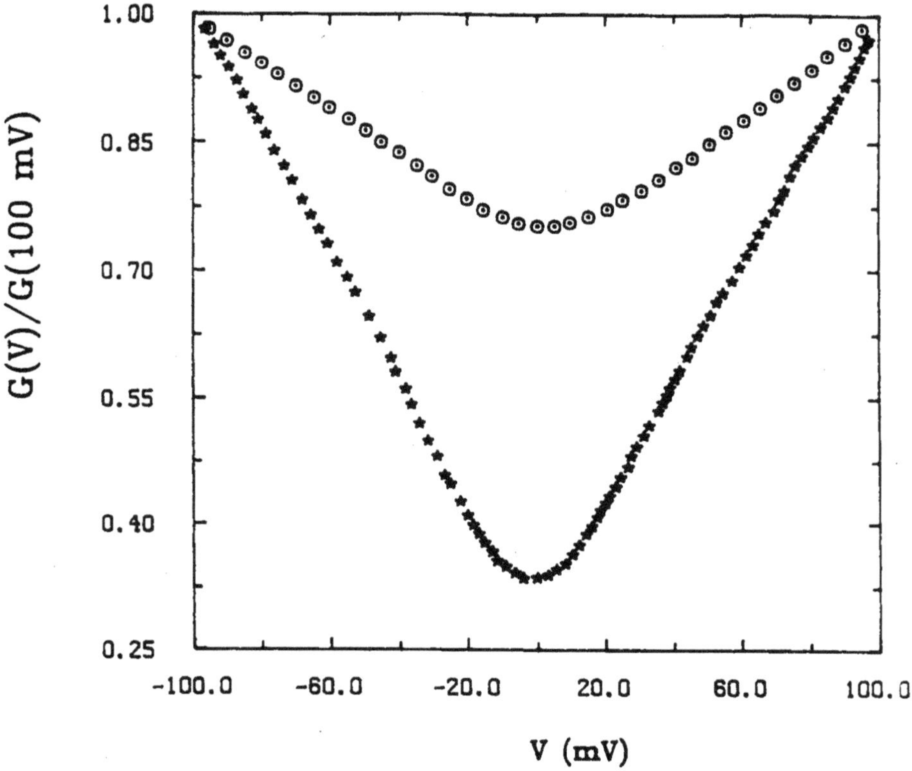

Fig. 3 Normalized conductance vs. voltage at $T>T_c$ for the same junctions of Fig. 1 .

5. Discussion and comments

The experiments here reported cannot be interpreted in terms of conventional tunneling results. At $T<T_c$, the nice and compact Eqs. (1), (2) do not describe the results of Fig. 1 and the exponential behaviour of the gap opening indicated by the BCS theory is not found in Fig.2. For normal tunneling, the V^2 dependence of the $G_n(V)$ is not observed.

All these discrepancies can develop due to anomalous densities of states and/or to unusual barrier shapes. In literature these hypotheses have been extensively discussed and it is clear that the anomalous normal properties of the high-T_C compound have to be taken into account in building the superconducting state. The linear background conductances, measured both above and below T_C, have been analyzed in terms of low and asymmetric barriers[7], inelastic tunneling[8], and density of states effect[9,10]. Unfortunately, at the moment, no experiment can unambiguously confirm one of the proposed mechanisms. However, the fact that this feature, independently from the junction fabrication method, has been found in junctions fabricated on Y-Ba-Cu-O[3,11], LSCCO[8], TBCCO[12], as well as in BPBO[13] and BKBO[13,14] strongly support the idea that this behaviour is intrinsic in this compound.

The results found in this paper are consistent with what has been found by NRM spectroscopy and specific heat measurements on theY-Ba-Cu-O compound. NRM studies[15] have revealed a temperature dependent spin susceptibility along the a-b planes and the discontinuity observed at $T=T_C$ in fully oxygenated samples does not appear in the smooth dependence found in oxygen deficient samples. Similar contrasts have been reported by specific heat measurements[16] in which the lower T_C compound exhibits, at $T=T_C$, a diminished peak compared to that of the 90K phase.

Due do these considerations, the tunneling characteristics of high-T_C superconductors has been recently modeled by superimposing BCS structures on linear background conductances[17].

Acknowledgements

We wish to thank R.C. Dynes, M. Gurvitch, J.M. Valles for their contribution to this work, L.F. Schneemeyer for growing the single-crystals and R. Di Florum for the near zero-field microwave measurements.

Referances

1. R.J. Cava et al., Physica C165 (1990) 419.
2. L.F. Schneemeyer et al. , Nature 328 (1987) 601.
3. M. Gurvitch et al., Phys. Rev. Lett. 63 (1989) 1008.
4. J.M. Valley et al., Phys. Rev. B 44 (1991) 11986
5. A.M. Cucolo et al., Phys. Rev. B 44 (1991) 2857.
6. A.M. Cucolo et al., Physica C 179 (1991) 69.
7. J.R. Kirtley, Int J. Mod. Phys. B 4 (1990) 201.
8. J.R. Kirtley et al., Phys. Rev. Lett. 65 (1990) 798.
9. P.W. Anderson et al., Phys. Rev. Lett. 60 (1988) 132.
10. C. M. Varma et al., Phys. Rev. Lett. 63 (1989) 1996.
11. M. Lee et al., Solid. State Comun. 70 (1898) 449.
12. I. Takeuchi et al., Physica C 158 (1989) 83.
13. F. Shafiri et al., Phys. Rev. B submitted.
14. J.F. Zasadinski et al., IEEE Trans. Magn. 27 (1991) 833.
15. R.E. Walstedt et al., Phys. Rev B 41 (1990) 9574.
16. A. Junod et al., Physica C 162-164 (1989) 1401.
17. A.M. Cucolo et al., Phys Letts, in press.

AN INTERPRETATION OF EXPERIMENTAL DATA ON HIGH T_C JUNCTIONS

A. DI CHIARA[†•], F. FONTANA[‡•], G. PELUSO[†•] and F. TAFURI[†•]

[†] University of Naples - *Federico II* - Napoli (ITALY)
[‡] University of Molise - Campobasso (ITALY)
[•] Istituto Nazionale Fisica della Materia

Abstract

The point contact junctions can be very likely described in terms of the phenomenological Bloder, Thinkam and Klapwijk model. In such a framework, we propose that the oder parameter at the interface of junction made by a high T_c superconducting base electrode can be described through the Mc Millan proximity approach. To show the reliability of our assumptions, experimental results are compared to the model outcomes.

1. Introduction

The depression of surface superconducting properties usually meet in high T_c superconducting electrodes plays a relevant role in the definition of electronic transport properties of devices based on those electrodes.

As already shown in the current literature[1,2] it is possible to model such a feature in terms of proximity effect. On this basis a phenomenological model which explains some among the main experimental tunnel characteristics of high critical temperature superconductors has been devised. In particular in a recent paper[3,4] the conductivity vs. voltage function for tunnel junctions was desumed in the framework of conventional theory of superconductivity. The validation of the model was made comparing numerical results to quite reliable and reproducible experimental measurements available in the current literature.

In this work we discuss the possible extension of the model to point contact junctions. The comparison between the computed outcomes and experimental data obtained in our laboratory shows that even if the configuration analised is quite different from the case of tunnel junctions, the model reproduces the main features of the experimental behavior.

2. Description of the model

The extension of the model we proposed for tunnel junctions to point contact junctions (PCJ) is quite easy. It founds on the following steps, both of them belonging to the framework of the "classical theory" of supercondutivity: a) the modellization of the electronic properties of the base electrode through the McMillan proximity model; b) the description of the features of the contact between the needle and the base electrode using the Blonder, Thinkam and Klapwjik model.

Following our assumptions, the basic electrode in the junction is viewed as a S/N structure and the whole junction is made by a "base electrode-constriction-(S or N)" assembly. The physics of the base electrode is described by the dependence of the order parameter on the voltage. For such a configuration, it is supplied by the well known Mc Millan proximity model[5]. Thermal smearing can be accounted in the usual way introducing the thermally smeared Fermi functions. Moreover, non-uniformities in the transmission efficiency between the superconducting and the normal layer of the base electrode, if needles with large sections compared to the surface disuniformities are concerned, could be easily introduced in the model by mean of a suitable "transparency coefficient"[3,4,5,6,7].

As far as the base electrode is concerned, according to the Mc Millan model, the order parameter for the two slabs (normal and superconducting) in the configuration above introduced ($\Delta_{N,S}(E,T)$) basically depends on the order parameters $\Delta^{ph}_{N,S}(T)$, characteristic of the "bulk" materials, and on two parameters $\Gamma_{N,S}$. These last are proportional to the ratios $\alpha/d_{N,S}$, where α is the trasmission probability coefficient between the normal and the superconducting sides of the junction and $d_{N,S}$ are the thicknesses of the normal and superconductive slabs respectively[5].

The model basic equations of the model are:

$$\Delta_{N,S}(E,T) = \left(\Delta^{ph}_{N,S}(T) + \frac{\imath\Gamma_{N,S}\cdot\Delta_{S,N}(E,T)}{\sqrt{E^2-\Delta^2_{S,N}(E,T)}}\right)\times \\ \times\left(1+\frac{\imath\Gamma_{N,S}}{\sqrt{E^2-\Delta^2_{S,N}(E,T)}}\right)^{-1} \tag{1}$$

The conductance $\sigma(E,T)$ between the base electrode and the needle, is defined by the equation:

$$\sigma(E,T) = \int_{-\infty}^{+\infty} N(V,T)\frac{\partial f(E,V,T)}{\partial E}\,dV \tag{2}$$

where $N(V,T)$ is a function depending on the order parameter and $f(E,V,T)$ is the difference between the Fermi functions at the energies $V-E$ and V.

The theoretical aspects of typical phenomena occuring in superconductor-constriction-(normal metal/superconductor) configurations (S-c-N, S-c-S') have been fully developed for classical superconductors[2,8]. As stated before, attempts to apply the theory also to high T_c structures has given interesting results. In this framework, to find the form of the function $N(V,T)$ we will apply a "classical model" to high T_c point contact junctions.

According to the phenomenological model proposed by Blonder, Tinkham and Klapwijk (BTK)[9], we assume the possibility that incident electrons may produce both reflected electrons and reflected holes. The conductivity σ_{NS} in the N/S configuration is supplied by the expression:

$$\sigma_{NS}(E,T) = \int_{-\infty}^{+\infty} [1+A(V,T)-B(V,T)]\,\frac{\partial f(E,V,T)}{\partial E}\,dV \tag{3}$$

where $A(V,T)$ and $B(V,T)$ are the probability of Andreev and ordinary reflection, respectively. It is clear that, in this framework:

$$N(V,T) = 1 + A(V,T) - B(V,T) \tag{4}$$

The following equations hold for $A(V,T)$ and $B(V,T)$[10]:

$$A(V,T) = \frac{\Delta_N^2(V,T)}{V^2 + (\Delta_N^2(V,T) - V^2)(1+2Z^2)^2} \tag{5.1}$$

$$B(V,T) = 1 - A(V,T) \tag{5.2}$$

for V lower than the phononic gap. Whereas, in the opposite case it results[9]:

$$A(V,T) = \frac{v_0(1-v_0)}{[(1-v_0) + Z^2(1-2v_0)]^2} \tag{6.1}$$

$$B(V,T) = \frac{(1-2v_0)Z^2(1+Z^2)}{[(1-v_0) + Z^2(1-2v_0)]^2} \tag{6.2}$$

where: v_0 is given by the following:

$$v_0 = \frac{1}{2}\left[1 - \left(\frac{V^2 - \Delta(V,T)^2}{V^2}\right)^{\frac{1}{2}}\right] \tag{7}$$

In the previous equations: Z is a not negative numerical parameter which is proportional to the eigth of the potential barrier at the N/S interface; the value of the phononic gap can be estimated in a self consistent way using either the Mc Millan model or the above equations.

3. Numerical results

The PCJ we used for the experimental measurements were obtained pressing Nb or Cu needles of different diameters (from .1 to 2 mm) on BISCCO pellets used as base electrode. The pellets were prepared in our laboratory following the standard procedure, their T_c was 90 K. A detailed description of the preparation procedure is given in elsewhere[10].

The contact pressure between the needle and the base electrode was controlled by a mechanical device suitable for micrometric adjustability while the assembly was in a cryostatic bath. To improve the electrical contact between the pellet surface and the needle, both of them were vacuum cleaned by mean of a fast atom gun.

The measurements of the I *vs.* V characteristics were performed in a LHe cryostat in the range 4.2 up to 300 K. The conductance vs V curves were obtained by the standard modulation technique. Depending on the needle pressure, its thickness and the temperature, the contact resistance varied from $\approx 1.\ \Omega$ (high contact pressure, $V = 30$mV, $T = 4.2$ K) up to 2KΩ (low contact pressure, $V = 30$ mV, $T = 4.2$ K).

In figure 1, the experimental curve of a $PCJ\ Nb - BiSCCO$ obtained at $T = 13K$ is compared with the computed one according to the model previously introduced. The values of the used parameters are: $\Delta_S^p h$=16 meV, $\Gamma_N = 1.$ meV, $\Gamma_S = .0001$ meV, $Z = .05$.

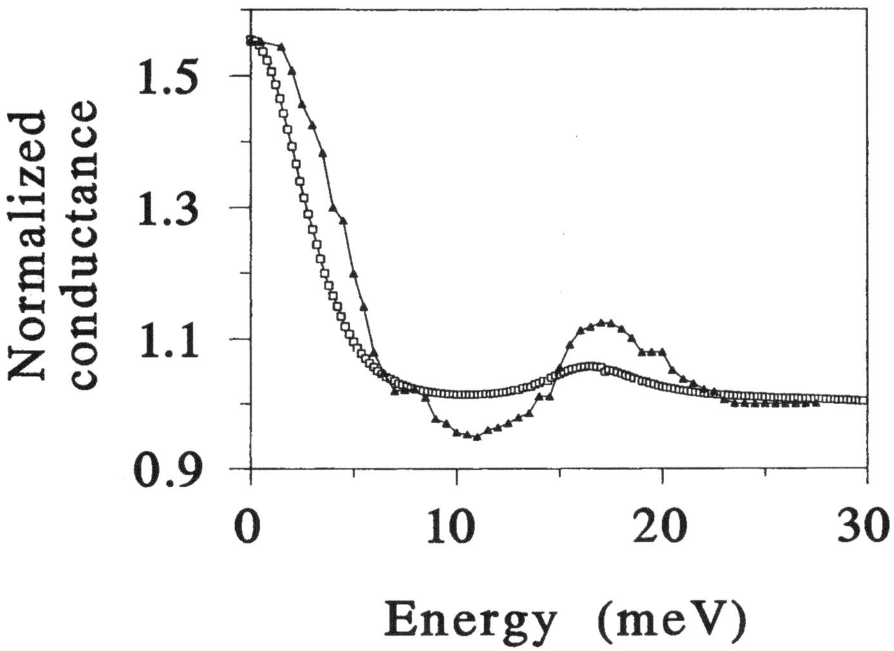

fig. 1: *Comparison between experimental data from a PCJ Nb-BiSCCO (full triangles) and the fitting curve (squares) obtained from the model described in the text.*

As shown by the figure the model reproduces the main features of the experimental data. The most remarkable issue is that, also for this configuration, the numerical values obtained for the fitting parameters are quite reasonable from the physical point of view.

Acknowledgements

Many thanks are due to Dr. A. Andreone, Dr. G. Pepe and Dr. U. Scotti di Uccio for their fundamental contribution to the experimental part of this work and to Mr. A. Maggio for technical aid. The work has been partially supported by the National Research Council of Italy (C.N.R.) under the: "Progetto Finalizzato: Cryogenic and Superconductive Technologies".

References

[1] V.Z. Kresin, in *"Novel Superconductivity"*, edited by S.A. Wolf and V.Z. Kresin (Plenum Press), New York (1987).

[2] I.O. Kulik, A.N. Omel'yanchuk and S.I. Beloborod'ko, in *Weak Superconductivity*, A. Barone editor, World Scientific, Singapore, 321 (1987).

[3] A. Di Chiara, F. Fontana, G. Peluso and F. Tafuri, *Journ. Supercond.*, 4, 35, 1991.

[4] A. Di Chiara, F. Fontana, G. Peluso and F. Tafuri, *Phys. Rev.*, **44**, 12026, 1991.

[5] W.L. Mc Millan, *Phys. Rev.* **175**, 537 (1968).

[6] J. Vrba and S.B. Woods, *Phys. Rev. B* **3**, 2243 (1971).

[7] J. Vrba and S.B. Woods, *Can.J.Phys.***49**,3133 (1971).

[8] A.F. Andreev *Zh. Eksp. Teor. Fiz.*, **48**, 1823 (1964), *Sov. Phys. JETP*, **19**, 1228 (1964).

[9] G.E. Blonder, M. Tinkham and T.M. Klapwijk, *Phys. Rev.*, B, **25**, 4515 (1982).

[10] A. Barone, A. Di Chiara, F. Fontana, G. Paternò, L. Maritato, G. Peluso, G. Pepe and U. Scotti di Uccio, *IEEE Trans. on Magnetics*, **25**, 993 (1989).

TUNNELING SPECTROSCOPY OF HIGH-T_c SUPERCONDUCTORS: THE POSSIBLE ROLE OF PROXIMITY EFFECT AND THE DETERMINATION OF ELIASHBERG FUNCTION

R.S. GONNELLI
Dipartimento di Fisica, Politecnico di Torino, Torino, Italy

D. ANDREONE
Istituto Elettrotecnico Nazionale Galileo Ferraris, Torino, Italy

P. SAMUELY
Hochfeld-Magnetlabor, Max-Planck-Institut fur Festkorperforschung, Grenoble, France

S.I. VEDENEEV
P. N. Lebedev Institute, Russian Academy of Sciences, Moscow, Russia

ABSTRACT

Point-contact tunneling measurements in a number of ceramic monophasic $Y_1Ba_2Cu_3O_{7-x}$ (YBCO) samples with different oxygen content and in $Bi_2(Sr,Ca)_3Cu_2O_{8-x}$ (BSCCO) single crystals are presented. A deconvolution procedure for unfolding the thermal broadening was applied to the experimental data and then the resulting conductance curves were fitted by a model of the total tunneling current containing a background "normal" tunneling current in parallel to a quasiparticle lifetime-broadened one. The significant deviations from the ideal BCS behaviour of the temperature dependences of the energy gap observed both in Cu/YBCO and in Nb/BSCCO junctions were analysed in the framework of the McMillan model of the superconducting proximity effect and a good agreement has been found between theoretical and experimental derived Δ vs T dependences. The ratios $2\Delta/kT_c$ deduced using this approach vary between 7.5 and 8.3.

Symmetric phonon-like deviations have been observed at energies greater than Δ in two Nb/BSCCO junctions. A good reproducibility is present in the peak positions of the $-d^2I/dV^2$ of experimental data. Two different models for extracting the normal state conductance from the tunneling data are presented. The application of these two models to the tunneling data exhibiting sharp gap-like features yielded to the solution of Eliashberg equations and to the determination of similar $\alpha^2(\omega)F(\omega)$ spectra as well as almost identical values for λ and μ^*. The Eliashberg functions obtained from these data and from recent tunneling measurement in $Nb/Bi_2Sr_2CaCu_2O_x$ point contact junctions present in literature are compared to the phonon density of states $F(\omega)$.

1. Introduction

In the past five years a very large number of papers on tunneling spectroscopy in high-T_c superconductors (HTS) appeared in literature. A very extensive review paper covering the research work up to the present year has been recently published [1]. The reason for this large interest in performing tunneling measurements is connected with the well known capability of these experiments to give information on the energy gap and on the density of states of the superconductor as well as on their temperature and magnetic field dependences [2]. In low-T_c strong coupling superconductors this technique was an extremely powerful tool for the confirmation of the phonon-mediated origin of the coupling between Cooper's pairs through the solution of the Eliashberg equations.

A large number of the puzzling effects and the non-ideal features very often observed in tunneling in HTS can now be explained in terms of the conventional theories for low-T_c superconductors [1]. A very important role is played, for example, by the anisotropic properties of these materials and by the complexity of the surfaces and interfaces that very often are quite different from the bulk material. But, limiting our analysis to YBCO and BSCCO compounds, there are some important properties that are still controversial. Among these properties, we are particularly interested in the temperature dependence of the energy gap and in the possibility to extract the phonon spectral density $\alpha^2(\omega)F(\omega)$ from tunneling experiments.

The determination of the Δ versus T dependence in YBCO has demonstrated very problematic. Even taking into account the thermal smearing, the temperature dependence was not observed at all [3] or observed, but not up to the gap closing, in reproducible point-contact experiments [4] and in artificial barrier junctions [5]. In their point-contact measurements, Ekino and Akimitsu [6] evidenced two gap structures that follow the BCS curve up to 0.7 T/T_c and they claimed to be due to the gap anisotropy. A BCS-like Δ vs T dependence for YBCO has been reported in edge junctions [7], but, apparently, the very important effect of thermal broadening has been ignored. The situation for BSCCO compounds is not very different even if it seems more simple to consistently observe a temperature dependence of the gap. Once again some papers that claim a BCS-like dependence use only the peak-to-peak values for Δ [8] or neglect the thermal smearing suggesting it can be done to a good approximation if the amount of lifetime broadening is large in experimental data [9]. Only very recently, some results in artificial barrier junctions on BSCCO single crystals have shown a good BCS dependence for the gap corrected for the thermal broadening [1]. In point-contact measurements on BSCCO single crystals we observed large deviations of the thermally unfolded Δ vs T dependence from BCS curve, which can be ascribed to the possible presence of proximity effect at the sample surface [10].

In the first part of the paper we analyze the significant deviations from the BCS behaviour of the temperature dependences of the gap obtained, after the thermal unfolding procedure, from measurements on BSCCO single crystals and on ceramic monophasic YBCO samples, in terms of the McMillan model of the superconducting proximity effect.

Concerning the determination of the phonon spectral density, only a few tunneling results are present in literature for YBCO and BSCCO systems. To our knowledge only one paper has claimed the determination of the $\alpha^2(\omega)F(\omega)$ spectrum from point-contact tunneling data on $EuBa_2Cu_3O_y$ [11]. In the BSCCO 2212 system, Vedeneev et al. [12-14] published some papers concerning the determination of the Eliashberg function where the agreement with the phonon spectrum obtained by neutron scattering [15] was rather good. But, at the same time, the inversion procedure determined negative values for the Coulomb pseudo-potential μ^* and relatively small values for the coupling coefficient λ [11] and, in some cases, these values were not estimated at all due to the uncertainty in the determination of the normal state conductance. More recently we analyzed some tunneling data already present in literature [14] by using a new phenomenological model for the estimation of the normal state conductance [16]. We were able to obtain a good agreement between $\alpha^2(\omega)F(\omega)$ and $F(\omega)$ as well as a positive value for the Coulomb pseudo-potential μ^* and a λ rather large but not incompatible with the high value of $2\Delta/kT_c$ obtained for this material.

In the second part of the paper we will analyze the tunneling conductances of two different Nb/BSCCO point-contact junctions by using two different models for inferring the normal state conductance and we will compare the results of the inversion of the gap equations to recent results in the determination of the Eliashberg function from tunneling data in BSCCO single crystals [17].

2. Proximity effect in point-contact junctions

Monophasic polycrystalline YBCO samples were prepared using a standard procedure described in literature [18] and then they were annealed in flowing oxygen at four different temperatures: 400, 500, 580 and 650 °C inducing variations in their oxygen content. The resulting bulk critical temperatures, measured using the standard four-contact resistive method, were (90 ± 1), (89 ± 1), (85 ± 3) and (46 ± 3) K for the different annealing temperatures respectively [4]. Monocrystalline BSCCO samples were prepared using a rapid melting process, in which powders of the appropriate composition are heated by means of the light of a 10 kW Xenon lamp. Solid drops are obtained which contain a large number of small single crystals. Bulk critical temperatures of these samples were in the interval 74-87 K [10] while the x-ray diffraction analysis showed the presence of two different superconducting phases, corresponding to the compositions $Bi_2(Sr,Ca)_{n+1}Cu_nO_{4+2n+x}$ with n=1 and n=2.

Point-contact tunnel junctions were made on the *fresh* surface of the material by cleaving the sample at 4.2 K by means of the needle used as counterelectrode. Cu counterelectrodes with large tip diameters (50-200 μm) were used in YBCO junctions while single crystal Nb tips of small diameter (≤ 5 μm) were pressed perpendicularly to the *c* axis of the single crystals of BSCCO. The differences in the nature of the samples and in the form and material of the counterelectrodes produced differences in the resulting tunneling characteristics. In Cu/YBCO junctions the dI/dV curves are typical of an averaging of the tunneling properties over different grains and crystallographic directions showing broadened gap-like features at about 20 mV, while in Nb/BSCCO point-contact junctions more evident sharp gap-like features are observed. In all the cases the thermal broadening was removed from the original data by a numerical unfolding procedure. Fig. 1 (a) shows the deconvoluted dI/dV curves of a Cu/YBCO junction at different temperatures while Fig. 1

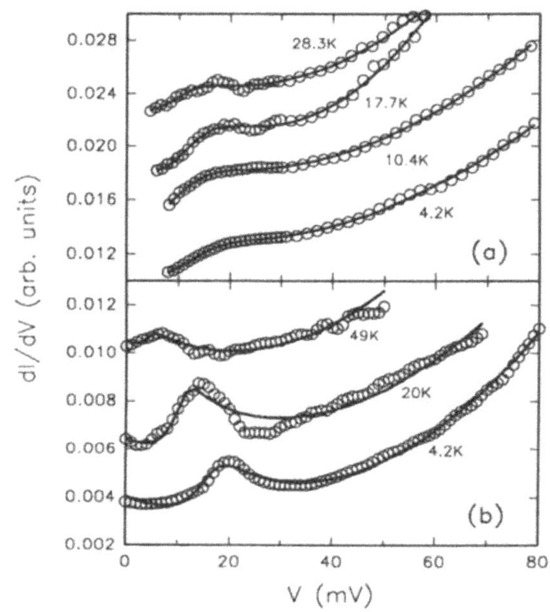

Fig.1 Deconvoluted differential conductances as a function of applied voltage in a Cu/YBCO junction (a) and in a Nb/BSCCO junction (b) at various temperatures (open circles). Solid lines are the fitted *dI/dV* curves obtained by the phenomenological model described in the text.

(b) shows the deconvoluted conductances of a Nb/BSCCO junction (open circles). The solid lines are the fitted conductances obtained by a phenomenological model that will be discussed later.

Some common features can be observed in the data of Fig.1: a) the positions of the gap-like maxima or plateaus shift toward lower voltages as the temperature increases (more clear in BSCCO data); b) a parabolic background conductance is present up to 50-100 mV; c) because of large zero-bias tunneling conductances the ratios $\sigma_s(0)/\sigma_n(0)$ are of the order of 0.5 at 4.2K. A complete closing of the gap was observed at a temperature of 64K in Nb/BSCCO junctions, while the gap-like features completely vanished at temperatures never greater than 43K in tunnel junctions obtained in YBCO samples without showing a complete gap closing and, therefore, not allowing a precise determination of T_c at the point of contact.

A phenomenological model of the tunneling conductance that accounts for the main features of the experimental data previously described was used to analyze the conductance curves [4,10]. The total tunneling current flowing in the point-contact junction is modelled by a normal tunneling current due to the presence of normal electrons or normal regions at the sample surface even at $T \ll T_c$ in parallel to a superconducting tunneling current due to the quasi-particles having a gap. Their density of states was thought smeared out by the lifetime broadening introduced by Dynes some years ago [19].

When the tunneling current of the normal channel, represented by a Metal-Insulator-Metal (MIM) contribution with a cubic dependence on voltage, is added to the quasi-particle tunneling current and the derivative with respect to the voltage is taken, we obtain the dynamic conductance dI/dV of the junction that at low temperatures ($T/T_c \ll 0.5$) simplifies as follows:

$$\frac{dI}{dV} = (1-\alpha)(a+2bV+3cV^2) + \alpha(b+2cV)e^{-1}\int_0^{eV} N_S(E)dE + \alpha(a+bV+cV^2)N_S(eV). \tag{1}$$

a, b and c are the model parameters that describe the parabolic and asymmetric behaviour of the normal background conductance with voltage and α quantifies the amount of quasi-particle tunneling present in the process. According to the lifetime model the density of states is written as [19]:

$$N_S(E) = Re\{[E - i\Gamma]/[(E - i\Gamma)^2 - \Delta^2]^{1/2}\} \tag{2}$$

where Δ and Γ are the energy gap and the lifetime broadening parameter of the density of states respectively.

The theoretical conductance of Eq. (1) can be used to fit the experimental data and the 6 parameters of the model $a, b, c, \alpha, \Gamma, \Delta$ can be determined by minimizing the standard deviation of the fit [4,10]. On the contrary, at temperatures higher than 15-20K the simplified conductance of Eq. (1) cannot be used and, on the other hand, in presence of a Nb tip, it should contain the contribution of the density of states of the superconducting counter-electrode too. These two drawbacks have been overcome by deconvolving the experimental data and, after, by fitting the unfolded conductance using Eq. (1) valid at low temperatures and, in the case of superconducting tip, by subtracting the gap of the counterelectrode at that temperature from the Δ obtained by the fit [20].

Using this approach we determined the dependence of the gap Δ on the critical temperature T_c in YBCO samples with different oxygen content. This dependence is linear with a ratio $2\Delta/kT_c$ of 4.3 ± 0.3 using the bulk values of T_c. However, if we use the minimum T_c values that can be inferred from Δ vs T dependence (of the order of 48 K and 20 K for the different YBCO samples in absence of complete gap closing) we obtain $2\Delta/kT_c = 7.2 \pm 0.6$ [4].

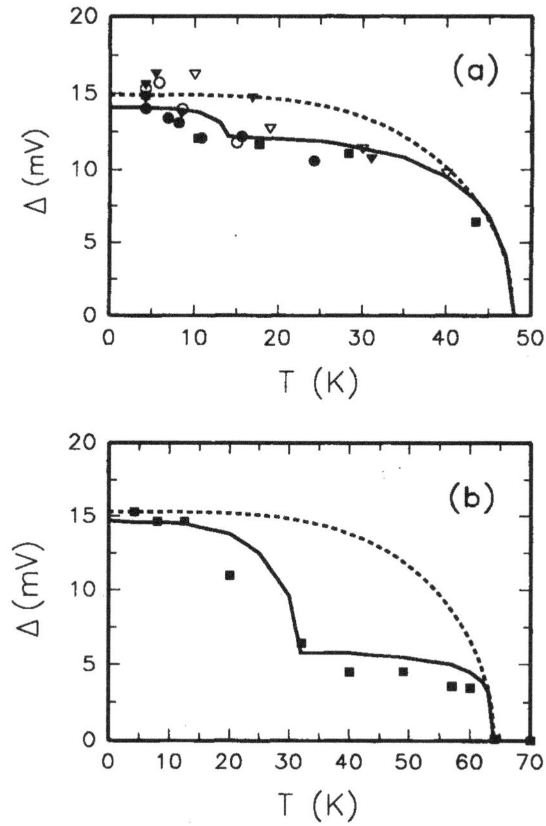

Fig. 2 (a) Dependence of the gap on temperature as obtained by the model described in the text in Cu/YBCO junctions. Different markers represent the results of measurements on different junctions; (b) $\Delta(T)$ dependence for the Nb/BSCCO junction. Short dash lines are the BCS behaviours for $\Delta(0) = 15$ mV and $T_c = 48$K (a) and for $\Delta(0) = 15.3$ mV and $T_c = 64$K (b). Solid lines represent the results of the proximity model described in the text.

Fig. 2 shows the Δ versus T dependences obtained from tunneling data by applying the described model to different Cu/YBCO junctions with approximately the same bulk T_c (a) and to Nb/BSCCO (single crystal) junctions (b). The short dash lines are the BCS behaviours for $T_c = 48$K (a) and $T_c = 64$K (b), while the solid lines represent the results of the application of the proximity model we will describe in the following. Reproducible deviations from the BCS curve can be observed in the graph (a) of Fig. 2, where a sharp decrease of Δ at $T \approx 12$K is followed by a sort of plateau up to 30-35K.

Deviations of the same kind, but much more evident, are present in the data of Nb/BSCCO junctions in Fig. 2 (b), where the plateau is now extending from 30 up to 60K with a ratio $\Delta(T)/\Delta(0) \approx 0.3$, and the gap closing is observed at 64K.

The features observed in Fig. 2 are not produced by the thermal unfolding procedure or by the method for the determination of the gap since they are already clearly present, for example, in the dependence of the conductance peak positions on temperature in original Nb/BSCCO tunneling data [20].

These anomalies are very similar to the behaviour of the gap on temperature in conventional superconductors when the surface is covered by a thin proximity layer showing a lower Δ and T_c [21]. As a consequence, we based our analysis of the Δ versus T data in both Cu/YBCO and Nb/BSCCO junctions on the McMillan tunneling model of the superconducting proximity effect [22]. The model states that the energy gaps in the superconducting (S) and in the proximized (N) layer are energy dependent complex quantities $\Delta_{N,S}(E)$ defined by two coupled equations. They depend on the constant bulk gaps in the same layers $\Delta_{N,S}^{phon}$ and on the two characteristic parameters $\Gamma_{N,S}$ which are function of the properties of the layers, in particular, of the Fermi velocities and of the inverse of the thicknesses. In this framework $\Gamma_{N,S}$ are considered as single parameters to be determined as a whole by the model. When the usual BCS dependence of $\Delta_{N,S}^{phon}$ on T is included in the coupled equations, it is possible to calculate the $\Delta_{N,S}(T)$ dependence. Tunneling experiments sample the gap of the proximized layer $\Delta_N(T)$, where the subscript N is conventionally written even if the layer could be superconducting. The coupled equations for the energy dependent gaps are easily solved if we can consider the thickness of the superconducting layer d_S very large in comparison with the thickness of proximized layer, i.e. $d_S \gg d_N$. In fact Γ_S becomes much smaller than Γ_N and can be neglected simplifying the equations as follows:

$$\Delta_N(E,T) = \frac{\left(\Delta_N^{phon}(T) + \Gamma_N \Delta_S(T) / [\Delta_S^2(T) - E^2]^{1/2}\right)}{(1 + \Gamma_N / [\Delta_S^2(T) - E^2]^{1/2})} \qquad (3a)$$

$$\Delta_S(T) = \Delta_S^{phon}(T) \qquad (3b)$$

By fitting the theoretical $\Delta_N(T)$ of Eqs. 3 to the Δ versus T dependences obtained in Cu/YBCO and Nb/BSCCO junctions we were able to determine the free parameters of the model, i.e. $\Delta_{N,S}^{phon}(0)$, Γ_N and the critical temperatures $T_c^{N,S}$ in the proximized and in the superconducting layer respectively. The $\Delta_N(T)$ curves, obtained by optimizing the model parameters, are shown in solid lines in Fig. 2 (a) and 2 (b) and evidence a good agreement with gap versus temperature dependences deduced from experimental data, particularly in Nb/BSCCO junctions. The fitted values of the parameters for the two different kinds of junctions are the following:

i) Cu/YBCO junctions: $\Delta_N^{phon}(0) = 5$ mV, $T_c^N = 14$ K, $\Delta_S^{phon}(0) = 17.1$ mV, $T_c^S = 48$ K, $\Gamma_N = 30$ mV;

ii) Nb/BSCCO junctions: $\Delta_N^{phon}(0) = 11.4$ mV, $T_c^N = 32$ K, $\Delta_S^{phon}(0) = 20.6$ mV, $T_c^S = 64$ K, $\Gamma_N = 8$ mV.

The ratios $2\Delta/kT_c$, calculated using for T_c and for Δ the new values obtained in the N layer by the proximity model, exhibit similar values in the two different materials varying between 7.5 and 8.3.

In summary, to include the possible effect of superconducting proximity in our tunneling data, we firstly extracted the Δ versus T dependence by using a model that contains only the effect of thermal broadening and of pair-breaking phenomena and then we analyzed the obtained dependence in terms of proximity effect. The proximity effect should have been included in the model from the beginning, but, in this case, the complex and integral form of the model equations would have prevented the inverse solution. The procedure just described is not the application of two different models to the same data but, on the contrary, it is a single complex model applied in two steps to the measured data. The applicability of this two-step procedure is somehow confirmed by the results of direct calculations for Cu/YBCO and for Nb/BSCCO junctions. The theoretical conductances obtained by the proximity model using the fitted parameters and including thermal and lifetime broadening as well as the background conductances, show a good agreement with the experimental dI/dV curves [20].

3. The role of phonon-like anomalies and the determination of $\alpha^2(\omega)F(\omega)$

The Eliashberg function $\alpha^2(\omega)F(\omega)$ can be calculated by means of usual inversion procedures [23,24] once the reduced density of states (RDS) is obtained from the tunneling measurements and, for this, one has to know the normal state conductance $G_N(V)$, which can be determined either from tunneling data measured in such a high magnetic field to suppress the superconductivity, or from data at $T > T_c$ supposing that $G_N(V)$ is not changed with temperature on the scale of T_c [11].

In low-T_c strong-coupling superconductors, RDS shows clear deviations from zero at voltages greater than the gap. Also in low temperature tunneling data from point-contact junctions between Nb (single crystal) and $Bi_2(Sr,Ca)_3Cu_2O_{8-x}$ (single crystal) symmetric deviations from the background conductance can be observed at $V > \Delta/e$ [14]. Fig. 3 (a) shows the part at positive bias of the experimental conductances of two Nb/BSCCO point-contact junctions at 4.2K. The curve 1 in Fig. 3 (a) represents old tunneling data [14] while the data of curve 2 have been obtained more recently. The second derivatives of the tunneling currents $-d^2I/dV^2$ numerically computed from the data of Fig. 3 (a) are shown in Fig. 3 (b). The bulk critical temperatures of the single crystals were 82K (curve 1) and 87K (curve

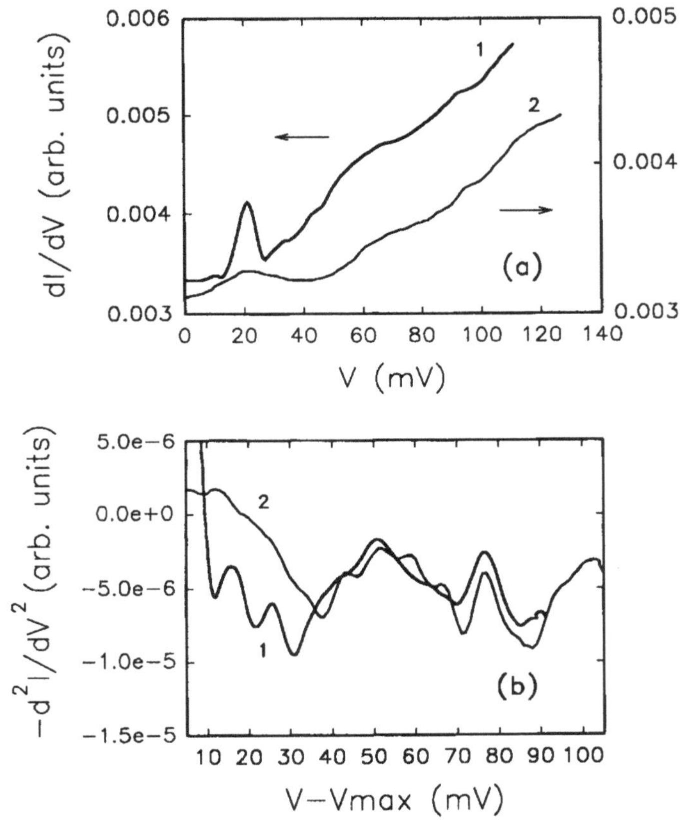

Fig. 3 (a) Experimental conductances dI/dV of two different Nb/$Bi_2(Sr,Ca)_3Cu_2O_8$ (single crystal) point-contact junction at 4.2K; (b) Second derivatives of the tunneling current $-d^2I/dV^2$ numerically calculated from the data of Fig. 3 (a).

2) while T_c determined by tunneling measurements ($\Delta = 0$) is 65K for junction 1. It is clear from these data that reproducible phonon-like anomalies are present in the experimental conductances of the two different junctions at V >30 mV which produce characteristic peaks at the same voltages in the $-d^2I/dV^2$. These peaks resemble the structures present in the phonon spectral density $F(\omega)$ [15]. We believe they are representative of the electron-phonon interaction in this high-T_c superconductor.

The high value of the upper critical field H_{c2} in high-T_c superconductors (T_c >25-30K) prevents the determination of $G_N(V)$. On the other hand, $G_N(V)$ varies remarkably with temperature because of the high values of T_c. A possible solution consists in trying to infer the normal state conductance from tunneling data at low temperature by using appropriate models for fitting the total experimental conductance.

Taking into account the common features observed in all our Nb/BSCCO tunneling data and, in particular, the dI/dV which does not vanish at $V \rightarrow 0$ and the linear or parabolic background conductance at high voltages, we used a modified version of the model discussed in the previous section for fitting the data [16]. For the tunneling current due to normal electrons $I_N(V)$ and for the normal conductance $G_N(V)$ due to the quasi-particles behaving as normal at $E > \Delta$ we used the following expressions respectively:

$$I_N(V) = aV + bV^2 + cV^3 \tag{4a}$$

$$G_N(V) = a' + b'V + c'V^2 \tag{4b}$$

where the coefficients a', b' and c' differ from a, b and c in order to account the possible different normal tunneling process that occurs in quasi-particles at $E > \Delta$ and in electrons already normal at $T \ll T_c$. Assuming that one electrode of the junction is a normal metal and calculating the total differential conductance for $T/T_c \ll 0.5$ ($T < 10$K for HTS) we obtain:

$$\frac{dI}{dV} = a + 2bV + 3cV^2 + (b' + 2c'V)e^{-1}\int_0^{eV} N_S(E)dE + (a' + b'V + c'V^2)N_S(eV) \tag{5}$$

where $N_S(E)$ is the usual lifetime broadening density of states of Eq. (2), and Eq. (5) can be used also if the counterelectrode is a low-T_c superconductor since we can assume its density of states is constant, for example, at $V > 10$ mV for niobium. Neglecting the phonon anomalies in the model density of states (in fact they are very small in comparison with the other features) and fitting the model conductance obtained from expressions (5) to the experimental one, it is possible to obtain the eight free parameters of the model: $a, b, c, a', b', c', \Delta, \Gamma$. The fitted values of Δ and Γ for the data of curve 1 in Fig. 3 (a) were 18.8 mV and 2.5 mV respectively. Finally, from the experimental data of dI/dV and from the obtained functions $I_N(V)$ and $G_N(V)$ we were able to get the RDS $[\overline{N}_S(E)/N_S(E)] - 1$ where $\overline{N}_S(E)$ is the density of states extracted from the experimental data by the model and containing the phonon-like anomalies. The reduced density of states so calculated allows to determine the $\alpha^2(\omega)F(\omega)$ spectrum by direct inversion procedure [24].

The results of this analysis are shown in Fig. 4 (b) where the Eliashberg function obtained from the data of Fig. 3 (a) (curve 1) is presented. It is compared to the phonon density of states $F(\omega)$ obtained from inelastic neutron scattering data on BSCCO 2212 samples [15] shown in Fig. 4 (a). We obtained a good agreement between the calculated $\alpha^2(\omega)F(\omega)$ and the measured phonon density $F(\omega)$ and, for the first time, a positive value

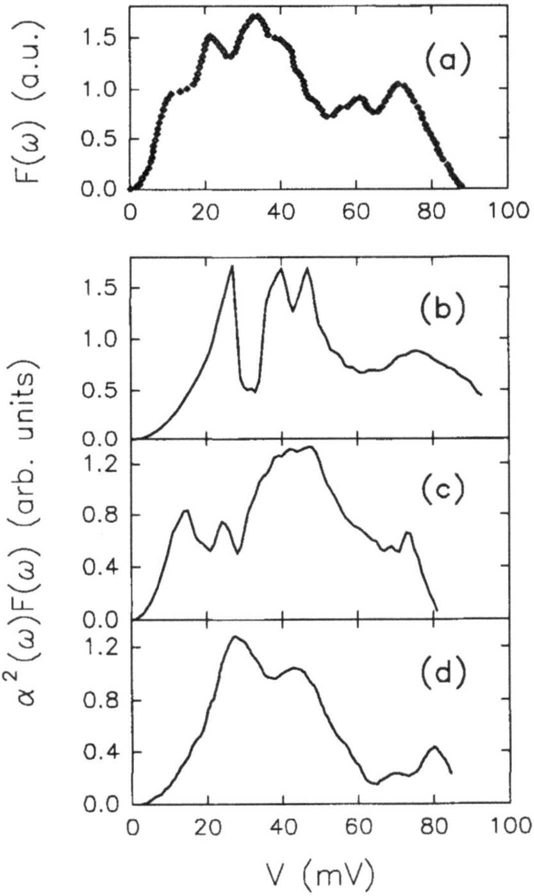

Fig. 4 Comparison between the phonon spectral density $F(\omega)$ of $Bi_2Sr_2CaCu_2O_x$ determined using neutron scattering data (Ref. 15) and appropriately scaled (a) and the electron-phonon spectral functions $\alpha^2(\omega)F(\omega)$ determined from tunneling data on the same material (curves (b), (c) and (d)). Curve (b) has been taken from Ref. 16 and curve (d) from Ref. 17. For details see the text.

for the Coulomb pseudo-potential μ^*=0.20 together with a large McMillan coupling parameter λ=3.95 which should be compared to the high ratio $2\Delta/kT_c$= 6.7 determined by the model. Details on this analysis can be found in literature [16].

Some common objections to these results concern their reproducibility and their supposed dependence on the particular model used for fitting the conductance data. In order to investigate this point we analyzed the same data of curve 1 in Fig. 3 (a) by using

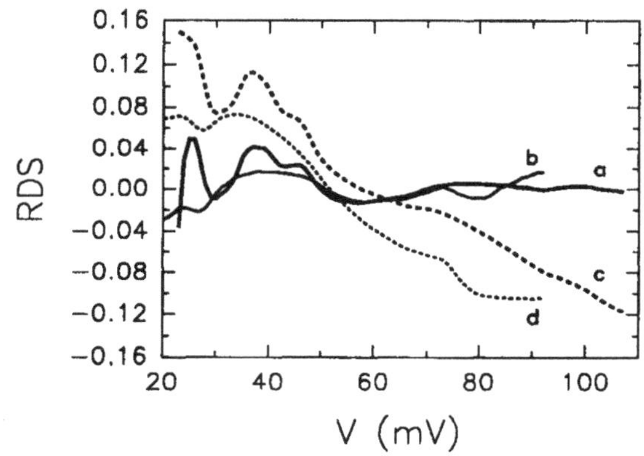

Fig. 5 Reduced densities of states deduced from data of Fig. 3 (a) (curve 1) by using the first model (curve a) and the second method (curve b) described in the text. Short dash lines (curve c and d) are the corresponding reduced densities of states calculated by the program for the inversion of the gap equations. See text for details.

a different procedure. The experimental conductance at low temperature was approximated, using a direct calculation, by the theoretical conductance expressed in Eq. (1). The normalized conductance $(dI/dV)_S/(dI/dV)_N$ containing the phonon-like deviations was then extracted from the experimental data and the approximating function. At low temperature this normalized conductance is a good approximation for the experimental derived superconducting density of states $\overline{N}_S(E)$ and, therefore, the reduced density of states can be evaluated in the same way as previously described. The result of the successive inversion procedure is shown in Fig. 4 (c). The reproducibility of the peaks and, in particular, of the valley positions in the two $\alpha^2(\omega)F(\omega)$ curves is rather good, apart the peak at about 15 mV that is not present in Fig. 4 (b). The new values of the Coulomb pseudo-potential μ^*=0.24 and of the coupling parameter λ=3.85 are close to the values previously obtained and, in this case, the critical temperature of the material calculated by the program was 82 K. It is an indication that the model has a minimal influence on the results if the phonon-like anomalies are clear and the gap-like peaks are sharp and not very broadened.

Fig. 5 shows a comparison between the reduced densities of states (RDS) deduced from the experimental conductance of Fig. 3 (a) (curve 1) by using the two models just described (curve a and b) and the corresponding RDS obtained as output of the program for the determination of the Eliashberg function (curve c and d). The curve a, obtained by using the first phenomenological model, corresponds to the calculated RDS shown in the curve c while the curve b, obtained by using the second procedure, corresponds to the curve d. A general agreement between the RDS extracted from the measurements and the theoretical ones is present for what concerns the shape of the curves, but a background roughly linear with voltage appears in the calculated RDS. Such a situation, which could be produced by the influence of the proximity effect on the density of states we have ignored in this framework up to now, suggests the use of the modified McMillan-Rowell Stage III program for the determination of $\alpha^2(\omega)F(\omega)$ in presence of proximity effect.

We analyzed the data of the other Nb/BSCCO junction (curve 2 in Fig. 3 (a)) by using the same two models but, in spite of all our efforts and of the presence of peaks at similar energies in the $-d^2I/dV^2$, it was impossible to obtain reasonable results for the shape of the $\alpha^2(\omega)F(\omega)$ function and for the λ and μ^* values. This is probably due to the rather large broadening present in the gap-like features of these data that superimposes to the phonon-like anomalies between 30 and 45 mV.

Recent tunneling data collected by two of the authors of the present paper give a direct confirmation of the results shown in Fig. 4 (b) and (c) [17]. Sharp gap-like peaks and clear and symmetric phonon-like anomalies of the same kind of those shown in the curves of Fig. 3 have been observed in the measured conductance of a Nb/$Bi_2Sr_2CaCu_2O_x$ (single crystal) point-contact junction at 4.2K. The tunneling conductance at zero bias voltage showed non-zero values producing a $(dI/dV)_S/(dI/dV)_N$ at V=0 of the order of 0.4. A magnetic field up to 10 T has been applied in a direction perpendicular to the c-axis of the crystal i.e. the same direction along which the point-contact on the fresh surface of the sample was made. This field broadened the observed gap-like features but, practically, had no influence on the peak positions. The tunneling conductances and the second derivative data d^2I/dV^2 show additional structures at $V > \Delta/e$ which are rather similar for positive and negative bias polarities. A smoothing procedure was applied to these curves in order to take into account the increase of the conductance above the gap and to infer the normal state conductance. The lifetime broadened density of states of Eq. (2) was then fitted to the normalized conductance yielding Δ=20 mV, Γ=1.5 mV and a ratio $2\Delta/kT_c$=5.5. The successive solution of the Eliashberg equations and the determination of the electron-phonon spectra for different voltage polarities and in presence or absence of the applied field gave the averaged $\alpha^2(\omega)F(\omega)$ function shown in Fig. 4 (d). The averaged values for the Coulomb pseudo-potential μ^*=0.1 $\pm$ 0.1 and for the coupling parameter λ=3.5 $\pm$ 0.5 are in good agreement with the values relative to the curves (b) and (c) of Fig. 4 and give the calculated T_c=90K.

A good agreement between the peak and valley positions in the Eliashberg functions obtained from different samples and with different models and the structures of $F(\omega)$ obtained from neutron scattering data is also evidenced in Fig. 4.

4. Conclusions

In the first part of the paper we have presented the application of the McMillan tunneling model of the proximity effect to the $\Delta(T)$ curves deduced from tunneling conductance measurements in Cu/YBCO (polycrystalline) and Nb/BSCCO (monocrystalline) point-contact junctions.

An unfolding procedure for removing the thermal broadening was firstly applied to the experimental dI/dV curves. We believe this step cannot be neglected if one is interested in the temperature dependence of the energy gap, since, in the great part of the cases for HTS, the peak positions in the tunneling curves result nearly independent of temperature up to at least 0.9 T_c because of the combined effect of the Fermi function and of the ideal BCS $\Delta(T)$ curve. The deconvoluted conductances were then analysed by means of a phenomenological model of the tunneling current which accounts for the coexisting presence of superconducting and normal tunneling channels in the contact and includes the smeared nature of gap-like features by means of a lifetime broadened density of states. Finally the proximity model has been used to fit the obtained Δ vs T dependences that, in our case, presented deviations from the theoretical BCS curve. We consider this method as a single model which is applied in two steps to the experimental tunneling conductances. A confirmation of the applicability of this approach is given by the direct calculation of the tunneling conductance starting from the values of the parameters obtained by the proximity model and including thermal and lifetime broadening. The computed dI/dV curves are in good agreement with the measured ones [20]. The proximity model fits well the Δ vs T curves inferred from our tunneling data (Fig. 2) and the parameters of the bulk superconductor $\Delta_S^{phon}(0)$ and T_c^S, obtained in independent way in the two materials, are in the range of the values present in literature for the same superconductors. Moreover the parameters of the surface proximized layer in BSCCO, and particularly the T_c^N, seem to agree with the x-ray analysis of our samples which evidenced the presence of the superconducting phase $Bi_2(Sr,Ca)_{n+1}Cu_nO_{4+2n+x}$ with n=1 and $T_c \approx 30$K together with the phase with n=2. The bulk and the proximized layers in YBCO and BSCCO show very similar values for the ratio $2\Delta/kT_c$ ranging from 7.5 to 8.3. These high ratios would suggest a very strong coupling mechanism in these materials and, for YBCO, are in agreement with experiments that sample the *bulk* value of Δ as, for example, infrared optical measurements [25].

These results suggest the influence of the proximity effect in tunneling experiments on high-T_c superconductors even if, at the present moment, a definitive answer concerning the nature and the origin of the proximized layer cannot be given. Both extrinsic reasons, connected with the composition and the oxygen content at the sample surface, and intrinsic ones, connected with a lowering of the pair potential at the surfaces and interfaces of the sample [26], can be adduced to justify the presence of proximized layers. In our samples, a non-ideal oxygen stoichiometry at the surface of YBCO and the presence of the phase with n=1 in BSCCO seem to suggest an extrinsic explanation of the phenomenon. The presence of the proximity effect, together with the probable differences of the surface values of T_c with respect to the bulk ones, can explain part of the discrepancies in the gap values observed in tunneling experiments.

In the second part of the paper we have described reproducible structures clearly observed in the dI/dV vs V characteristics of two Nb/$Bi_2Sr_2CaCu_2O_x$ (single crystal) point-contact junctions at $E > \Delta$. These features have been interpreted as the contributions of the electron-phonon interaction to the density of states of the superconducting material in the framework of the Eliashberg theory of strong coupling superconductors. In order to invert the gap equations and to determine the electron-phonon spectral function $\alpha^2(\omega)F(\omega)$ it is necessary to know the normal state conductance of the material at low temperature which cannot be measured in high-T_c superconductors. A proper model is therefore needed, for trying to infer from the tunneling conductance an appropriate form of the $(dI/dV)_N$. Apparently this fact is a weak point of the method since it seems that the results are strongly model-dependent. But, as a matter of fact, it is not true.

In the paper we have presented two different models for extracting a form of the normalized conductance from tunneling data. The application of these two models to the experimental conductances of the two Nb/$Bi_2Sr_2CaCu_2O_x$ junctions, which exhibited similar structures at energies above the gap in the $-d^2I/dV^2$, has yielded to different results according to the amount of broadening. When the two models are applied to the tunneling data which exhibit rather sharp gap-like features and a small amount of broadening, as curve 1 in Fig. 3 (a), it is possible to solve the Eliashberg equations using the standard numerical methods obtaining similar $\alpha^2(\omega)F(\omega)$ functions (Fig. 4 (b) and (c)). The values of λ and μ^* obtained from the inversion procedure are also very similar (λ=3.95 and 3.85, μ^*=0.20 and 0.24 respectively) giving a further indication that, in this case, the kind of model for extracting the normalized conductance has not a large influence on the results. On the contrary, it was not possible to obtain any solution of the gap equations when the two models were applied to the data of curve 2 in Fig. 3 (a) where a larger amount of broadening is present.

Recent tunneling measurements in Nb/$Bi_2Sr_2CaCu_2O_x$ point contact junctions [17] have shown very sharp gap features and clear additional structures at $V > \Delta/e$. The method used for deducing the $(dI/dV)_N$, consisting in a smoothing procedure applied to the data in the superconducting state, was, in some way, similar to the second method presented here and also the results of the inversion procedure were very similar both in the shape of $\alpha^2(\omega)F(\omega)$ spectrum (Fig. 4 (d)) and in the values of λ, μ^* and calculated T_c.

As a consequence we can draw the following conclusions:

i) The $\alpha^2(\omega)F(\omega)$ spectra can be successfully obtained only from tunneling conductances which show evident and symmetric phonon-like anomalies at $V > \Delta/e$ and exhibit sharp gap-like peaks, i.e., are affected by a small amount of broadening.

ii) In the case that condition i) is fulfilled, the use of different models for approximating the experimental conductance curves and extracting the normalized conductance seems not to be very influential on the overall results of the inversion procedure.

iii) The observed conductance anomalies at $E > \Delta$ and the $\alpha^2(\omega)F(\omega)$ spectra, obtained from different junctions using different models for inferring the $(dI/dV)_N$ and compared to the phonon density of states $F(\omega)$, give evidence in the favour of a strong electron-phonon coupling in $Bi_2Sr_2CaCu_2O_x$ and reveal the important role played by the electron-phonon interaction in the pairing mechanism of this high-T_c superconductor.

Acknowledgements

We are deeply indebted to L.N. Bulaevskii, O.V. Dolgov and G.F. Durin for model calculations and fitting procedures, to I.P. Kazakov and M. Vallino for the production of the samples and to C. Appino, V. Lacquaniti and V. A. Stepanov for the help in tunneling measurements. This work has been partially supported by the National Research Council of Italy, CNR, under the Progetto Finalizzato "Superconductive and Cryogenic Technologies".

References

1. T. Hasegawa, H. Ikuta and K. Kitazawa, in *Physical Properties of High Temperature Superconductors III*, ed. D.M. Ginsberg, (World Scientific Publishing, 1992).
2. E.L. Wolf, *Principles of Electron Tunneling Spectroscopy*, (Oxford Univ. Press, 1985), chapters 3 and 4.
3. J. Geerk, X.X. Xi, and G. Linker, *Zeit. Phys.* **B73** (1988) 329.
4. R.S. Gonnelli, G.F. Durin, D. Andreone, V. Lacquaniti, C. Appino and M. Vallino, *Int. J. Mod. Phys.* **B5** (1991) 1899.
5. J. Takada, T. Terashima, Y. Bando, H. Mazaki, K. Iijima, K. Yamamoto and K. Hirata, *Phys. Rev.* **B40** (1989) 4478.
6. T. Ekino and J. Akimitsu, *Phys. Rev.* **B42** (1990) 8049.
7. J.S. Tsai, I. Takeuchi, J. Fujita, S. Miura, T. Terashima, Y. Bando, K. Iijima and K. Yamamoto, *Physica* **C157** (1989) 537.
8. M. Lee, D.B. Mitzi, A. Kapitulnik and M.R. Beasley, *Phys. Rev.* **B39** (1989) 801.
9. E.L. Wolf, H.J. Tao and B. Susla, *Solid State Commun.* **77** (1991) 519.
10. R.S. Gonnelli, S.I. Vedeneev and V.A. Stepanov, in *High Temperature Superconductivity*, eds. C. Ferdeghini and A.S. Siri (World Scientific, 1990), p. 188.

11. L.N. Bulaevskii, O.V. Dolgov, I.P. Kazakov, S.N. Maksimovskii, M.O. Ptitsyn, V.A. Stepanov and S.I. Vedeneev, *Supercond. Sci. Technol.* **1** (1988) 205.
12. S.I. Vedeneev, I.P. Kazakov, S.N. Maksimovskii and V.A. Stepanov, *JETP Lett.* **47** (1988) 679
13. S.I. Vedeneev and V.A. Stepanov, *Pis'ma Zh. Eksp. Teor. Fiz.* **49 (9)** (1989) 510.
14. S.I. Vedeneev and V.A. Stepanov, *Physica C* **162-164** (1989) 1131.
15. B. Renker, F. Gompf, D. Ewert, P. Adelmann, H. Schmidt, E. Gering and H. Mutka, *Z. Phys.* **B77** (1989) 65.
16. R.S. Gonnelli, L.N. Bulaevskii, O.V. Dolgov and S.I. Vedeneev, in *High Temperature Superconductivity,* eds. C. Ferdeghini and A.S. Siri (World Scientific, 1990), p. 194.
17. S.I. Vedeneev, P. Samuely, S.V. Meshkov, G.M. Eliashberg, A.G.M. Jansen and P. Wyder, *Physica C* **198** (1992) 47.
18. F. Abbattista, M. Vallino, D. Mazza, P. Brovetto, and D. Delunas, in *Proceedings of the First European Workshop on High-Tc Superconductors and Potential Applications,* eds. J. Vilain and S. Gregoli (Commission of the European Communities, 1987), p. 279.
19. R. C. Dynes, V. Narayanamurti, and J. P. Garno, *Phys Rev. Lett.* **41** (1978) 1509.
20. D. Andreone, R.S. Gonnelli and S.I. Vedeneev, *Proceedings of the International Workshop on "Tunneling Phenomena in High and Low Tc Superconductors"*, Capri, Italy, October 1-4, 1991 (World Scientific), in press.
21. A. Gilabert, J. P. Romagnan, and E. Guyon, *Solid State Commun.* **9** (1971) 1295.
22. W. L. McMillan, *Phys. Rev.* **175** (1968) 537.
23. W.L. McMillan and J.M. Rowell, *Phys. Rev. Lett.* **14** (1965) 108.
24. A.A. Galkin, A.I. D'yachenko and V.M. Svistunov, *Sov. Phys. JETP* **39** (1974) 1115
25. Z. Schlesinger, R.T. Collins, F. Holtzberg, C. Feild, S.H. Blanton, U. Welp, G.W. Crabtree, Y. Fang and J.Z. Liu, *Phys. Rev. Lett.* **65** (1990) 801.
26. G. Deutscher, and K. A. Müller, *Phys. Rev. Lett.* **59** (1987) 1745.

AUTHOR INDEX

www.ingramcontent.com/pod-product-compliance
Lightning Source LLC
La Vergne TN
LVHW060813200926
848329LV00040B/44